Engineering Materials and Processes

Series editor

Brian Derby, Manchester, UK

More information about this series at http://www.springer.com/series/4604

Jan Kośny

PCM-Enhanced Building Components

An Application of Phase Change Materials in Building Envelopes and Internal Structures

Springer

Jan Kośny
Fraunhofer Center for Sustainable Energy
Systems CSE
Boston, MA
USA

ISSN 1619-0181
Engineering Materials and Processes
ISBN 978-3-319-14285-2 ISBN 978-3-319-14286-9 (eBook)
DOI 10.1007/978-3-319-14286-9

Library of Congress Control Number: 2015934911

Springer Cham Heidelberg New York Dordrecht London

Printed on acid-free paper

Springer International Publishing AG Switzerland is part of Springer Science+Business Media (www.springer.com)

Preface

Thermal mass plays an important role in building energy conservation and in control of internal thermal comfort. It has been observed that it can be also greatly assisted by the incorporation of building components with latent heat storage capabilities. Phase change materials (PCMs) are one of the thermal control means used today in building envelopes and in internal construction components. PCMs in buildings can be utilized for many different purposes including reduction of space conditioning energy consumption, thermal peak load shaving and shifting, local temperature control in building envelope components, or improvement of overall system durability. The scope of this book is to summarize and explain the most important basics of PCM applications in building structures. Despite wide interest in PCM-enhanced building technologies by researchers from industry and academia, engineers, architects, building developers, energy policy makers, code officials, and home owners, there is still a shortage of publications supporting design and analysis in this field. In addition, the industry lacks sufficient technical data and performance information for performance comparisons and development of new technologies. At the same time, industry and government code bodies call for adequate performance testing and rating standards.

Note that there are a large number of engineering and research publications focus on PCMs as a major topic. However, even though PCM-enhanced building materials represent today the major market share for the PCM industry, there is still very little engineering literature dedicated to this subject. Most recent publications treat PCM-enhanced building components more from the material perspective (i.e., PCM types, PCM packaging and encapsulation, PCM manufacturing processes, and experimental analysis of PCMs from the chemical and thermal engineering points of view), rather than focusing on the building component scale. As a result, analysis of the PCM-enhanced building components is most often based on the very basic material scale (only the PCM's performance is examined), or a relatively inaccurate whole building scale analysis is performed without taking into consideration that PCM-enhanced envelopes are distinct building systems with their own properties and performance characteristics.

This work is almost exclusively focused on PCM applications as parts of building envelopes and internal building fabric components. A variety of PCM building products and applications are presented here, followed by subsequent thermal and energy performance data. This publication also presents state-of-the-art testing methods to enable thermal performance analysis of building envelope systems containing PCMs. In addition, numerical methods for dynamic thermal analysis of PCM-enhanced building envelopes and whole buildings containing PCM building envelope components are presented here.

This work was motivated by my desire to further the evolution and widespread application of PCM-based building technologies. Since my goal was to reach a wide audience, I organized this book so that it could be easily understood by advanced undergraduate mechanical engineering students and first-year graduate students of architecture and different engineering disciplines with sufficient thermal/energy analysis and material engineering backgrounds. However, this publication also offers an inclusive collection of references leading to more detailed technology descriptions, performance data, and advanced analytical methods, which may be helpful in research work. In my opinion, this publication is mainly intended for:

- Architects, building designers, home owners, and architectural students, who I trust, will benefit from learning about the history of PCM applications in building envelopes and will be able to study most common material configurations, and PCM locations within a building.
- Building materials and systems developers, engineers, and researchers will find in this book an overview of different types of PCMs, their physical characteristics, commonly used PCM carriers, and a selection of commercially available building products containing PCMs. This group of potential readers may also benefit from the patent list associated with PCM-enhanced building products.
- Researchers, engineers, and code officials will learn from information presented here about performance characteristics of the PCM-based building technologies and descriptions of experimental methods used worldwide for testing of PCMs and PCM-enhanced building products.
- Students, engineers, researchers, product developers, designers, home owners, and finally, energy policy government officials should find the field performance data generated during various whole-system and whole-building field experiments worldwide very helpful.
- Lastly, students, engineers, researchers, and energy modelers should find useful the chapter dedicated to numerical performance analysis of the PCM-enhanced building envelopes and whole buildings utilizing these technologies.

Please bear in mind that publications of this type inevitably reflect the opinions and prejudices of their authors. Hence, some readers may inevitably disagree with my opinions, book structure, and of course the choice of presented material.

In my opinion, such disagreements usually represent healthy reflections coming from the diversity of technology under discussion and are essential for its evolution. Nevertheless, I hope that all future readers will find something of interest here.

Boston

Jan Kośny

Contents

About the author

Dr. Jan Kośny is a leading building envelope researcher with over 30 years experiences in the building sciences. Dr. Kośny leads Building Enclosure Research Program at Fraunhofer Center for Sustainable Energy Systems (CSE) in Boston, MA, USA. The main objectives of this R&D program are Building Physics, High Performance Thermal Control Materials, and Building Integrated PV Technologies. Dr. Kośny holds a Ph.D. degree in Building Physics from the Polish Academy of Sciences. His doctoral research was in the area of passive solar systems. Prior to joining Fraunhofer CSE, Dr. Kośny spent 18 years at Oak Ridge National Laboratory where he developed several high-performance building envelope and BIPV concepts. Since 1981, he has held a number of faculty positions in Europe and North America and has published more than 120 technical papers and numerous patents related to advanced building concepts. Dr. Kośny actively works for the ASTM C16, ASHRAE TC 4.4 and TC 4.7 committees. He has been representing the United States at many international organizations and standards bodies including the International Energy Agency. Dr. Kośny, is a winner of the 2009 R&D 100 Award for the development of novel phase-change materials.

Chapter 1
Introduction

Thermal storage plays an important role in energy conservation, which is greatly assisted by the incorporation of latent heat storage in different products. It has been observed that building components, which store heat during peak power operation, can reduce at the same time space-conditioning energy consumption. Phase-change materials (PCMs) are one of the thermal control devices used today in building envelopes. PCMs have been tested as a thermal mass component in buildings for at least six decades. A key goal in low-energy building research is to find ways to manage differences in time between energy sources and energy consumption (i.e., building loads). According to the International Energy Agency (IEA) Annex 24/42 Energy Conservation through Energy Storage/Solar Heating Cooling, thermal energy storage, including PCM technologies, is essential for energy efficiency.[1] PCMs are emerging materials with the potential to dramatically reduce cooling energy consumption and peak cooling demand in buildings, when applied properly.

Published research data have shown that PCMs can notably enhance building energy performance. A large variety of research studies demonstrated that an application of thermal mass in well-insulated structures could generate heating and cooling energy savings between 5 and 30 % in residential buildings (Feustel 1995; Tomlinson et al. 1992; Yamaha 2006; Schossig et al. 2005; Castellón et al. 2007; Kośny et al. 2006, 2012a, b). Considering that new PCM-enhanced building envelope components could be installed in about 10 % of both new and existing US residential homes, the potential for energy savings would be between 0.6×10^{11} and 1.5×10^{11} KWh/year (Kośny et al. 2008).

There is large number of engineering publications focused on PCM as a major subject. However, even though building envelopes represent major PCM applications, there is very little literature dedicated to this topic. Most of recent publications give overviews of PCMs performance, manufacturing processes, and testing from chemical and thermal engineering points of view. Building applications are usually presented more generally as a combination of wall applications,

[1] http://www.iea-eces.org/files/annex_24_extension_work_plan_v6_20120917.pdf.

J. Kośny, *PCM-Enhanced Building Components*, Engineering Materials and Processes, DOI 10.1007/978-3-319-14286-9_1

space-conditioning systems containing PCMs, active solar heating systems, and the large-scale whole building seasonal thermal storage applications. This publication is focused almost exclusively on PCM applications in building envelope components. A variety of PCM-based building products and PCM blends with building materials and insulations are presented here. Performance data for most major PCM building envelope applications are also described here. Furthermore, this publication presents state-of-the-art testing methods enabling thermal performance analysis of building envelope systems containing PCMs. In addition, numerical methods for dynamic thermal analysis of PCM-enhanced building envelopes and whole buildings containing PCM building envelope components are discussed here.

PCMs in building envelopes can be used for many different purposes including reduction of space-conditioning energy consumption, thermal peak load shaving and shifting, local temperature control in building envelope components, or improvement of overall system durability. This publication mainly discusses several strategies for mitigation of thermal loads generated in residential and commercial buildings. Reduction of the space-conditioning energy consumption can be achieved either by improvements in building thermal shell components (i.e., through increasing the thickness of the insulation), or by adding PCM to the walls, roof, attic floor insulation, floors and ceilings, and fenestration components.

In a traditional, simplified understanding, the thermal performance of insulation is directly proportional to the insulation thickness, when isolating the exterior surface from the rest of the building. However, reported results of field testing, energy simulations, and cost analysis demonstrated that conventional thermal insulations, due to relatively high-cost and diminishing energy benefits, cannot be considered as the only means to achieve improved thermal performance of a building shell in low-energy buildings (Kośny et al. 2012a, b). From the whole system perspective, the impact of thermal insulation thickness on overall energy efficiency is significantly more complex. Thermal analysis becomes more difficult when adding the effects of structural members, local thermal bridging caused by imperfections in insulation installation, air leakage, moisture content impact, deterioration of material properties caused by aging, etc. This shows that taking into account only thickness and thermal conductivity of insulation is not sufficient while analyzing the overall thermal performance of building envelopes. Several alternative building envelope systems have been developed during last decades to assist thermal insulation with control of building thermal loads and in the reduction of the building space-conditioning energy consumption. These systems can be grouped into the following basic areas:

- Exterior radiation control technologies—cool-roof and cool wall coatings (Miller et al. 2008);
- Radiant barriers and foil-faced insulations (Medina 2012);
- Conventional thermal mass and building components utilizing thermal inertia for whole building energy consumption mitigation (Kośny et al. 1998; Kossecka and Kośny 2002);
- Phase-change materials (Mehling and Cabeza 2008);

- Airspaces and naturally ventilated cavities (Sedlbauer and Künzel 1999; Miller et al. 2010);
- Roofs with above the deck inclined airspaces (Miller et al. 2007).

Listed above, recent improvements in building envelope technologies suggest that in the near future, residences can be routinely constructed to operate with very low heating and cooling loads. Optimized design of building envelopes using PCMs can be one of the engineering means to meet low-energy consumption targets in the future. It may result in notable savings in energy consumption, reductions in peak-hour power loads, shifting of building thermal loads, and sometimes, in long-term system durability improvements. An application of PCM thermal mass in buildings has been a main building research topic for the last 60 years. Although the information is quantitatively enormous, it is also spread widely in the literature, and not easy to come across. In general, a material that uses its phase-changing ability for the purposes of heating, cooling, or temperature stabilization is defined as a PCM. PCMs have found applications in many fields, including thermal energy storage, building energy efficiency, cooling of food products, packaging and transportation, spacecraft thermal systems, solar power plants, microelectronics, thermal protection of military installations, waste heat recovery, etc. In most current building applications, PCMs continue to absorb heat without a significant rise in temperature until all the material is transformed into the liquid phase. When PCMs reach the phase transition temperature, they absorb large amounts of heat at an almost constant temperature. Then, when the ambient temperature around a liquid material falls, the PCMs solidify releasing stored latent heat.

Concepts of latent heat and specific heat were discovered by Scottish scientist Joseph Black in the mid-eighteenth century. He was a professor of Medicine and Chemistry at University of Glasgow, UK. Prof. Black assisted James Watt in the development of the steam engine. They also collaborated in a project to manufacture sodium hydroxide.

Ice is the best-known PCM used by humans for food preparation, food conditioning, cold drinks, and space cooling. The knowledge of preserving snow for cooling drinks and preserving food during the summer in warm countries has endured since ancient times. Greeks and Romans bought snow and ice transported on donkey trains from mountains. Most urban residents bought it at snow shops, although few could afford private ice houses. Roman refrigeration techniques involved digging deep pits that were then filled with snow and covered with straw. Ice was also preserved for similar purposes in the northern regions of the world.

For generations, inhabitants of northern arctic regions have been using ice for thermal stabilization of their dwellings. Igloos are the first-known application of the phase-change latent heat in building structures. Derived from "igdlu," the Inuit word for "house," igloos have been the traditional dwellings of the natives of the frozen northern reaches of Europe, Canada, and Greenland for thousands of years (Fig. 1.1). The igloo is an ingenious invention, very effective in keeping arctic people warm. Igloos are relatively easy to construct and made from materials found

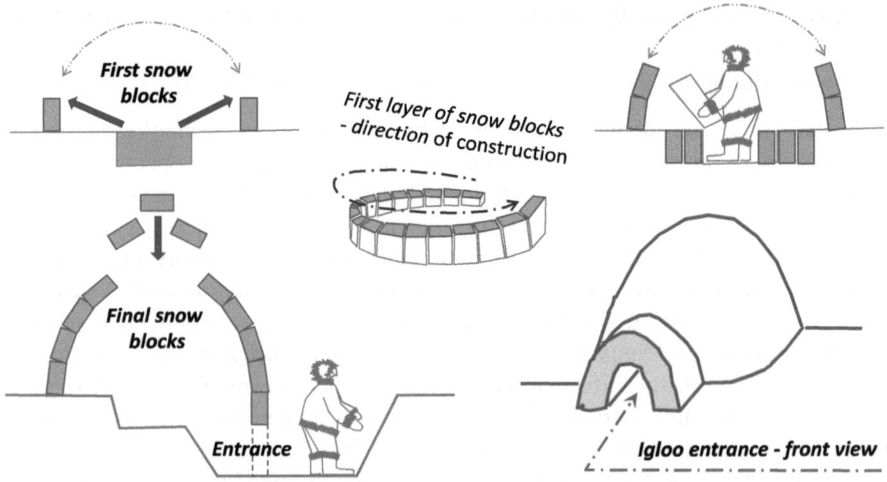

Fig. 1.1 Igloo—a first human construction utilizing PCM—erection details

in abundance: snow and ice which serve simultaneously as building structural components, thermal insulation, thermal radiation shield, and energy storage. Blocks of ice are formed into the dome shape, joined together by snow. To prevent excessive amount of snow and cold wind from coming in, a sunken entrance is typically constructed, along with a raised sleeping platform covered with fur for comfort and warmth.[2] According to González-Espada et al. (2001), internal igloo temperature circulates between 9 and 15 °C, when occupied, even during harsh arctic winters where outside temperatures can drop to −45 °C.

Since the mid-twentieth century, latent heat of ice and a large group of materials with capability of storing heat through the phase transition have been successfully used in buildings. The first documented use of a PCM was by Massachusetts Institute of Technology (MIT) researcher Dr. Maria Telkes in 1948 in a house located in Dover, Massachusetts, USA. In this building, Dr. Telkes followed passive solar wall design, which was first explored by Edward S. Morse and patented by him in 1881.[3] In her house, metal drums filled with Glauber's salt were used as a part of a passive solar heating system. The five-room 135-m^2 house comprised of two bedrooms. Solar energy was collected by the galvanized steel absorber plates, painted black and placed behind the double glazing. Heat generated by this passive solar system passed along a duct via a fan to three heat storage bins situated inside the rooms (see Frysinger and Sliwkowski 1987).

During two and a half winters, in the Northern US coastal climate of Massachusetts, the Dover house was exclusively heated by the sun, before the experiment

[2]http://www.dspace.library.cornell.edu/bitstream/1813/125/2/Igloo.pdf.

[3]http://www.dailykos.com/story/2007/05/17/335503/-Old-Solar-1881#.

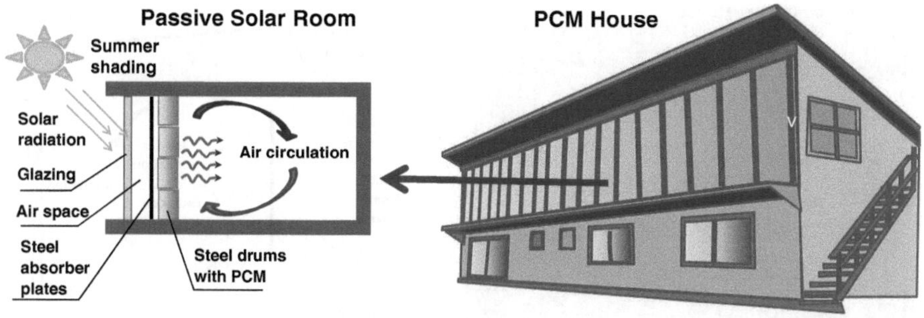

Fig. 1.2 PCM house built in Dover, Massachusetts, USA, by Dr. Maria Telkes and schematic of its passive solar heating operation

ended.[4] On cloudy days, when no solar energy was entering the system, a fan system blew heat from the PCM storage, recrystallizing the salt. Dr. Telkes had analyzed climatic data from the National Weather Bureau and found that, for a period of 65 years, Boston had not gone more than nine days without the sun. She estimated that 21 tons of salt would be enough to heat her house through a 10-day sunless period (see Telkes 1978, 1980) (Fig. 1.2).

A large number of similar residential houses were designed and built in the 1950–1980s worldwide using similar solar design principals to Dr. Maria Telkes approach. One of the first follow-up houses built by Lawrence Gardshire in New Mexico, USA, consisted of a two-story 102-m^2 floor area with collector glazing located in the roof structure (see Ghoneim and Klein 1991). Unlike Telkes' design, comprising of steel barrels with PCM located at the ground floor level; smaller 5-gallon steel storage cans were located within the roof space. The major difference in configuration of the PCM containers was the addition of Borex to the Gluaber's Salts to act as a nucleation enhancing agent. In the 1980s, a large number of solar houses using PCM were built in the USA following very promising field testing results—such as performance data published by Balcomb and McFarland (1978) of Los Alamos National Laboratory, New Mexico, USA, followed with numerous publications describing design principles of passive solar houses using thermal mass components (see Collier and Grimmer 1979; Wilson 1979; Cook 1980; Balcomb et al. 1983; Garg et al. 1985; Neeper 1986), etc. In that time, in experimental passive solar buildings utilizing PCM, different types of hydrated salts were often used for heat storage (see Balcomb and McFarland 1978; Farouk and Guceri 1979; Ghoneim et al. 1991; Crosbie 1997). For example, in late 1970s, Bordeau tested a passive solar collector containing $CaCl_2 \times 6H_2O$, finding that a 8.1-cm-thick PCM wall had an appreciable thermal accumulation (see Bordeau 1980). Unfortunately, these substances frequently created serious durability and corrosion problems, which later affected a number of similar constructions.

[4]http://www.doverhistoricalsociety.org/documents/DoverDays_new/bicentennial/fifties.html.

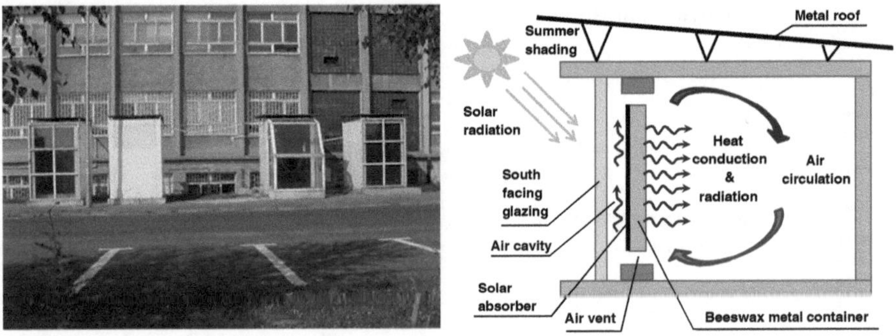

Fig. 1.3 Metal container with beeswax used as PCM heat storage in the test Trombe Wall—Technical University of Rzeszow, Poland (Kośny and Starakiewicz 1987)

In Europe, during the second half of the twentieth century, PCM applications in buildings naturally evolved from a rich tradition of building thermal mass applications, combined with new findings in area of passive solar architecture. Generations of building thermal designers and solar architects often followed an early idea of passive solar walls, which were the developments of French engineer Félix Trombe and architect Jacques Michel http://solarhousehistory.com/book/. Figure 1.3 shows one of the first applications of the bio-based PCM used as a thermal storage in the passive solar Trombe wall. This field experiment was performed in the early 1980s by the Rzeszow University of Technology, Rzeszow, Poland (see Kośny and Starakiewicz 1987). A solar wall contained double glazing and about a 6-cm layer of beeswax packed into the metal container, serving also as a solar absorber. In 1980s, however, PCMs were most often incorporated into more conventional construction materials including PCMs mixed with structural concrete, gypsum, or stucco. New manufacturing techniques had been developed for impregnating concrete, gypsum, or ceramic masonry with PCM. It is important to mention that in early applications, a great variety of PCMs had been tried, including inorganic salt hydrates, organic fatty acids and eutectic mixtures, fatty alcohols, neopentyl glycol, and paraffinic hydrocarbons. An overview of the main PCMs is given by Feldman et al. (1989), Hawes et al. (1993), and Demirbas (2006), whereas other reviews on PCMs may be found in works by Hasnain (1998), Farid et al. (2004), Khudhair and Farid (2007), Mehling and Cabeza (2008), and by Baetens et al. (2010). Most of these studies found that PCMs improved building energy performance by reducing peak-hour cooling loads and by shifting peak-demand time.

In the 1980s, a wide application of organic paraffinic PCMs in gypsum boards took place based on pioneering work of Shapiro et al. (1987) from Concordia University, Canada, Salyer and Sircar (1989) of University of Dayton, USA, and Tomlinson et al. (1992) of Oak Ridge National Laboratory, USA. In Europe, similar work was initiated by development of encapsulated paraffinic PCM products. PCM blends with building structural and board materials have been studied by several researchers (see Banu et al. 1998; Hawlader et al. 2003; Zalba et al. 2003;

Cabeza et al. 2007), and new PCM building products were developed by large chemical companies[5] (Schossig et al. 2003; Fitz et al. 2006). Similar developments took place in Asiatic countries (Sharma and Sagara 2005; Pasupathy et al. 2008) and in Australia and New Zealand (see Khudhair and Farid 2004). Various paraffinic PCM products are commercially available today, but flammability issues combined with a low thermal conductivity (Farid et al. 2004) and a volume change during phase transition (Hasnain 1998) limit their building application.

In large number of demonstration projects worldwide, paraffinic hydrocarbon PCMs generally performed well, which made them a number one line of the PCM building products today. Kissock et al. (1998) reported that wallboard including a paraffin mixture made up mostly of *n*-octadecane, which has a mean melting temperature of 24 °C and a latent heat of fusion of 143 kJ/kg, "was easy to handle and did not possess a waxy or slick surface. Limited surface discoloration was observed. Its unpainted color changed from white to gray. The dry wall with PCM required no special surface preparation for painting." Even though gypsum boards were just impregnated by liquid PCM (no micro-encapsulation was used), the PCM leakage was not a major issue at that time. Salyer and Sircar (1989) reported that during testing of 1.22 × 2.44 m wallboard with PCM, there was an insignificant loss of PCM after 3 months of exposure to continuously cycled 37 °C air. The capability of PCMs to reduce peak loads is also well documented. For example, Kissock et al. (1998) measured peak temperature reductions of up to 10 °C in side-by-side testing of unconditioned experimental houses with and without paraffinic PCM wallboard. Similarly, Zhang et al. (2005) found peak cooling load reductions of 35–40 % in side-by-side testing of conditioned small houses with and without paraffinic PCM inside the walls.

During first decade of the twenty-first century, several PCM blends with thermal insulations were introduced by different research centers. They were mostly used as wall cavity insulation or attic floor insulation, providing reductions in overall cooling loads, shaving and shifting of peak-hour cooling loads, or just local temperature control.[6] In 2003, the first wall assembly utilizing PCM-enhanced foam was designed and constructed by the author. In 2004, the first dynamic hot-box test was performed on a wall containing two types of PCM-impregnated foam by the research team from Guardian, USA (insulation manufacture), Outlast, USA (manufacture of PCM-enhanced fabrics and foams), and Oak Ridge National Laboratory (ORNL) (Kośny et al. 2007; see Figs. 1.4a and 1.5c). In the following years, a similar wall assembly containing PCM and reflective insulation was analyzed as a retrofit solution for concrete masonry walls (Kośny et al. 2010a; see Fig. 1.4b).

In 2004, the author and the research team from ORNL and Advanced Fiber Technologies (AFT), USA, introduced the first PCM-enhanced fiber insulation by mixing cellulose fibers with micro-encapsulated PCM (see Fig. 1.5a). For the 25 % PCM blend, Kośny et al. (2006) and Kośny (2008) reported that PCM-enhanced

---

[5] http://www.basf.com/group/pressrelease/P-09-390.

[6] http://web.ornl.gov/sci/roofs+walls/AWT/HotboxTest/PCMCellulose/index.htm.

Fig. 1.4 Wall application of PCM-impregnated foam. **a** Fabrication of the PCM-enhanced foam and reflective insulation assembly for the test wall, **b** Masonry wall retrofits with PCM-enhanced foam and reflective insulation—Oak Ridge, TN, USA

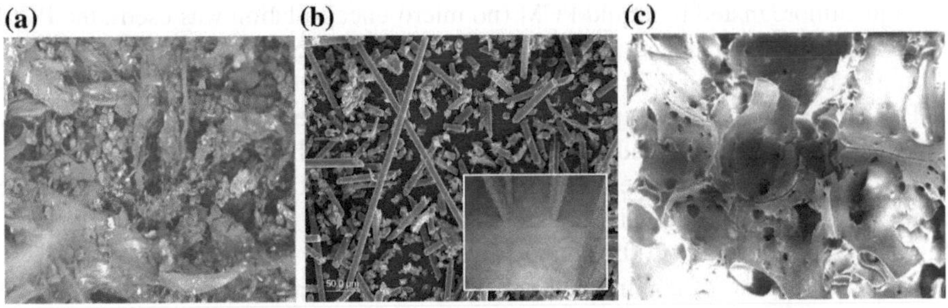

Fig. 1.5 Electron microscope images of thermal insulations mixed with microencapsulated organic PCM. **a** PCM mixed with cellulose fiber insulation. **b** Fiberglass and a photo of PCM mixed with blown fiberglass in the corner. **c** PCM mixed with polyurethane foam

cellulose insulation can reduce wall-generated peak-hour cooling loads by about 40 %. In former applications, the chosen locations for flammable paraffinic PCMs were the interior surfaces of walls, ceilings, or floors. In the following works, however, the PCM-enhanced were positioned inside the wall cavity or installed as a part of the attic insulation system. Placement in these locations is expected to significantly reduce flammability issues that were common in earlier applications of the technology. Also, detailed optimizations performed for PCM applications showed a significant reduction of initial costs with a corresponding reduction in payback time. In 2008, the same research team, in collaboration with Johns Manville (US insulation manufacture), developed a PCM blend with blown fiberglass insulation, as presented in Figs. 1.5 and 1.6. This new dynamic insulation was field tested as an attic floor insulation reducing peak-hour cooling loads generated by the attic by about 70 % (Kośny et al. 2010b). In addition, field application tests took place and a dynamic hot-box test was performed on the wood-framed wall insulated with PCM-enhanced fiberglass. Following these developments, PCM micro-pellets were successfully blended with polyurethane foam for building applications (see Fig. 1.4c).

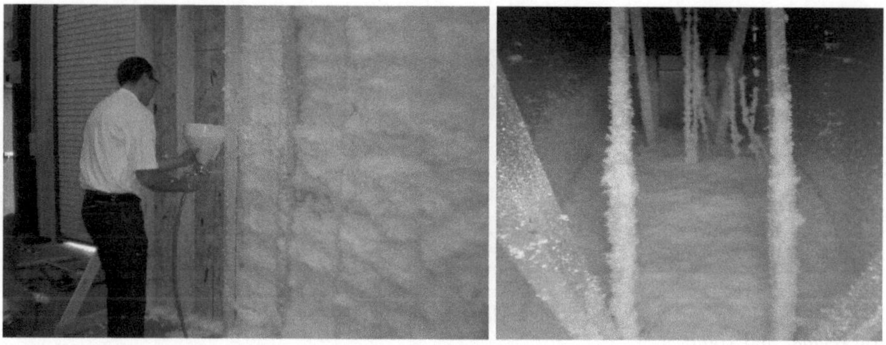

Fig. 1.6 Microencapsulated PCM blended with fiberglass insulation: installation on the test wall —*left*, view of the attic with installed PCM-enhanced fiberglass—*right*

Since a significant amount of the solar heat gain in buildings comes through the windows, one of the most effective ways to control it is to either use windows containing PCM or conventional thermal storage, or to install exterior shading devices, such as window shutters with incorporated PCM. [7, 8]

In conventional applications, thermal performance of windows can be improved by the use of absorbing gases filling the gap between the glass sheets, radiation controlling glass surface treatments, and by the application of thermally broken window frames. The other thermal improvement options may incorporate translucent filling materials such as silica aerogel or a semi-transparent PCM. Though, both of these options require serious consideration of optical properties and window functionality. The objective of using PCM in the window glazing or window attachments is to utilize its high latent heat of fusion to reduce the cooling loads by absorbing the solar-generated heat wave before it reaches the indoor space.

Starting from 2010, the European Union (EU) took the lead in the development of a new generation of PCM-enhanced building insulations including PCM-enhanced plastic foams and PCM blends with nano-insulations. The "New Advanced Insulation Phase Change Materials" project is one out of six funded research topics in the FP7-2010-NMP-EeB research program launched by the EU E2B-PPP initiative—see: http://www.e2b-ei.eu/default.php. This research is expected to bring a new innovative and highly technical solution to increase in a cost-effective manner the thermal inertia of the building's envelope without increasing building mass or space. It is anticipated that through the incorporation of new PCM building technologies, improved with the use of nano-materials, the new building envelopes will be provided in the future with an extra thermal storage capacity based on latent heat. Similarly, starting from 2010 in the USA, Fraunhofer CSE has been involved in several projects focused on development of nano-pore insulations, radiant barriers,

[7]http://www.mendeley.com/catalog/comparison-between-pcm-filled-glass-windows-absorbing-gas-filled-windows/.

[8]http://www.sciencedirect.com/science/article/pii/S0378778811006190.

Fig. 1.7 PCM-enhanced aerogel blanket (*left*), aerogel insulation laminated with different types of foil (*center* and *right*)

and reflective insulations, all thermally enhanced with PCM. Figure 1.7 shows photos of newly developed aerogel insulation products, which were tested during one of the projects funded by the US Department of Defense in collaboration with Aspen Aerogels.

The addition of different forms of PCMs to walls, floors, roofs, or attic floor insulations is widely considered today as a promising thermal improvement method. Similar material options with anticipated high chance for building applications include PCM enhancements of structural boards, stucco, gypsum finish, or blending PCMs with concrete or composite elements. PCMs used for building envelopes are inorganic or organic substances with a high heat of fusion, which melt and solidify at a specific temperature range. The high heat of fusion gives PCMs the capability of storing and releasing large amounts of energy. PCM, energy is absorbed or released when the material changes from a solid to liquid state and vice versa.

In building envelopes, PCMs can be used and incorporated into different types of building products, capsules, packages, etc. New applications of PCM in the design of building envelope components require careful selection of materials, identification of PCM location, bounding thermal resistances, and evaluation of the quantity of PCM to be used. Concentrated PCM applications (such as gypsum boards or arrays of PCM containers) have been tested as a thermal mass component in Northern American and European buildings for the last forty years. Most of the published research data have demonstrated that PCMs can notably enhance energy performance of walls, roofs, and attics. Recent North American studies (Feustel 1995; Athienitis et al. 1997, 2005; Kissock et al. 1998; Zhang et al. 2005; Murugananthama et al. 2010; Kośny et al. 2006, 2010a, b) demonstrated that the application of concentrated PCMs could generate heating and cooling energy savings of up to 25–30 % in well-insulated residential buildings in the southern USA. Childs and Stovall (2012) performed numerical analysis of thermal performance of different configurations of layered wall cavity insulation using micro-encapsulated PCM.

Keep in mind that roofs and attics are the most dynamic working building thermal envelopes. Initial experimental studies of residential systems utilizing latent energy storage have demonstrated very promising performance characteristics. For

Fig. 1.8 Construction of the experimental PCM-PV roof—Oak Ridge, TN, USA. *Left* Arrays of PCM cells and reflective insulation, *center* ventilation air cavities on top of reflective insulation and between PCM cells, *right* top view of the test PCM-PV roof during construction

example, several generations of novel roof designs developed and tested by Oak Ridge National Laboratory, USA, showed about 95 % cut of peak-hour cooling loads and over 30 % overall cooling load reduction compared with traditional roofs design of the same insulation *R*-value (see Kośny et al. 2010a). In addition to conventional insulation, these attics contained over-the-deck ventilation, reflective insulation, and a PCM heat sink. Figure 1.8 shows a roof cross-sectional view of a similar metal roofing system containing arrays of PCM pouches, over the roof ventilation, reflective insulation, and building integrated PV (BIPV) laminate (Kośny et al. 2012a). Correspondingly, the parametric performance comparisons of roofs and attics utilizing dispersed and concentrated PCM systems has been performed by Kośny et al. (2014a).

Majority of the North American experimental studies of dispersed PCM applications have been focused on wood-framed walls or attics (see Petrie et al. 1997; Tabares-Velasco et al. 2012; Kośny et al. 2006, 2012b). PCM-enhanced cellulose insulation was most often used. Advantages of cellulose originate the fact that natural cellulose fibers can efficiently mix with micro-encapsulated PCM, while glass or composite fibers have difficulties in structural support of the weight of the PCM micro-pellets. In early North American experiments, paraffinic PCM was utilized (Kośny et al. 2008) and it was later replaced by less flammable bio-based PCM fabricated from waste fatty oils. In 2009, this technology was awarded the prestigious North American R&D 100 Award,[9] and in 2010, a first residential house using this technology in its walls and attic was constructed in Oak Ridge, TN, USA (Miller et al. 2010; Kośny et al. 2014b; see Fig. 1.9). In 2012/2013, new PCM cellulose blends were tested by Fraunhofer CSE and ORNL in the USA. These mixtures used extruded shape stabilized bio-PCM, currently developed by Syntroleum Corporation (see Fig. 1.10). In addition to the 2012 thermal testing, a series of flammability tests was performed to ensure ignition resistance of these new products (see Shukla et al. 2014).

---

[9] http://www.rdmag.com/award-winners/2009/07/organic-thermal-material-changes-phase-stalls-fire.

Fig. 1.9 Development team of ignition resistant PCM-enhanced cellulose insulation and the first house with PCM-enhanced cellulose insulation in walls and attic. *Left side* Development team from *left* D. Yarbrough and J. Kośny—ORNL, D. Leuthold—AFT, T. Riazzi—Microtek Labs, *right side* PCM house during construction—Oak Ridge, TN, USA

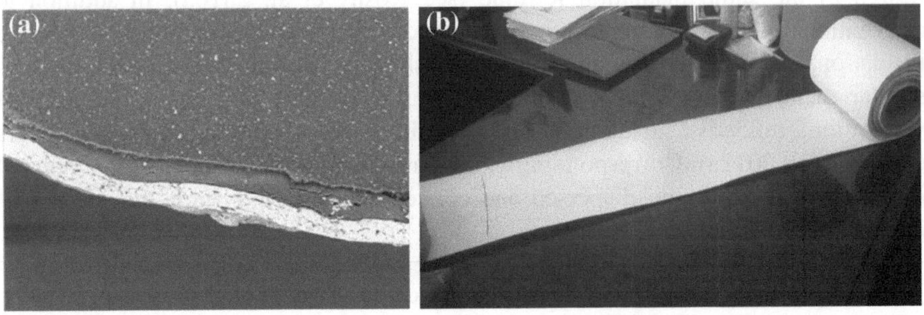

Fig. 1.10 Shape stabilized bio-PCM—Syntroleum Corporation, USA, **a** cross-sectional SEM image of sliced polymer-coated PCM pellet (256 × magnification), **b** a roll of the shape stabilized PCM membrane

Earlier findings in areas of thermal mass, building thermal inertia, and building envelope integrated PCMs help in defining a new dynamic building design paradigm for today. The conventional concept of "statically" designed building envelopes is replaced with a novel, dynamically working, and fully integrated envelope system. For this purpose, several major dynamic thermal effects, which can be generated either by thick layers of conventional insulations, PCM-enhanced insulations, or by different sequences of insulation and massive components, are considered.

- Traditional approach of using thermal mass for stabilization of the internal air space temperature.
- Reduction of the heat flow across building enclosure components through an application of PCM.
- Use of PCM for reduction of peak thermal loads.
- Utilization of PCM for time shifting of dynamic thermal loads generated by building enclosure components.

- Use of PCMs to assist floor, wall, or ceiling hydronic space conditioning systems.

Thermal response shift and an application of materials causing reverse heat flow may be the future low-cost means for energy performance improvements in small residential and commercial buildings (Kośny and Kossecka 2013; Kośny et al. 2014a). Time shift can be controlled by the form of PCM application (dispersed or concentrated), the PCM location in the case of condensed PCM application, and the total R-value of the building component insulation (in most cases a function of its thickness). In short, in low-energy buildings, in order to optimize performance of the building envelope as a thermal system, the following design goals need to be considered:

- Design of thermal insulation of high *R*-value and low amount of penetrating thermal bridges.
- Selection of dispersed or concentrated PCM thermal mass.
- Optimized location of thermal mass within the building envelope component.
- Proper selection of the PCM phase transition temperature.
- Finally, a wider usage of synchronized dynamic thermal response characteristics of the building structure, instead of adding costly thermal insulations or oversized space-conditioning systems.

In addition, several current studies have demonstrated that PCMs can also be very useful in local temperature control in different building envelope designs. In several PCM-stucco applications, slightly higher surface temperatures during the night and early morning hours eliminated the risk of condensation, reduced mold grow, and improved overall lifespan of the exterior finish (Künzel 2007). In typical North American shingle roofs, the under-the-deck PCM applications can reduce roof surface peak temperatures and increase roof lifespan.

Heat storage materials used in thermal engineering to enhance energy efficiency, control temperatures, and shift thermal loads are invariably subjected to dynamic thermal excitations. In order to incorporate dynamic thermal characteristics in thermal analysis or design work, information about dynamic thermal performance of materials or composites containing PCMs is needed to supplement the material physical properties measured under the steady-state conditions.

In the past, the most frequently used method for PCM thermal evaluations utilized the differential scanning calorimeter (DSC). The accuracy of DSC measurements of standard materials is discussed in detail by Richardson (1997) and Rudtsch (2002). Concerning DSC measurements of materials with strong phase transition, the accuracy of the measurement is dominated by the heating rate and sample size. Unfortunately, the DSC method requires small and relatively uniform test samples. This requirement is unrealistic in the case of many PCM-enhanced building envelope products. Small samples are not representative of PCM-based blends, since these materials are not homogeneous. Considering the limitations of the DSC testing (when applied to more complex building envelope applications), a combined experimental–analytical protocol has been developed. This method

Fig. 1.11 Modified heat flowmeter apparatus and hot-box testers used by the author for dynamic testing PCM-enhanced building envelopes. **a** Heat flowmeter apparatus with a complex PCM assembly inside, **b** ORNL hot-box apparatus used for dynamic testing of walls containing PCMs, **c** ORNL horizontal hot-box tester used for dynamic testing of attics containing PCMs

utilizes a conventional ASTM C518 heat flowmeter apparatus (Kośny et al. 2009; Shukla et al. 2014) to analyze major thermal characteristics and to determine the amount of phase-change energy available to provide thermal storage (see Fig. 1.11a). Another testing alternative for non-uniform PCMs is the T-history method, which was proposed by Zhang (1999). This method is a relatively inexpensive and easy way to measure the phase-change enthalpy of PCM products using samples significantly larger from what is required for DSC testing.

Dynamic heat flow meter apparatus testing and the T-history method make possible the material and system-scale analyses of thermal characteristics of PCM blends or complex arrays of PCM containers. It is important to remember that thermal characteristics of final PCM-enhanced products can be significantly different from initial characteristics of pure PCMs used for their production. Finally, building envelopes containing PCMs can be also tested in a full scale using dynamic hot-box testing method (see Kośny et al. 2007, 2010a). Figure 1.11b, c shows two hot-box testing apparatus used in the past by the author for dynamic hot-box testing of walls and attics containing PCMs at Oak Ridge National Laboratory, USA.

In order to assess and optimize the PCMs included in building walls, floors, roof, or attic, transient numerical simulations are necessary. An extensive literature review of the available numerical models for PCM simulation in building envelopes is presented in this publication. As for all building envelope technologies, numerical analysis of PCM-enhanced building systems can be performed both on a system scale and in a whole building scale. The application of materials with phase transition for lightweight construction (e.g., a wood or steel frame buildings, lightweight masonry buildings) makes it possible to modify thermal inertia, improve thermal comfort, and reduce energy consumption. Simulation models are useful tools due to large number of variables and input parameters usually required to estimate energy consumption and demand for buildings with PCM products. There are many state-of-the-art models capable of performing analysis of PCM products, building envelopes containing PCM, as well as for whole buildings containing PCM-enhanced building envelopes. However, in most cases, the calculation accuracy has not been fully evaluated and many of these models have not yet been experimentally validated.

The most interesting models used by engineers and researches worldwide have been selected and analyzed in more detail. Their weak and strong points have been highlighted. All of the models have been considered, regardless if they were implemented solely for thermal analysis of building components, or in whole building energy simulation programs. The mathematical algorithms used to approach the phase transition, numerical methods, modeling accuracy, the simulation time, and the need for experimental data (previous measurements are necessary) have been studied as well.

There are different commercially available programs today which are useful for the thermal and energy performance analysis of whole buildings and for building envelope components: DOE-2, Energy 10, BLAST, EnergyPlus, TRNSYS, ESP-r, RADCOOL, IDA-ICE3, HVACSim, CLIM2000, WUFI, and many others. However, only a few of them have the full capability to analyze the phase transition processes. In these cases, the effect of phase transition has been added to the energy balance equation as a latent heat generation term. Most commonly, the phase change is modeled using temperature-dependent specific heat.

The evaluation of PCM products for building envelope applications requires careful determination of their energy-efficiency potential. Very often, comparisons need to be drawn to determine mass equivalents between PCM and conventional thermal mass including stone, concrete, and brick. In addition, detailed thermal inertia characteristics, temperature, and sensible cooling profiles need to be developed for a prototypical building based on energy simulations. It is also important to determine which climate zones are most appropriate for the considered PCM installations. In most cases of building envelopes containing PCMs, good thermal performance depends on regular day-after-day phase transition ability, which is most often a function of night local temperature variations required to regenerate PCM. It is important to mention that a great majority of early-developed numerical models used for whole building energy analysis were based on research findings from projects using simple paraffinic PCMs, thereby using simplified enthalpy curves and ignoring hysteresis and subcooling effects. A good representation of the subcooling effect and complexity of the enthalpy curve is critical for performance analysis of PCM-enhanced building envelope systems. It is essential for proper estimation of PCM's solidification capability during the discharging time (in most cases overnight cooling), which is a key energy performance factor for the PCM-enhanced building envelopes.

The main goal of this publication is to improve the understanding of how PCM-enhanced building envelopes are constructed, what their thermal behavior is, and what could be their energy impact on the whole building energy balance. The author hopes that this publication will help in bringing PCM-enhance building envelopes to the design mainstream of future residential and commercial constructions.

HAPPY READING!

References

Athienitis AK, Liu C, Hawes D, Banu D, Feldman D (1997) Investigation of the thermal performance of a passive solar test-room with wall latent heat storage. Build Environ 32 (5):405–410

Athienitis AK, Zhang J, Feldman D (2005) A study of double facades with phase-change storage and photovoltaics. In: Proceedings of international conference "passive and low energy cooling for the built environment", Santorini

Baetens R, Jelle BP, Gustavsen A (2010) Phase change materials for building applications: a state-of-the-art review. Energy Build 42(9):1361–1368

Balcomb JD, McFarland RD (1978) Passive testing in Los Alamos. Los Alamos National Laboratory. Report LA-UR-7B-1178

Balcomb JD, Kosiewicz CE, Lazarus GS, McFarland RD, Wray WO (1983) Passive solar design handbook, vol 3. Los Alamos National Laboratory. ISBN 0-89553-106-2

Banu D, Feldman D, Haghighat F, Paris J, Hawes D (1998) Energy-storing wallboard: flammability tests. J Mater Civ Eng 10(1998):98–105

Bourdeau LE (1980) Study of two passive solar systems containing phase change materials for thermal storage. In: Hayes J, Snyder R (eds) Proceedings of the fifth national passive solar conference. American Solar Energy Society, Amherst, pp 297–301

Cabeza LF, Castellón C, Nogués M, Medrano M, Leppers R, Zubillaga O (2007) Use of microencapsulated PCM in concrete walls for energy savings. Energy Build 39:113–119

Castellón C, Medrano M, Roca J, Nogués A, Castell M, Cabeza LF (2007) Use of microencapsulated phase change materials in building applications. In: Proceedings of X conference—thermal performance of the exterior envelopes of buildings. Clearwater, Florida

Childs K, Stovall T (2012) Potential energy savings due to phase change material in a building wall assembly: an examination of two climates. Report ORNL/TM-2012/6. Oak Ridge National Laboratory, USA

Collier RK, Grimmer DP (1979) The experimental evaluation of phase change material building walls using small passive test boxes. In: Proceedings of 3rd national passive solar conference. San Jose, Jan 11–13 1979

Cook J (1980) Passive design for desert houses. Arizona Solar Energy Commission. ASIN: B00070NZ7S

Crosbie MJ (ed) (1997) The passive solar design and construction handbook. Steven Winter Associates, Inc., Wiley. ISBN 978-0-471-18308-2

Demirbas FM (2006) Thermal energy storage and phase change materials: an overview. Energy Sour Part B 1:85–95

Farid MM, Khudhair AM, Ali K, Razack S, Al-Hallaj F (2004) A review on phase change energy storage: materials and applications. Energy Convers Manag 45(2004):1597–1615

Farouk B, Guceri SI (1979) Tromb–Michal wall using a phase change material. In: Proceedings of the second Miami international conference on alternative energy, Miami

Feldman D, Shapiro M, Banu D, Fuks CJ (1989) Fatty acids and their mixtures as phase change materials for thermal energy storage. Sol Energy Mater 18:201–16

Feustel HE (1995) Simplified numerical description of latent storage characteristics for phase change wallboard. Indoor Environmental Program, Energy and Environment Division. Lawrence Berkeley Laboratory. University of California, Berkeley

Fitz C, Hofbauer W, Sedlbauer K, Krus M, Breuer K (2006) Prognoseverfahren zum biologischen befall durch algen pilze und flechten an bauteiloberflächen auf basis bauphysikalischer und mikrobieller untersuchungen. Bauforschung für die Praxis Band 77, Fraunhofer IRB Verlag, 304 Seiten

Frysinger GA, Sliwkowski J (1987) Phase change material storage assisted heating systems. University of Delaware, USA, PH-87-5, No. 4, p 516

Garg HP, Mullick SC, Bhargava AK (1985) Energy storage in building materials In: Solar thermal energy storage. Springer, pp 495–546. ISBN 978-94-010-8841-1

Ghoneim AA, Klein SA (1991) Phase change materials analysis. Sol Energy 47(3):237–242
Ghoneim AA, Kllein SA, Duffie JA (1991) Analysis of collector—storage building walls using phase change materials. Sol Energy 47(1):237–42
González-Espada WJ, Bryan LA, Kang N-H (2001) The intriguing physics inside an Igloo. Phys Educ 36(4):290
Hasnain SM (1998) Review on sustainable thermal energy storage technologies, Part 1: heat storage materials and techniques. Energy Convers Manag 397(11):1127–1138
Hawes DW, Feldman D, Banu D (1993) Latent heat storage in building materials. Energy Build 20:77–86
Hawlader MNA, Uddin MS, Khin MM (2003) Microencapsulated PCM thermal-energy storage systems. Appl Energy 74:195–202
Khudhair AM, Farid MM (2004) A review on energy conservation in building applications with thermal storage by latent heat using phase change materials. En Conv Manag 45:263–275
Khudhair AM, Farid MM (2007) Use of phase change materials for thermal comfort and electrical energy peak load shifting: experimental investigations. In: Goswami DY, Zhao Y (eds) Solar World Congress 2007, Beijing, 18–21 Sept 2007
Kissock JK, Hannig JM, Thomas I (1998) Testing and simulation of phase change wallboard for thermal storage in buildings. In: Morehouse JM, Hogan RE (eds) Proceedings of 1998 international solar energy conference, Albuquerque, American Society of Mechanical Engineers, New York
Kośny J (2008) 2006/07 field testing of cellulose fiber insulation enhanced with phase change material. Oak Ridge National Laboratory report ORNL/TM-2007/186
Kośny J, Kossecka E (2013) Understanding a potential for application of phase-change materials (PCMs) in building envelopes. ASHRAE Trans 119(Part 1):DA-13-001
Kośny J, Starakiewicz A (1987) Ocena Efektow Stosowania w Budownictwie Niewentylowanych Przegrod Trombe'go. PZITB Lodz, Konferencja Naukowo—Techniczna, Fizyka Budowli w Teorii i Praktyce, Lodz, Poland, 11–13 Listopad, 1987
Kośny J, Kossecka E, Desjarlais AO, Christian JE (1998) A dynamic thermal performance of concrete and masonry walls. In: Proceedings of DOE, ASHRAE, ORNL conference—thermal envelopes VII, Clearwater
Kośny J, Yarbrough DW, Wilkes KE, Leuthold D, Mohiuddin SA (2006) PCM-enhanced cellulose insulation thermal mass in lightweight natural fibers—ECOSTOCK—IEA international conference on thermal energy storage. Stockton College of New Jersey, May 31–June 02
Kośny J, Yarbrough DW, Miller W, Childs P, Syed AM (2007) Thermal performance of PCM-enhanced building envelope systems. In: Proceedings of X conference—thermal performance of the exterior envelopes of buildings, Clearwater
Kośny J, Yarbrough DW, Miller W (2008) Use of PCM-enhanced insulations in the building envelope. J Build Enclos Des (Summer/Fall 2008)
Kośny J, Kossecka E, Yarbrough DW (2009) Use of a heat flow meter to determine active PCM content in an insulation. In: Proceedings of the 2009 international thermal conductivity conference (ITCC) and the international thermal expansion symposium (ITES). Pittsburgh, August 29–September 2, 2009
Kośny J, Miller W, Zaltash A (2010a) Dynamic thermally-disconnected building envelopes—a new paradigm for walls and roofs in low energy buildings. In: Proceedings of DOE, ASHRAE, ORNL conference—Thermal envelopes XI. Thermal performance of the exterior envelopes of buildings, Clearwater
Kośny J, Yarbrough D, Kossecka E, Miller W, Bianchi M, Smith J, Lee E (2010b) Theoretical and experimental thermal performance analysis of building shell components containing blown fiber glass insulation enhanced with phase change material (PCM). In: Proceedings of DOE, ASHRAE, ORNL conference—Thermal envelopes XI. Thermal performance of the exterior envelopes of buildings, Clearwater
Kośny J, Biswas K, Miller WA, Kriner S (2012a) Field thermal performance of naturally ventilated solar roof with PCM heat sink. Sol Energy 86(9):2504–2514

Kośny J, Kossecka E, Brzezinski B, Tleoubaev A, Yarbrough D (2012b) Dynamic thermal performance analysis of fiber insulations containing bio-based phase change materials (PCMs). Energy Build 52:122–131

Kośny J, Fallahi A, Shukla N, Kossecka E, Ahbari R (2014a) Thermal load mitigation and passive cooling in residential attics containing PCM-enhanced insulations. Sol Energy 108:164–177

Kośny J, Asiz A, Smith I, Shrestha S, Fallahi A (2014b) A review of high r-value wood framed and composite wood wall technologies using advanced insulation techniques. Energy Build 72 (2014):441–456

Kossecka E, Kośny J (2002) Influence of insulation configuration on heating and cooling loads in continuously used building. Energy Build 34:321–331

Künzel H (2007) Factors determining surface moisture on external walls. In: Proceedings of DOE, ASHRAE, ORNL conference—Thermal envelopes X. Thermal performance of the exterior envelopes of buildings, Clearwater

Medina M (2012) A comprehensive review of radiant barrier research including laboratory and field experiments. ASHRAE Trans 1:2012

Mehling H, Cabeza LF (2008) Heat and cold storage with PCM. Springer, Berlin, 15 Aug 2008. ISBN 978 3 540 68556 2

Miller WA, Keyhani M, Stovall T, Youngquist A (2007) Natural convection heat transfer in roofs with above-sheathing ventilation. In: Proceedings of X conference. Thermal performance of the exterior envelopes of buildings, Clearwater

Miller WA, Childs PW, Desjarlais AO, Haun L, Jacobs J, Karagiozis AN, Kośny J, Petrie TW, Syed AM, Wilson J, Yarbrough DW, Youngquist A (2008) Next generation roofs and attics for homes. ACEEE summer conference, Pacific Grove

Miller WA, Kośny J, Shrestha S, Christian J, Karagiozis A, Kohler C, Dinse D (2010) Advanced residential envelopes for two pair of energy-saver homes. In: Proceedings of 2010 ACEEE summer study on energy efficiency in building. Pacific Grove, 15–20 Aug 2010

Muruganantham K, Phelan P, Horwath P, Ludlam D, McDonald T (2010) Experimental investigation of a bio-based phase-change material to improve building energy performance. In: Proceedings of ASME 2010 4th international conference on energy sustainability, ES2010, Phoenix. http://proceedings.asmedigitalcollection.asme.org/volume.aspx?volumeid=15077, 17–22 May 2010

Neeper DA (1986) Solar buildings research: what are the best directions? Passive Sol J 3:213–219

Pasupathy A, Velraj R, Seeniraj RV (2008) Phase change material-based building architecture for thermal management in residential and commercial establishments. Renew Sust Energy Rev 12 (1):39–64

Petrie TW, Childs KW, Childs PW, Christian JE, Shramo DJ (1997) Thermal behavior of mixtures of perlite and phase change material in a simulated climate. In: Graves RS, Zarr RR (eds) Insulation materials: testing and applications, vol 3. American Society for Testing and Materials, West Conshohocken, ASTM STP 1320, pp 180–194

Richardson MJ (1997) Quantitative aspects of differential scanning calorimetry. Thermochim Acta 300:15–28

Rudtsch S (2002) Uncertainty of heat capacity measurements with differential scanning calorimeters. Thermochim Acta 382:17–25

Salyer I, Sircar A (1989) Development of PCM wallboard for heating and cooling of residential buildings. In: Thermal energy storage research activities review. U.S. Department of Energy, New Orleans, 15–17 Mar 1989, pp 97–123

Schossig P, Henning HM, Haussman T, Raicu A (2003) Encapsulated phase-change materials integrated into construction materials. In: Proceedings of the 9th international conference on thermal energy storage. Futurestock'2003, Warsaw

Schossig P, Henning H, Gschwander S, Haussmann T (2005) Microencapsulated phase change materials integrated into construction materials. Sol Energy Mater Sol Cells 89:297–306

Sedlbauer K, Künzel HM (1999) Luftkonvektionseinflüsse auf den Wärmedurchgang von belüfteten Fassaden mit Mineralwolledämmung. Zeitschrift für Wärmeschutz, Kälteschutz, Schallschutz, Brandschutz 42(43):53–59

Shapiro M, Feldman D, Hawes D, Banu D (1987) P.C.M. thermal storage in drywall using organic phase-change material. Passive Sol J 4(4):419–438

Sharma SD, Sagara K (2005) Latent heat storage materials and systems: a review. Int J Green Energy 2:1–56. ISSN 0197-1522

Shukla N, Cao P, Abhari R, Kośny J (2014) Lab-scale dynamic thermal testing of PCM-enhanced building materials. In: ASTM STP1574, Symposium on next-generation thermal insulation challenges and opportunities, Jacksonville, 23–24 Oct 2013. http://www.astm.org/DIGITAL_LIBRARY/STP/PAGES/STP157420130090.htm

Tabares-Velasco PC, Christensen C, Bianchi M (2012) Verification and validation of energyplus phase change material model for opaque wall assemblies. Build Environ 54(2012):186–196

Telkes M (1978) Trombe wall with phase change storage material. In: Proceedings of the 2nd national passive solar conference, Philadelphia

Telkes M (1980) Thermal energy storage in salt hydrates. Sol Energy Mater 2:381–393

Tomlinson J, Jotshi C, Goswami D (1992) Solar thermal energy storage in phase change materials. In: Proceedings of solar'92: the American solar energy society annual conference, Cocoa Beach

Willson A (1979) Thermal storage wall design manual. New Mexico solar energy association. Modern Press, Albuquerque

Zalba B, Marín JM, Cabeza LF, Mehling H (2003) Review on thermal energy storage with phase change: materials, heat transfer analysis and applications. Appl Therm Eng 23(2003):251–283

Zhang Y (1999) A simple method, the T-history method, of determining the heat of fusion, specific heat and thermal conductivity of phase-change materials. Meas Sci Technol 10(3):201–205

Zhang M, Medina MA, King J (2005) Development of a thermally enhanced frame wall with phase-change materials for on-peak air conditioning demand reduction and energy savings in residential buildings. Int J Energy Res 29(9):795–809

Chapter 2
Short History of PCM Applications in Building Envelopes

2.1 PCM for Building Applications—Background

Ongoing improvements in building envelope technologies suggest that residences will soon be routinely constructed with low heating and cooling loads. The use of novel building materials containing active thermal components (e.g., PCM, sub-venting, radiant barriers, and integrated hydronic systems) would be an ultimate step in achieving significant heating and cooling energy savings from technological building envelope improvements. The key benefit of using PCM is that it affords structures improved thermal storage capabilities with minimal change to the existing building design (Sharma 2013). The main methods for incorporating PCM into building materials include the use of gypsum plaster boards and other structural boards, blending PCM with thermal insulations, and by macro-packaging. The thermal energy storage property of PCM is based on its latent heat storage capacity, given that large amounts of energy can be stored in a small volume. Therefore, the material containing PCM can absorb and release heat more effectively than conventional building materials. However, the selection of the PCM locations, PCM transition temperature range, and the amount of used PCM are essential for effective and durable use of the PCM-enhanced technologies, considering a relatively long life span of building envelopes.

Present-day residential and commercial buildings are becoming more structurally lightweight, and concerns are been raised over indoor thermal comfort due to reduced thermal storage potential. A strong tension exists between the drive to build better efficient structures with less impact on the environment and the tendency to add more mass to the structure for thermal storage. These issues are further heightened by global climate change and the continual rise in energy cost. Without counting traditional arctic ice igloo constructions, the use of PCM in buildings began in the mid-1940s and this passive solar system with thermal storage capability was one of the first applications studied at that time (Telkes 1978; Frysinger and Sliwkowski 1987; Ghoneim and Klein 1991). During the 1980s, bulk

J. Kośny, *PCM-Enhanced Building Components*, Engineering Materials and Processes, DOI 10.1007/978-3-319-14286-9_2

encapsulated paraffinic PCM had been manufactured for building applications. When phase changing heat storage materials are incorporated into the building envelope or internal building structural components, during the day they absorb heat from the glazing and opaque enclosure. As the PCM melts, they stabilize the indoor temperature. At night, when the interior space temperatures decrease, in passive cooling scenarios, the PCM releases the stored energy, thus preventing the temperature in the room from getting excessively cold (Abhat 1983; Lane 1983; Shapiro et al. 1987; Peippo et al. 1991; Schossig et al. 2005; Sharma and Sagara 2005; Mehling and Cabeza 2008). In active cooling scenario when air-conditioning is used, at least part of the PCM's energy is discharged by the space conditioning systems—so, this is not happening without associated energy cost. In some buildings, this cyclic process may result in a reduction of heat flow from the outdoor to the indoor enclosures, which in turn shifts peak cooling loads to off-peak periods, thereby evenly distributing the demand for electricity and avoiding energy shortages often encountered during the peak periods. PCM-enhanced building envelopes offer higher per unit heat storage capacity than conventional building materials and provide lightweight structures the benefit of increased thermal mass (see Kośny and Kossecka 2013). The capabilities of incorporating PCM in structures to thermally stabilize interior space and shift peak-hour cooling loads are highlighted in publications of Castellón et al. (2007), Zhuang et al. (2010), Kośny et al. (Kośny et al. 2010a, 2014), Tardieu et al. (2011), and Childs and Stovall (2012). Since available surface area for heat transfer is significantly greater in building envelopes, in comparison with encapsulated local heat storage applications, large amounts of energy can be stored with minimal fluctuations in the transition temperature. As a result of the improved thermal performance gained from PCM incorporation, lighter and thinner building envelopes can be designed and constructed to take full advantage of the performance.

2.2 First PCM Application for Passive Solar Heating

The use of PCM as thermal storage systems for buildings has been of interest throughout the second half of the twentieth century. Most often latent heat storage materials are used to stabilize interior building temperatures. In building envelope applications, PCM stores latent heat as the ambient temperature rises to the melting point (most PCMs change from a solid to a liquid state). As the temperature cools down, the PCM returns to a solid phase and the latent heat is released. This absorption and release of heat takes place at a constant temperature, which is ideal to smooth external temperature fluctuations.

The first documented use of a PCM for passive solar heating of the residential building took place in 1948 by Dr. Maria Telkes (December 12, 1900–December 2, 1995), a Hungarian born researcher at the Massachusetts Institute of Technology (MIT—Cambridge, MA, USA). By her friends and co-workers, she was called the Sun Queen because of her long fascination by the possibilities of solar heating, since the 1920s. Unable to get a research grant from MIT, Telkes collaborated with sculptor Amelia Peabody, the client, who personally funded the project, and architect Eleanor Raymond.[1]

The first PCM residential house was constructed in Dover, Massachusetts, USA (Telkes 1947). It contained approximately 4 m^3 of Glauber's salts, which were packed in steel drums located in the southern glazed sun spaces that were ventilated with fans to move the warm air into the living space during the winter. In summer months, PCM thermal storage was able to cool surrounding rooms as well (see Telkes 1978, 1980). The Dover house worked very well for two and half seasons. Unfortunately, Glauber's salt disintegrates during a short-time period and loses its phase transition capability, if not sealed and chemically enhanced. During the third winter season in the Dover house, containers with Glauber's salts permanently stopped working. As a result, the experiment was terminated. However, Dr. Telkes stayed for log time optimistic for PCM application in solar heating applications. In 1951 she wrote—"Sunlight will be used as a source of energy sooner or later anyway. Why wait?"[2]

Since 1948, more resources have been invested in the development of PCM-based heat storage technologies. These systems have been extensively studied over decades, notably in the late 1980s and 1990s (see Mancini 1980; Feldman and Shapiro 1989; Peippo et al. 1991; Hawes and Feldman 1993; Bromley and McKay 1994; Salyer and Sircar 1989; Stetiu and Feustel 1996). The next, portion of rapid development PCM technologies experienced during the first decade of twenty-first century—to be discussed on the following sections. Yet, despite numerous experimentally proven performance pluses and the fact that a large portion of difficulties associated with designing of PCM-enhanced systems and their long-term durability have been already resolved, the technology is still moving through the bumpy road of development and market acceptance. In 2014, the assumption could be drawn that for many PCM applications the cost-to-benefit ratio is simply still too high for many PCM applications, to become a common design option of choice.

2.3 PCM Solar Thermal Storage Walls

Applying a PCM in building construction can utilize both the heat from external solar energy gains and thermal loads produced by mechanical heating and cooling systems. Effective storage of this energy is then required in order to match the energy demand of the building sufficiently and at the appropriate time. PCM solar walls are typically used in mixed and cold climates. Generations of passive solar walls containing PCM have been studied for decades as a way of heating buildings from a renewable energy source. A key ingredient of these walls is their heat

[1]http://www.technologyreview.com/article/419445/the-house-of-the-day-after-tomorrow/.

[2]http://www.uh.edu/engines/epi2608.htm.

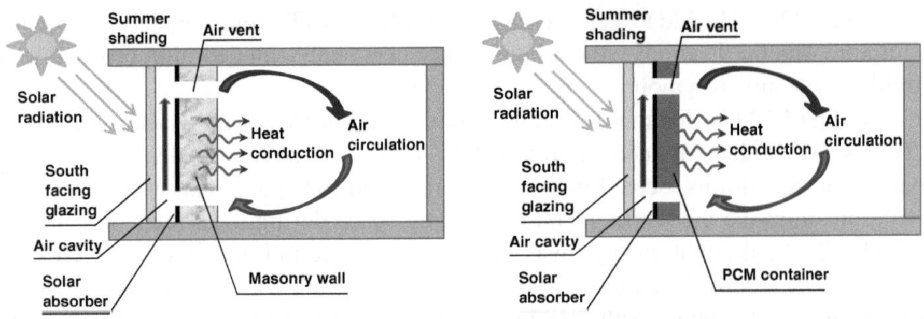

Fig. 2.1 Schematics of conventional Trombe Wall and PCM-enhanced Trombe Wall

storage capacity. In conventional non-PCM applications, the storage capacity increases weight and volume of passive solar systems, which makes difficult their merge with, common today, lightweight construction methods and limits integration into the existing buildings. To alleviate this problem, conventional heavy-weight thermal mass is replaced by PCM.

PCM solar walls are designed to trap and transmit solar energy efficiently into a building. They are sometimes called solar thermal storage walls with PCM. A single- or double-pane glazing is used as an outer thermal barrier of the wall and to provide a greenhouse effect. Research has been carried out on various building applications to determine the effectiveness of PCM latent heat storage. Generations of building thermal designers and solar architects pursued very often an early idea of passive solar walls, which were developed by French engineer Félix Trombe and architect Jacques Michel.[3] The Trombe wall is an example of an indirect gains approach to passive heating which has been tested both experimentally and theoretically with the integration of PCM. The working principal of a Trombe wall is based on sensible heat storage and can be seen in Fig. 2.1. Single or double glazing is usually placed on the exterior surface of the massive heat storage module, with a thin air gap separating these two materials. The exterior surface of the heat storage part is usually painted black so as to absorb the solar energy, which is then stored and conducted through the wall over the period of the day. In winter solar heating scenarios, when the evening and night heating energy demands approach and the internal space temperature drops, heat from the wall storage radiates into the building from the Trombe wall over several hours.

During the last decades, several modifications have been developed from the basic design of a classical Trombe wall and composite Trombe–Michell wall (Zalewski et al. 1997, 2002; Sharma et al. 1989; Jie et al. 2007). Originally, Trombe walls have been built using either masonry or water to provide the appropriate sensible heat storage capacity. This requires large areas of space in order to construct the walls. For this reason, the greater heat storage per unit mass provided by

[3]http://solarhousehistory.com/book/.

PCM is seen as the ideal property which can be incorporated into the existing systems and optimize the space for other practical uses. Thinner PCM walls are also much lighter in weight in comparison with the traditional concrete and masonry materials. These factors offer convenience and attractive reductions in the construction costs associated both with construction of new buildings and their retrofit applications (see Tyagi and Buddhi 2007).

A large number of experimental and theoretical assessments have been conducted to investigate the energy performance and long-term reliability of the PCM-based Trombe wall heat storage components. Initially, hydrated slats have been sampled for this purpose. Telkes (1978, 1980) worked on a construction similar to the Trombe wall, using Glauber's salt behind a polyhedral glazing. Her work was only a first-order theoretical analysis demonstrating the potential for energy and space savings. In experiments performed by Swet (1980), Ghoneim et al. (1991), and Chandra et al. (1985), a Glauber's salt was utilized as well (sodium sulfate decahydrate with melting point 32.1 °C) as a phase-change material in a south-facing Trombe wall. Bourdeau and Jaffrin (1979) and Bourdeau et al. (1980) simulated and tested a Trombe wall using chliarolithe ($CaCl_2 \cdot 6H_2O$) as a PCM heat storage. A numerical model demonstrated that a 3.5-cm wall using PCM could replace a 15-cm-thick conventional wall made of concrete. In a following project, Bourdeau (1982) studied the behavior of Trombe wall made of polyethylene containers placed on a wood shelf behind a double glazing. A series of field tests was carried out at Los Alamos National Laboratory, USA (Balcomb and McFarland 1978; Balcomb et al. 1983). These experimental results were used to validate the numerical model, which demonstrated that a Trombe wall with latent heat storage was more efficient than conventional concrete walls. This research indicated that the optimum thickness of a PCM wall was of a factor 4 thinner than an equivalent concrete wall.

In the following work, varieties of natural- and petroleum-based hydrocarbons have been used as thermal storage. Knowles (1983) used paraffin mixed with metal shavings for increasing the overall conductivity and efficiency in the Trombe wall. Using numerical analysis, Knowles developed a guidelines for the design of low-mass, high-efficiency walls. One conclusion from this experiment was that thermal resistance of the solar thermal wall should be as low as possible. Compared with concrete, paraffin–metal mixtures were found to offer a 90 % reduction in storage mass and a 20 % increase in thermal efficiency. Stritih and Novak (1996) investigated a passive solar wall, which absorbed solar energy into a black paraffin wax (with a melting point between 25 and 30 °C). The stored heat was used for the space heating with efficiency close to 79 %. The results of numerical analysis showed that the optimum thickness of paraffinic heat storage should be close to 5 cm and the melting point of PCM should be few degrees higher from the room set point temperature. In early 1980s, Kośny and Starakiewicz (1987) tested PCM Trombe wall using as a thermal storage a 5-cm-thick metal container with beeswax. Benard et al. (1985) performed a 3-year field experiment in test cells on composite Trombe walls containing 40-cm-thick layer of concrete, 8-cm hard paraffin, and 8-cm soft paraffin. These test cells allowed the analysis of various degrees of thermal coupling between the wall and the conditioned space behind.

The subsequent development included an application of fatty acids and esters for thermal storage. For example, Buddhi and Sharma (1999) measured the transmittance of solar radiation through a solar storage wall containing stearic acid. This parametric analysis was performed at different melting temperatures and wall thicknesses. In addition, Benson et al. (1985) carried an analysis on polyalcohols used as PCM. They also performed numerical analysis on PCM-enhanced Trombe walls compared to conventional concrete structures. They found an optimum melting temperature for PCM which was close to 27 °C. Numerical analysis demonstrated that an increase in thermal diffusivity can be beneficial to the thermal performances of PCM solar walls. Accordingly, laboratory tests demonstrated that diffusivity can be increased by a factor of five through the addition of 2 % of graphite, which should lead to about 30 % improvement in performance. Similar to the Los Alamos test findings discussed earlier, Benson et al. (1985) concluded that a Trombe wall containing PCM could be four times thinner and a factor nine lighter than its equivalence made of concrete.

2.4 Impregnated Concrete Blocks and Ceramic Masonry

The passive use of PCM-enhanced construction and finish materials is relatively not a new concept. As discussed earlier, there were several moderately successful attempts in the 1970s and 1980s to utilize different types of organic and inorganic PCMs to reduce peak loads and heating and cooling energy consumptions (Balcomb 1979; Balcomb et al. 1983; Salyer and Sircar 1990; Hawes and Feldman 1993). These investigations focused on impregnating concrete, gypsum, or ceramic masonry with salt hydrates or paraffinic hydrocarbons.

For decades, masonry blocks or other building materials impregnated with a PCM have been tried in building construction, resulting in successful applications in structures with enhanced thermal inertia and without the heavy-weight mass associated with it. However, an incorporation of PCM in concrete masonry materials is a complex technological process. Some of the concerns include volume changes during melting and freezing, slow heat transfer rates of inorganic PCM products, problems of PCM leakage, and adverse effects on the physical properties of the PCM carrier materials. One of the simplest PCM-enhancement methods consists of impregnation of the concrete block with PCM in a constant volume liquid PCM (see Lee et al. 2000). This is a flexible method which can be applied to different PCM transition temperatures. Concrete blocks can be impregnated as a part of continuous process during their manufacturing.

Concrete is a common construction material made of four components: cement, water, aggregates, and additives. PCM can be either introduced to concrete as an additive or during the impregnation process. Cement is the bounding concrete element that is activated by water. Frequently used Portland cement usually consists of the following four major mineral ingredients: tricalciumsilicaat ($3CaO \cdot SiO_2$), dicalciumsilicaat ($2CaO \cdot SiO_2$), tricalciumaluminaat ($3CaO \cdot Al_2O_3$), and tetracalciumaluminaatferriet ($Al_2O_3 \cdot Fe_2O_3$). Occasionally, blast furnace cements or fly ash cement can be used in concrete mixtures instead of, or in addition to the portland cement. Sometimes lime and clay are added to the concrete mixture as well. A typical concrete hydration process can be described as the following development of C–S–H (Calcium–Silicate–Hydrates):

$$2(3CaO \cdot SiO_2) + 6H_2O \rightarrow 3CaO \cdot 2SiO_2 \cdot 3H_2O + 3Ca(OH)_2$$

$$2(2CaO \cdot SiO_2) + 4H_2O \rightarrow 3CaO \cdot 2SiO_2 \cdot 3H_2O + Ca(OH)_2$$

In 1980s and 1990s, impregnation was used as the major method of introduction of melted PCM to the carrier material. Incorporation of PCM into concrete masonry products was carried out through an immersion process into a liquid PCM bath. Hawes et al. (1990, 1992) have studied latent heat storage of concrete with different types of PCM in different types of concrete blocks. Innovative methods of PCM containment in hollow-core building blocks were studied by Salyer et al. (1995). One significant technical innovation from this work was development of a new PCM composite that could be made by blending of PCM with high-density polyethylene (HDPE)/ethylene-vinyl acetate (EVA) and silica in defined proportions. PCM-hydrophobic silica dry powder was incorporated into the concrete wet mix to provide an effective thermal storage material. Based on this discovery, Salyer and his team have developed different methods of PCM incorporation to building masonry products: by imbibing the PCM into porous materials, PCM absorption into silica or incorporation of PCM into polymeric carriers. Silica fume and fly ash were used often to reduce the alkalinity of concrete and to improve the compatibility with alkaline sensitive PCM. Romanowska et al. (1991) have analyzed the thermal performance of organic PCM's added by impregnation to different types of ceramic masonry. Lee et al. (2000) have studied and presented the results of macro-scale tests that compare the thermal storage performance of ordinary concrete blocks with those that have been impregnated with two types of PCM, butyl stearate, and commercial paraffin. Hawes et al. (1990, 1992) presented the thermal performance of PCM's (butyl stearate, dodecanol, paraffin, and tetradecanol) in different types of concrete blocks. This analysis has covered the effects of concrete alkalinity, temperature, immersion time, and PCM dilution on PCM absorption during the impregnation process. Hadjieva et al. (2000) have applied the same impregnation technique for concrete but with sodium thiosulphate penta-hydrate ($Na_2S_2O_3 \cdot 5H_2O$) as a PCM. They concluded that the large absorption area of cellular concrete may serve as a good carrier material of an incongruently melting inorganic PCM and improving its structure stability during thermal cycling.

Another PCM application method incorporating concrete products has been highlighted in more recent work performed by Zhang et al. (2004), Bentz and Turpin (2007), and van Haaren (2012). In these experiments, lightweight aggregates with high porosity were used as the PCM carriers to achieve adequate heat storage capacity. In some types of PCM-enhanced masonry products, porous aggregates are surrounded by tight cement-based paste, which reduces a potential for PCM leakage. In early 2000, Bentz and Turpin (2007) investigated the effectiveness of thermal storage mortar with lightweight expanded shale aggregates impregnated with paraffin and polyethylene glycols.

In addition to early applications of salt hydrates, Hawes et al. (1990) and Hawes and Feldman (1993) studied the thermal performance of several organic PCMs, such as butyl stearate, dodecanol, tetradecanol, and paraffin, in different types of concrete masonry products. However, it was found that most of the above methods of PCM introduction to the masonry products may have drawbacks of interacting with building structure and change the material matrix, causing possible leakage over its lifetime, among other issues (see Schossig et al. 2005). Despite many benefits, the problem of PCM leakage has been observed in many building applications after repeated thermal cycling. To reduce potential durability problems, in research work performed by van Haaren (2012), the introduction of PCMs into the concrete was achieved in two different ways: (1) The sand from a standard mixture was replaced by PCM particles and (2) porous lightweight aggregates were impregnated with PCM and later a concrete mix was made with use of impregnated and not impregnated lightweight aggregates (see Fig. 2.2).

Encapsulation of a solid–liquid PCM during its phase transition is crucial in most cases to hold the liquid phase of the PCM and to reduce PCM leakage and the reactivity of PCM with the outer environment (see Hawlader et al. 2003). In Microencapsulation, micronized materials (both liquids and solids) are packaged in the form of microcapsules, which range in size from less than 1 μm to more than 300 μm. The outer skin of the capsule can be made by using natural and synthetic polymers which provides a hard shell. Depending on the size of the final products, encapsulation can be classified as follows: macro-encapsulation or microencapsulation. Keep in mind that in early PCM building research, Collier and Grimmer (1979) already showed that a macro-encapsulated PCM material cemented within

Fig. 2.2 Microscopic view of the concrete—inorganic PCM mixture (*left*) and test samples of PCM—enhanced concrete (*right*) (van Haaren 2012)

masonry building blocks results in significant increase in the system performance over an equivalent volume of concrete. In Spain, the results from a field study of concrete blocks using microencapsulated PCM were presented by Cabeza et al. (2007). Improved thermal inertia was observed for PCM-enhanced products, which indicated a potential for energy savings in buildings. Other types of masonry products can be used as well as PCM carriers. For example, the application of PCM-treated ceramic bricks was investigated by Romanowska et al. (1991) and Lai and Chiang (2006). In addition, Mehling et al. (2002) found that PCM can be combined with lightweight wood concrete (holz-beton) and that the mechanical properties of the structure do not change significantly.

2.5 PCM-Enhanced Gypsum Board and Interior Plaster Products

As mentioned in the previous section, PCM can be easily blended with many construction materials, including gypsum building products. This makes it possible to increase the latent heat capacity of lightweight constructions by applying it in the form of interior plaster or finish gypsum boards. In natural form, gypsum is a common soft sulfate mineral composed of calcium sulfate dihydrate, with the chemical formula $CaSO_4 \cdot 2H_2O$. Manufacturers of commercial gypsum remove purposely much of natural gypsum's inherent water by crushing the rock into powder and heating the powder to remove its water molecules. Dehydration of gypsum allows the gypsum to become rehydrated later. After addition of water, the rehydrated gypsum dries and sets, it becomes a rock-hard substance, which is commonly used in variety of building materials.

For decades, paraffinic PCM has been the most widely used latent heat storage material for thermal enhancement of gypsum boards and plasters. The reason for this is very simple; in interior environments, PCM with a melting temperature between 19 and 24 °C can be used with best results, since this is a temperature range, which is close to human comfort level. Keep in mind that paraffin waxes (such as *n*-hexadecane, *n*-heptadecane, and *n*-octadecane) match the necessary temperature range well. That is why a large number of building applications used today by construction industry largely use paraffin for heat storage. However, due to relatively high cost of paraffin-based PCM, its flammability, and origin in non-sustainable petrochemicals, other organic PCMs are being explored now, including fatty acids, coming from agricultural and food industry waste.

Interior building surfaces of walls, ceilings, or floors have been traditionally considered as the best locations for the PCM. In gypsum board and plaster applications, PCM is used to stabilize the temperature of the building interior. As shown in Fig. 2.3, PCM concentrated in gypsum boards interacts mostly with the interior of the building. The energy storage capacity of the PCM-enhanced gypsum is used to reduce interior space temperature swings and absorb solar gains coming through

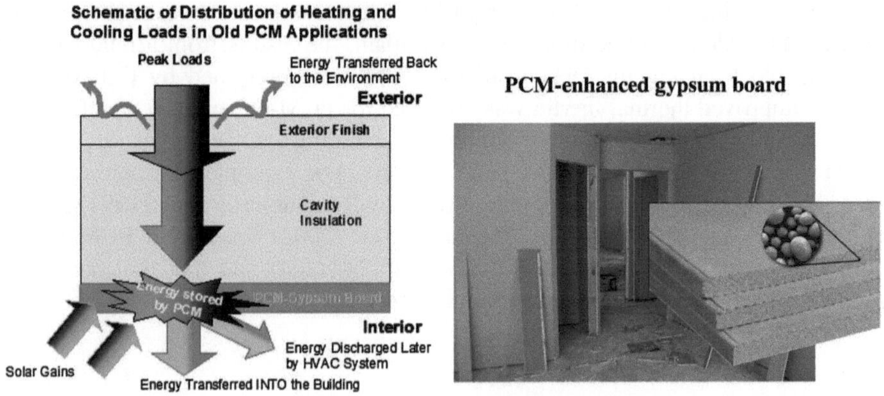

Fig. 2.3 PCM as part of the interior surface of the building envelope

the glazing. In this working scenario, heat stored in PCM needs to be discharged through, most-often, ventilation during the night or by the building's space conditioning systems. In the second case, when mechanical system is used, there may not be direct energy savings. The only energy benefits associated with this PCM setup can be peak-hour load savings and peak load shifting. In addition, in buildings using air-conditioning, due to the relatively small interior temperature fluctuations, PCM applications facing the interior of the building may require a long time to discharge the stored energy.

During the last two decades of twentieth century, interest has increased toward the energy saving potential achieved when combining PCM into wall or ceiling interior finish materials. PCM-enhanced wallboards are capable of stabilizing interior space temperature by capturing dynamic internal thermal loads and a large proportion of incident solar radiation falling on building surfaces. They are also relatively easy to install and are used in a wide variety of applications. Most of the earliest studies focused on wallboard that had been immersed into the molten paraffin. Gypsum, in wallboard, can absorb up to 30 weight percentage of PCM (Stovall and Tomlinson 1995; Neeper 2000). Initially, porous building materials such as gypsum were dipped into a molten PCM bath, absorbing the material into pores by capillary action. Next, the building material was removed from the paraffin immersion and was cooled, allowing the PCM to set. The great advantage of this method was that it enabled ordinary wallboard to be converted to PCM wallboard simply, inexpensively, and when it was required, i.e., imbibing the material either prior to installation or at the building site. The major disadvantage was potential leaking of PCM from the product causing aesthetical surface discolorations.

Non-encapsulated paraffins were the initial material of choice in early investigations of the PCM-impregnated gypsum boards. Today, they are still the most common PCM substances considered for thermal enhancement of gypsum boards, however, after microencapsulation. Research by Feldman and Shapiro (1989), Feldman et al. (1991), and Peippo et al. (1991) showed that gypsum–paraffin composites can be used to reduce room cooling loads. Paraffinic hydrocarbons generally perform well while impregnating gypsum boards with liquid material, but they increase the flammability of the building envelope (see Slayer and Sircar 1989; Tomlinson et al. 1992; Kissock et al. 1998). Salyer and Sircar reported that during the testing of wallboard with PCM, after 3 months of continuous exposure to cycled 40 °C air, there was no significant loss of PCM. In addition, Kissock and co-workers investigated the wallboard which contained a paraffin-based PCM mixture made up mostly of *n*-octadecane. This PCM had a mean melting temperature of 24 °C and a latent heat of fusion of 143 kJ/kg. They reported: "…it was easy to handle and did not possess a waxy or slick surface. It scored and fractured in a manner similar to regular wallboard. Its unpainted color changed from white to gray. The drywall with PCM required no special surface preparation for painting" (see Kissock et al. 1998).

In first PCM gypsum applications analyzed by Oak Ridge National Laboratory, USA, gypsum boards impregnated with paraffin were installed directly on the interior building surfaces (see Drake 1978; Tomlinson and Heberle 1990; Tomlinson et al. 1992). A following study focused on gypsum board impregnated with non-encapsulated PCM demonstrated a potential for significant cooling energy savings. However, this work indicated a need for greater convective heat exchange between PCM-enhanced gypsum boards and interior space air (see Stovall and Tomlinson 1995). In addition, Neeper (2000) examined gypsum wallboard infused with fatty acids and paraffin waxes, subjected to the diurnal variation of room temperature, but not directly illuminated by the sun. Kondo et al. (2000), with a research team from Kanagawa University and Tokyo University, Japan, researched the thermal storage of PCM-enhanced gypsum boards using the dual-PCM mixture containing 95 % octadecane and 5 % hexadecane.

Since, PCM must be able to cycle continuously day after day, it is critical to optimize its amount, phase transition temperature, and location within building envelopes. Numerical analysis is very helpful for this purpose. It has been used in many research projects worldwide in conjunction with laboratory or field experiments. For example, Athienitis et al. (1997) conducted extensive field experiments combined with one-dimensional nonlinear numerical simulation study in a full-scale outdoor test room with PCM gypsum board as inside wall lining. Their numerical results were compared with good agreement against the experimental data. Finite-difference heat transfer simulations were used by Lawrence Berkeley National Laboratory, USA, (Feustel 1995; Feustel and Stetiu 1997) and Oak Ridge National Laboratory, USA, (Stovall and Tomlinson 1995) to numerically evaluate the latent storage performance of PCM-enhanced wallboard. Simulation results for a living room with high internal loads and weather data in Sunnyvale, California, USA, showed significant reduction of room air temperature when heat was stored in PCM-treated wallboards. In the case of the prototype International Energy Agency (IEA) building located in California climate zone 4, it was estimated that PCM wallboard would reduce the peak cooling load by 28 % (see Feustel and Stetiu 1997).

According to many authors, the PCM used on wall surfaces may decrease overheating in the interior spaces and reduce energy consumption for space conditioning (see Kuznik and Virgone 2009; Schossig et al. 2005). In 2006, Kissock and Limas of University of Dayton, USA, investigated paraffinic PCM that can be added to the walls, to reduce the peak diurnal cooling and heating loads transmitted through the envelope (Kissock and Limas 2006). This work was a combined numerical–experimental study to quantify the effectiveness of PCM in reducing thermal loads through the building envelope components and to develop a design strategy for the placement of PCM within the massive walls. The PCM studied was paraffin octadecane, with an average melting temperature of 25.6 °C. For the climate of Dayton, OH, thermal loads through the PCM-enhanced wood-frame wall were simulated using an explicit finite-difference procedure with the indoor air temperature held constant. While comparing to the conventional wall, cooling load savings were close to 16 %. The simulation technique has been validated against the experimental work.

In addition to simple paraffinic PCM, a great variety of organic compounds, including fatty acids and esters, were also used for gypsum board impregnation. Shapiro (1989) investigated a potential of several methyl-esters PCM for introduction into gypsum wallboard with possible thermal storage applications for the Florida, USA, climate. These materials were blends of methyl palmitate, methyl stearate, and mixtures of short chain fatty acids (capric and lauric acids). Although these materials had relatively high latent heat capacity, the temperature ranges required in achieving the thermal storage were not fully sufficient for the internal wall location—PCM had problems with full cycling everyday. Similarly, (Feldman and Shapiro (1989), Feldman et al. (1991) performed extensive research on different organic compounds as candidates for PCM heat storage. This work included fatty acids (capric, lauric, palmitic, and stearic), butyl stearate, dodecanol, and polyethylene glycol 600. Various materials were considered as PCM carriers, including different types of concrete and gypsum products. The PCM gypsum board was made by soaking conventional gypsum board in liquid butyl stearate, a PCM with phase-change temperature range of 16–20.8 °C. The PCM gypsum board contained about 25 % of PCM by weight. Its thermal properties were measured with a differential scanning calorimeter (DSC). Wallboards with incorporated organic PCM were also analyzed by Shilei et al. (2006). Capric acid and lauric acid, as PCM, were applied in building wallboards for low-temperature latent heat storage due to their low carbon chains. It was found that they can automatically absorb indoor redundant heat, which can greatly reduce the building thermal loads and save electric energy for space conditioning. In full-scale experiments, a wallboard containing about 26 % PCM by weight was installed on top of the existing conventional wall. Compared with an ordinary room, it was found that the PCM wallboard room could greatly reduce the energy cost of HVAC systems and notable shift electric power peak load.

It is good to remember that PCM must be able to cycle continuously for years without loss of its reacting volume. In order to prevent PCM's loss in mass through leaking, while in melted stage or through evaporation, PCM must be enclosed, either by means of micro- or macro-encapsulation. Later studies of wallboards and plasters containing microencapsulated PCM have shown promising results in terms of thermal performance but also in the effectiveness of the manufacturing processes used to produce these types of components. The microencapsulation of PCM has allowed a PCM weight ratio to reach about 30 % in gypsum-based composites (see Fig. 2.4). In recent years, we have seen the advent of composite materials that can contain up to 60 % by weight of PCM (Athienitis et al. 1997). The opportunities presented by the microencapsulation of PCM in gypsum plaster were also investigated by Schossig et al. (2005). It has been found that, since the PCM micro-pellets are very small, the mechanical destruction of capsules during the production process is highly unlikely. The fine distribution of the PCM particles in the matrix provides larger surface area for heat transfer, so the heat transfer rate during melting and freezing cycle is enhanced significantly. It has been shown that microencapsulation of PCM results in easy application, improved heat transfer, and good compatibility with conventional construction materials. The PCM walls facilitate low fluctuations in the indoor air temperature.

An innovative method of incorporation of microencapsulated PCM into the plaster mix for indoor use was developed by Zamalloa et al. (2006). Thermal and mechanical properties of the developed new material and long-term PCM durability have been tested. Thermal simulations were performed to study the optimal distribution of this material inside the building. These simulations were validated in the field by the construction of two real-size concrete cubicles, using in one of them the PCM-enhanced plaster. The results of the study showed that the new plaster is effective to diminish the thermal oscillations inside the buildings and reduce the energy needs up to a 10–15 % in heating and about 30 % in cooling seasons.

In recent years, heat conduction boosting agents are added to PCM-enhanced building products to improve their overall thermal performances. Earlier ORNL research on PCM-impregnated gypsum boards showed that PCM boards required a significant increase of the surface convection coefficient (up to three times), to provide proper conditions for charging and discharging of PCM. It was related to

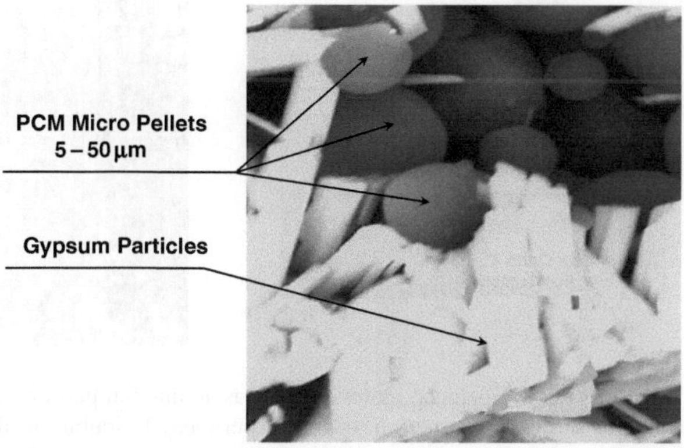

Fig. 2.4 Electron microscope image of gypsum-PCM composite

limited temperature fluctuations available inside the building space (see Stovall and Tomlinson 1995). It was also found that heat conduction enhancement of the interior wall surface layers may improve overall effectiveness of many PCM gypsum products. Adding high conductive materials such as metal (aluminum powder and sometimes silver), metal oxide, and expanded graphite into PCM produced effective thermal storage materials and enhanced heat transfer rates with the interior space (see Zhou et al. 2007, 2014; Sari et al. 2004; Sari and Karaipekli 2007; Kim and Drzal 2009). It was found that with the addition of 3 % by weight of expanded and exfoliated graphite, the thermal conductivity of PCM composites increased between 14 and 24 %. A series of thermal storage and release tests of pure hexadecane and aluminum/hexadecane composite demonstrated significant improvement of heat conduction. Darkwa and Zhou concluded that thermal response rate especially heat-flux rate was accelerated by introducing aluminum powder. Heat conductivity of composite was more than 8 times higher comparing to pure hexadecane (Darkwa and Zhou 2011).

2.6 Use of PCM-Enhanced Wall Cavity Insulation

Similarly to, earlier described, PCM applications in wall masonry units, air gaps in massive walls and framed wall cavities can be very convenient locations for PCM. As shown in Fig. 2.5, the application placing PCM inside the wall cavity takes advantage of the large temperature fluctuations that take place on the exterior building envelope surfaces. These energy fluctuations, which can be a significant part of the building cooling and heating loads, are largely absorbed by the PCM-enhanced insulation and later transferred to the environment without affecting the interior building energy balance. In this application, phase transition temperature

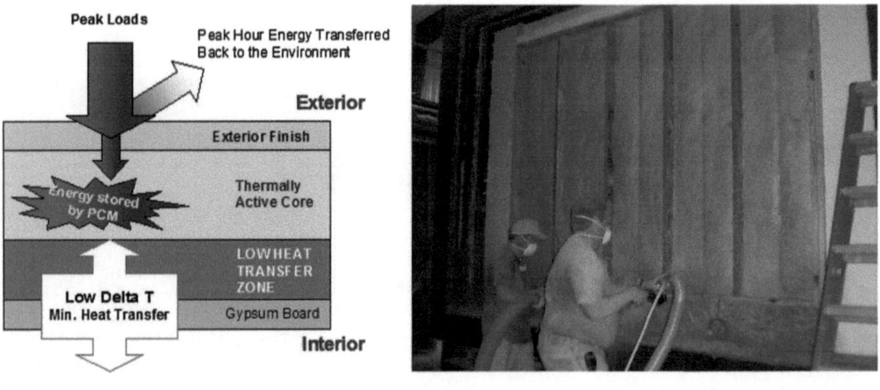

Fig. 2.5 PCM-enhanced materials used as an integral part of the building thermal envelope. *Right picture* shows construction of an experimental double wall with exterior layer of cavities containing PCM-cellulose insulation

range of PCM should be as close as possible to the interior space set point temperature. As a result, heat transfers between the core of the building envelope and the interior space is reduced. This simple change in material configuration means real space conditioning energy savings. It is also expected that this new placement method for PCM should significantly reduce flammability issues that were common in earlier technology developments. In addition, detailed optimizations performed for PCM applications showed significant potential for reduction of initial costs and a corresponding reduction in cost payback time (Kośny 2008; Kośny et al. 2009; Kośny et al. 2012a). A challenge for this application can be a need for significant latent heat storage capacity in place where PCM is located.

Ismail and Castro (1997) presented the results of a theoretical and experimental study of PCM-filled brick walls and attic floor insulation under real operational conditions to achieve passive thermal comfort. In theoretical analysis, Ismail and Castro used a one-dimensional finite-difference model for simulations of the phase transition problem in a wall constructed of two layers of brick with the PCM core. The results obtained were compared with field measurements. The experimental setup consists of a small room with movable roof and side wall. The test wall contains the PCM heat sink which was composed of two commercial grades of glycol in order to obtain the required fusion temperature range. Another wall, identical but without the PCM, was also used during comparative tests. Field tests demonstrated that in Brazil climate the PCM used was capable to maintain the indoor room temperature close to the thermal comfort levels. Economical analysis indicated that the PCM application concept may effectively help in reducing the electric energy consumption and improving the energy demand pattern in Brazilian buildings.

Another PCM application concept was proposed by Kośny and Yarbrough in 2002/2003 to the US Department of Energy (US DOE). The concept is based on incorporating PCM thermal insulation into the internal cavity of lightweight framed walls (see Kośny et al. 2006, 2007). This application is counter intuitive, because it deliberately restrains energy transport between the PCM and the conditioned space and the exterior environment (Khudhair and Farid 2004). However, this PCM location controls the temperature profile within the building envelope component, and thus influences the overall heat exchange. In some configurations, this can reduce the net energy transported through the interior envelope surfaces; in others it only changes the time when the peak energy crosses that boundary. In most air-conditioning operations, a night-time precooling is a well-established energy saving method. In the case of PCM application, the time delay can prove especially valuable, in economic terms, when utilities lower off-peak-time electricity rates. In energy consumption terms, the air-conditioning system can operate more efficiently (less costly) during the time of the shifted space conditioning loads.

Zhang and Medina of University of Kansas, USA, developed a thermally enhanced wood-frame wall that integrated a paraffinic PCM via macro-encapsulating (Zhang et al. 2005). Results from the field testing show that the PCM wall reduced wall peak heat fluxes by as much as 38 %. For a period of several days (in experiment that included walls facing different directions), the average wall peak heat-flux reduction was approximately 15 % for a 10 % concentration of PCM and approximately 9 % when a 20 % PCM concentration was used. The average space-cooling load was reduced by approximately 8.6 % when 10 % PCM was applied and 10.8 % when 20 % PCM was used.

In research work initiated by the US DOE on fiber insulations blended with PCM, in addition to microencapsulated paraffins, new PCM produced from organic fatty-acid esters was used (Kośny 2009). Compared to the paraffinic PCM, fatty-acid esters are less expensive and significantly less flammable. They are made from underused feedstocks such as soybean oils, coconut oils, palm oils, and beef tallow. Because these esters are fully hydrogenated, the fatty-acid ester PCM is expected to remain stable during thousands of phase-change cycles with no risk of oxidation.

PCM-enhanced cellulose was one of the first successful developments of PCM-enhanced thermal insulations in the building area (Kośny et al. 2006, see Fig. 2.6). Subsequently, PCM blended with blown fiberglass (Kośny et al. 2010a) and building plastic foam insulations were introduced (Kośny et al. 2007, 2010b; Mehling and Cabeza 2008). From 2006 to 2007, ORNL performed a series of dynamic hot-box tests of wall assemblies containing PCM-enhanced insulations. For the wood-framed wall containing PCM-enhanced foam insulation, Kośny reported that it can reduce wall-generated peak-hour cooling loads by about 40 % (Kośny et al. 2007, 2010a). The major advantage of PCM-enhanced insulations is their capability of significant lessening and shifting the peak-hour thermal loads generated by building envelopes. A detailed explanation of dynamic hot-box testing procedures is presented in the following chapters. Below, a short description is available for the dynamic hot-box test performed on the wood-frame wall insulated with PCM-enhanced cellulose.

In dynamic hot-box testing, nominal 2.4 × 2.4 m lightweight-frame wall specimens were used (see Fig. 2.7). One of the first test walls was constructed with 6 × 15.2 cm wood framing installed 40 cm on center. Three wall cavities were insulated with plain cellulose of a density of about 42 kg/m^3. Three remaining wall cavities were insulated with a cellulose–PCM blend of a density of about 42 kg/m^3 and containing about 22 % by weight of PCM. It is estimated that about 17 kg of

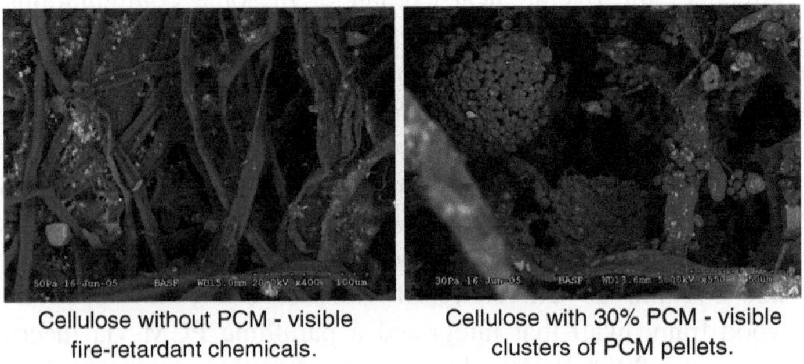

Fig. 2.6 Scanning electron microscope images of the PCM-cellulose blend. Clusters of PCM microcapsules are shown on the *right photograph*. BASF's Micronal PCM was used

Fig. 2.7 Installation of PCM-enhanced cellulose in the test walls

PCM-enhanced cellulose insulation was used for this dynamic experiment (Kośny 2008). Comparisons of measured heat flow rates on the wall surface, which was opposite the thermal excitation, enabled an estimate of the potential thermal load reduction generated by the PCM (3 test wall cavities contained about 3.6 kg of the microencapsulated PCM manufactured by BASF). In reality, most daily thermal excitations generated by solar irradiance are no longer than 5 h (peak-hour time). In this dynamic experiment, during the first 5 h after the thermal excitation, measured heat flux was reduced by about 40 % thanks to the PCM application.

In 2009, the dynamic hot-box experiment was performed on the wood-framed wall containing blown PCM-enhanced fiberglass insulation (see Fig. 2.8). This insulation was jointly developed by ORNL, Johns Manville (insulation manufacture), and Microtek Labs (bio-based PCM manufacture) (see Kośny et al. 2010a). Comparisons of measured heat flow rates on the wall surface opposite to the thermal excitation enabled estimation of the potential thermal load reduction generated by the PCM. On average, the PCM part of the wall demonstrated over 27 % total reduction of the heat flow during 8-1/2 h, and over 50 % during the first two

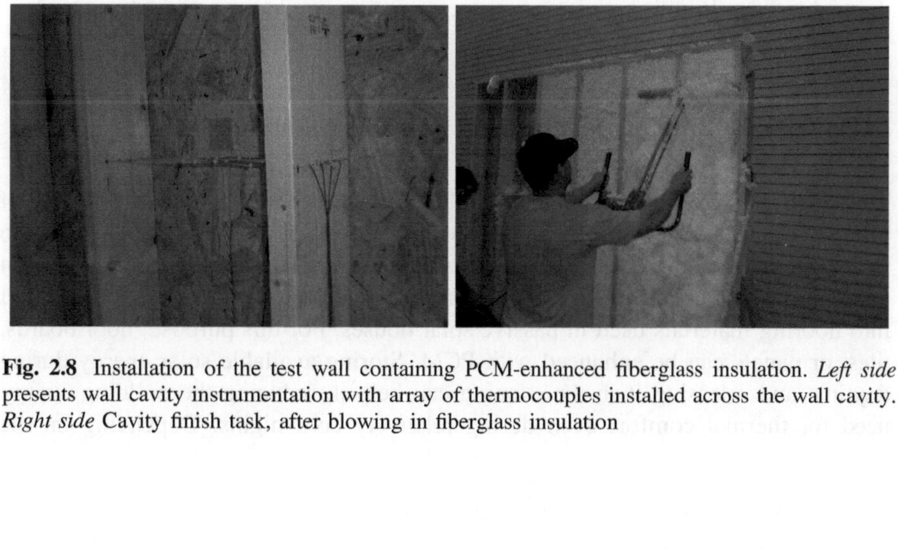

Fig. 2.8 Installation of the test wall containing PCM-enhanced fiberglass insulation. *Left side* presents wall cavity instrumentation with array of thermocouples installed across the wall cavity. *Right side* Cavity finish task, after blowing in fiberglass insulation

hours after the rapid heating process. It took about 15 h to fully charge the PCM-fiberglass wall. Recorded load reduction for the entire 15 h time period was close to 20 %. Thermal lag time for that heating process was between 7 and 8 h for the PCM part of the wall (Kośny et al. 2010a).

In addition to dynamic hot-box experiments, a series of full-scale field tests was performed on lightweight walls containing the PCM-enhanced cellulose insulation (Kośny 2008). A novel production method was developed and tested on a small-scale pilot line in the Advance Fiber Technologies production facility in Bucyrus, OH, USA. Then, cellulose/PCM material was produced in full-scale commercial plant conditions. Two field experiments were performed in test facilities located in Oak Ridge, TN and Charleston, SC, USA. Reduction of cooling loads averaged 42 % for PCM-insulated cavities at the southern oriented Oak Ridge test site. Heating loads during mixed season and winter were reduced by 16 % at the same location. A 5 % cooling load reduction was observed for the wall cavities insulated with PCM at the northwestern-oriented Charleston test facility. Peak-hour load reductions of 30 % were observed for PCM-insulated walls at the Charleston site during the summer months.

The incorporation of PCM-enhanced cellulose insulation for use as latent heat storage and for potential reduction of energy requirements in buildings was also analyzed by Evers et al. (2010). Two types of PCM, paraffin-based products and hydrated salt, were mixed into loose-fill cellulose insulation with 10 and 20 % by weight. A square 1.22 m × 1.22 m lightweight wood-frame wall was used in this experiment. The test walls containing PCM-enhanced cellulose insulation were heated and allowed to cool down in a dynamic wall simulator that replicated the sun's exposure to a wall of a building on a typical summer day. Peak heat fluxes, total "daily" heat flows, and surface and air temperatures were measured and recorded. Results show that the paraffin-based PCM-enhanced insulation reduced the average peak heat flux by up to 9.2 % and reduced the average total "daily" heat flow up to 1.2 %. At the same time, cellulose insulation blended with inorganic PCM did not work. Because of the hydroscopic behavior of non-encapsulated hydrated salt, the hydrated salt-based PCM-enhanced insulation did not provide any thermal storage benefit.

2.7 PCM-Enhanced Floors and Ceiling Systems

Traditional passive solar systems have relied for decades on sensible heat storage of internal walls and floors for energy savings. However, recent research has also investigated advantages of latent heat storage for additional energy savings in passive solar applications. This can be accomplished by the incorporation of PCM into flooring materials used in passive solar houses. For this purpose, floor boards, tiles, or panels can be enhanced with PCM. Storing available solar energy during daytime and releasing it in the evening can help in reducing the building energy need for thermal comfort even during relatively cold nights. Depending on the specific phase transition temperature, which needs to be around a comfortable room temperature between 21 and 23 °C, the PCM is expected to be effective during early and late winter as well as during spring and fall seasons. In these seasons, solar gains and peak ambient temperatures can be sufficient enough to melt PCM during daytime. During the evening or night, the ambient temperature can be low enough to discharge energy stored in PCM. An example of passive solar sunspace designed with PCM-enhanced floor components was the PCM floor system investigated by Stamatiadou et al. (2009) for the climate of Athens, Greece. It has been shown that the addition of paraffinic PCM (of specific heat capacity at 110 kJ/kg) to conventional floor tiles used for floor applications has a positive impact on the heating load, in cases of direct exposure to solar radiation. The PCM floor tiles in the sunspace, used for the numerical analysis, were compared to a typical marble floor tile. Two considered types of PCM-enhanced tiles contained 10 and 20 % of microencapsulated PCM by weight. The passive solar heating system yielded maximum solar savings of around 4 %, which can be directly translated to reduced building heating load requirements and can be expected when applying floor tiles containing 20 % of PCM by weight.

Research performed at the Colorado State University, USA, introduced a new flooring material that contained PCM (see Hittle 2002). An agglomerate floor tile containing 20 % by mass of encapsulated octadecane was manufactured. Peak melting transition temperature was determined to be 27.2 °C with a latent heat of 33.9 J/g of tile. It was determined that the maximum phase-change material content was 20 % by mass. Supplementary analysis of energy savings using this floor tile containing 20 % by mass of PCM was performed by A.S. Lee as a part of her PhD research at Colorado State University (Lee 2005). A series of dynamic whole building energy simulations was performed to enhance understanding of performance of PCM-enhanced floor tiles in residential applications (Barbour and Hittle 2006). The prototype tiles incorporating microencapsulated PCM have undergone preliminary testing indicating that the annual heating cost can be reduced by an average of 24 %. The prototype tiles were optimized to have similar physical properties as tiles without PCM.

In a similar research project performed by the University of Twente, the Netherlands, a series of field experiments was focused on PCM applications in concrete floors (Entrop et al. 2009, 2011). As shown in Fig. 2.9, the test containers used in this study had a similar configuration to what was originally used ten years earlier by the University of Dayton, OH, in the USA (see Kissock et al. 1998). The University of Twente study demonstrated that PCM-enhanced floors can significantly improve indoor thermal comfort in southern oriented rooms. It was recorded during the summer that in concrete floors containing PCM, maximum floor temperature was reduced by up to 16 °C ± 2 % comparing to similar assemblies without PCM. For the winter months, an increase of minimum floor temperatures up to 7 °C ± 3 % was reported.

In parallel work, Li et al. (2009) analyzed the usage of microencapsulated PCM in floor elements made out of a composite of high-density polyethylene and wood composites. Moreover, a research team from the Southeast University, Nanjing,

Fig. 2.9 Test containers used by University of Twente, the Netherlands, to test PCM floor tiles. Entrop et al. (2011)

China, investigated a new double-layer PCM floor system (Jin and Zhang 2011). The two layers of PCM had different phase transition temperatures. It was found that compared to the floor without PCM, the energy released by the floor with PCM in peak heating demand time period can be increased by around 40 %, depends on PCM temperature range. PCM of a total enthalpy about 150 kJ/kg was analyzed in this work.

From the design perspective, PCM cooling applications in ceilings are either passive (similar to PCM-enhanced wall gypsum boards or internal plasters) (see Fig. 2.10), or active, which are usually a part of more complex, dynamic air-conditioning systems using over-night precooling with often incorporated space conditioning components (i.e., hydronic systems, micro-tubing heat exchangers, and air channels) (see Fig. 2.11). Earlier research focused on conventional ceiling cooling systems demonstrated that they can offer significant advantages over traditional space air-conditioning technologies (see Kochendörfer 1996; Antonopoulos et al. 1997; Conroy and Mumma 2001). Briefly, considerable energy savings may be observed, primarily because thermal comfort is obtained with higher indoor temperatures, typically 28 °C. Additional energy savings are obtained because of the available large cooling surfaces, which enable higher cooling water temperatures. Comfort is improved due to minimization of air motion and surface temperature differences, elimination of noise associated with fan coils, and uniformity of indoor air temperature. Also, peak cooling loads may be reduced because of cool

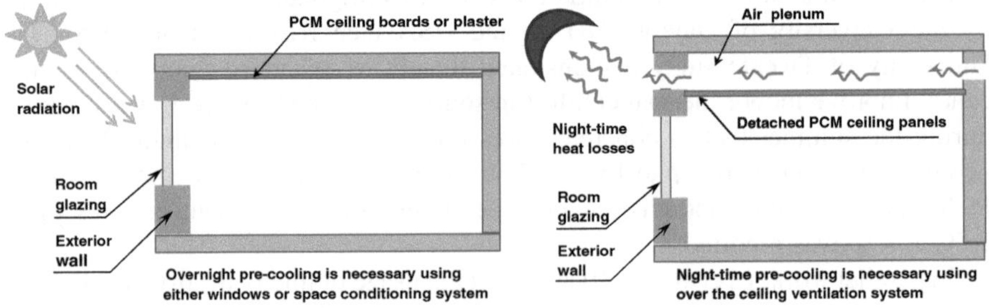

Fig. 2.10 Passive applications of PCM-enhanced ceiling systems

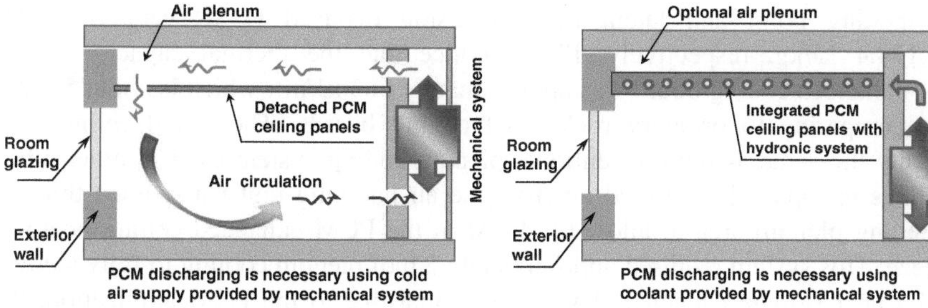

Fig. 2.11 Active applications of PCM-enhanced ceiling systems

storage within the ceiling and adjoining structural elements. Decades of testing and demonstrations worldwide have proven that adding PCM to the ceiling cooling systems can notable improve their energy performance and reduce a risk of moisture condensation. Due to a large variety of ceiling systems containing PCM described in engineering literature (see Fig. 2.12), this section is only limited to a small number of most typical, representative research projects, and applications.

One of the first passive solar heating systems that incorporate a PCM located in the ceiling was developed by Gutherz and Schiler (1991). Sun reflectors were used to direct the solar energy entering via the windows on to the PCM. The main advantage of the system was that it allowed a large area to be dedicated to heat storage without the need for large volumes of storage medium that would be required with sensible heat storage. It was shown that the use of such a system has the potential to recover 17–36 % of heat lost over the initial gains.

In an office building study performed by Kondo and Ibamoto (2006) from Kanagawa University, Japan, the PCM-enhanced ceiling rock wool boards were used, in order to reduce the cooling peak load of the air-conditioning system.

Fig. 2.12 Passively working delta cool ceiling panels containing inorganic PCM. Dörken GmbH & Co. KG, Germany

The melting point and latent heat of fusion of used PCM were 24.51 °C and 174.4 kJ/kg, respectively. Please notice that the thermal capacity of the PCM-enhanced ceiling board was approximately 663 kJ/m^2, which is about 5 times more comparing to ordinary rock wool ceiling boards. The tested ceiling was basically the same as the conventional plenum ceiling system used in other office buildings in Japan. The chilled air from the air-handling unit was passed through the ceiling plenum area to allow the PCM in the PCM-enhanced ceiling board to refreeze. This system required approximately 2 h of plenum cooling to fully recover PCM's heat storage capacity. It was found that due to the PCM application, the peak-hour cooling load was reduced by 14.8 %, compared to the conventional ceiling without PCM. There was a small increase in the night operation time for the space conditioning system. However, in Japan, off-peak electricity cost is significantly lower than in the middle of the day. As a result, the overall electricity cost was 8.4 % lower than that of using the rock wool ceiling boards. From these results, it can be concluded that the PCM ceiling system acts effectively to enable peak shaving and load shifting, bringing cost savings in electricity cost for space cooling.

The research team from the National Technical University of Athens, Greece, utilized the combined numerical–experimental methodology for the solution of the transient three-dimensional heat transfer problem for night cooling with use of the PCM-enhanced ceilings containing embedded piping (see Antonopoulos 1992; Antonopoulos and Democritou 1993; Antonopoulos et al. 1997; Antonopoulos and Tzivanidis 1997). This research demonstrated that thermal comfort requirements for the indoor space can be better satisfied during the day and night with an application of active PCM-enhanced ceiling systems, while compare to similar non-PCM concrete ceiling technologies. It was found that PCM applications yield significantly lower temperature fluctuations. For example, in case of the PCM-enhanced ceiling system, approximately 2.8 °C indoor temperature variations were observed, compared to about 4.4 °C in the case of the traditional concrete slab application. In addition, cycling of the space conditioning equipment was notable reduced during daily operation.

In the past, several research centers have considered an application of PCM-enhanced heat exchangers containing hydronic tubing or plastic micro-tubing systems. Very often, hydronic piping is not directly embedded in the concrete slab of the ceiling, as in many European ceiling applications. Instead, it is placed in a layer of the latent heat storage material located between the lower surface of the ceiling structure and the bottom, indoor-facing finish layer. Turnpenny et al. (2000) introduced a latent heat storage unit incorporating heat pipes embedded in phase-change material. The latent heat storage capacity in this dynamic system was restored during the night, to be available for cooling purposes in daytime. To allow sizing of an experimental mechanical system, Turnpenny also developed a one-dimensional mathematical heat transfer model of energy transfer from air to PCM.

At the same time, a large number of demonstration and commercial projects focused on novel chilled ceilings with integrated PCM have been studied by the Fraunhofer ISE, Germany (Schossig et al. 2003, 2005; Kalz et al. 2007; Haussmann et al. 2009). Different combinations of ceilings and mechanical systems were analyzed to demonstrate the main advantages of using PCM for space conditioning.

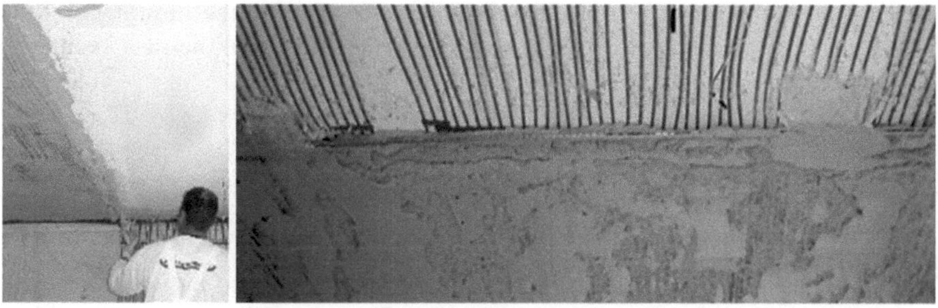

Fig. 2.13 Installation of the active-chilled ceiling system containing PCM-enhanced plaster and plastic micro-tubing. Fraunhofer ISE, Germany (http://www.ise.fraunhofer.de/en/areas-of-business-and-market-areas/energy-efficient-buildings/hot-and-cold-storage)

The aim of this research work was to develop new space conditioning technologies enabling reduction of the primary energy consumption associated with space conditioning. It was found that the best approach for these targets is adding PCM thermal mass to construction material which shifts building thermal loads and allows decoupling the cooling demand from the cold production. This allows moving cold production from day to night, where most space conditioning systems can work more efficient and cost-effective due to lower night air temperatures and lower energy prices. As shown in Fig. 2.13, a chilled ceiling with integrated PCM gypsum plaster and plastic micro-tubing was installed in five office rooms with an overall surface area of 100 m^2. The layer of the PCM plaster was about 3 cm thick with a density close to 950 kg/m^3. The overall system heat storage capacity in a 6 degree temperature range was nearly 162 Wh/m^2. For comparison, for the same ceiling without PCM, heat storage capacity would be just around 62 Wh/m^2. The experimental and analytical results show that the energy demand for cooling could be reduced by optimizing the control strategies. Shifting the energy demand from day to night by adding thermal mass to the building is a good solution to enhance the efficiency of most cold sources. Increasing the heat exchange area and the utilization capacity of a cold source is another important advantage achievable by using PCM in a chilled ceiling as dispersed storage. Measurements with chilled PCM ceilings also show that power output and response time are not negatively affected by the PCM, especially since the increased thermal mass only has an effect within the PCM operation temperature range. Outside the phase transition range, the PCM-enhanced ceilings react in a similar way as conventional non-PCM chilled ceilings.

According to Hasnain (1998), Khudhair and Farid (2004), and Farid et al. (2004), the ceiling structures containing PCM may store considerably greater amounts of heat than those stored in conventional concrete slabs. A properly designed application of PCM, under favorable conditions and the right temperature levels, may increase the storage capacity by more than five times or decrease storage space accordingly (Zalba et al. 2004). In hydronic ceiling applications, PCM solidification and further cooling by the cooling water take place during the night off-peak time at the reduced electricity price. During the day, melting and further heating of the PCM take place, following the daily thermal excitations generated within the conditioned indoor space.

2.8 PCM Used in Roofs and Attics

Currently, conventional roofs and attics are thermally designed based on a steady-state criteria with thermal resistance (*R*-value) being used by the majority of the building standards as a metric of thermal performance. At the same time, most of these building envelope components are subject to dynamic environmental conditions. Building loads are very often complex and dynamic combinations of convective, radiative, and conductive heat transfer mechanisms with frequently added air leakage, ventilation, or hygrothermal-related mass transport process. The steady-state principles used in design and code requirements for the roofs and attics and their dynamic operation are seriously conflicting. As a result, the dynamic operation combined with shortened thermal designs yields relatively low overall thermal efficiencies. The author believes that with proper thermal design incorporating heat storage, thermal breaks, and radiation control technologies, a significant part of these dynamic loads can be reduced, or even totally eliminated, thus improving overall building energy efficiency. Well-designed roof or attic thermal systems will not only improve overall thermal resistance, but also minimize transmission of dynamic thermal excitations (by shaving and shifting dynamic loads). Some advance thermal systems can also exhaust or absorb part of the dynamic environmental loads (depending on the climatic season).

Dynamic roofs containing PCM have been investigated in the USA for several decades. In 1994, blends of lightweight aggregates and salt hydrates were analyzed and tested by Oak Ridge National Laboratory, USA, as dynamic attic floor insulation (Petrie et al. 1997). This novel attic configuration contained inorganic PCM which was sandwiched between two layers of conventional insulation. The PCM heat storage that was used consisted of hydrated calcium chloride dispersed in perlite and contained in test cells. One cell had a PCM/perlite ratio of 2:1 by weight; while the other had a 6:1 mixture. Extruded polystyrene foam (XPS) was the insulation below and above the PCM. Heat-flux transducers on the top and bottom of each cell as well as arrays of thermocouples from the top to the bottom of each cell allowed to follow the progression of freezing and melting in the PCM as the project team subjected the cells to both steady and diurnally varying simulated outside temperatures. The above testing of the perlite granulates impregnated with PCM, demonstrated excellent capability for reduction of temperature fluctuations in comparison with traditional fiberglass insulation. For example, in the 2:1 PCM system, 22 % lower thermal loads were observed comparing to the conventional fiberglass system. Its peak heat flux was reduced by 42 % as well. In addition, a four-hour delay time was observed. The 6:1 PCM insulation system showed 32 %

lower total cooling loads. The thermal load profile was essentially flat at a level that was 21 % of the peak fiberglass heat flux (79 % cooling load reduction).

Very often, dynamic roofing and attic systems utilizing PCM were designed based on earlier experiences gained during similar projects focused on wall assemblies. Following findings from earlier wall performance investigations, in 1997, Ismail and Castro (1997) presented the results of a theoretical and experimental study of the PCM-enhanced roofing insulation system under real operational conditions to achieve passive thermal comfort in a test building. The experimental setup consists of a small room with movable roof and side walls. The roof was constructed in the traditional way using attic floor insulation with added PCM. Thermocouples were distributed across the cross section of the roof. Another roof, identical but without the PCM, was also used as a baseline. It was found that for Brazilian climate conditions PCM-enhanced envelopes can help in keeping the interior of the building within desired thermal comfort zone.

In 2006, Kissock and Limas investigated paraffinic PCM that can be added to the steel commercial roofs, to reduce the peak diurnal cooling and heating loads (see Kissock and Limas 2006). This work was a combined numerical–experimental study where the numerical algorithm was validated against the experimental data. The PCM studied was the paraffin octadecane, with an average melting temperature of 25.6 °C. The metal roof that was analyzed had two 1-inch-thick layers of polyisocyanurate foam. The bottom layer of the foam was enhanced with the paraffinic PCM. For the climate of Dayton, OH, USA, thermal loads through the PCM-enhanced polyisocyanurate board were simulated using an explicit finite-difference procedure while the indoor air temperature was held constant. When compared to a conventional roof (no PCM), cooling load savings were close to 14 %.

A prototype residential roof using a cool-roof surface, natural subventing, and PCM heat sink was designed and field tested by the author (Kośny et al. 2007; Miller and Kośny 2008). A multilayer configuration of PCM-enhanced polyurethane foams, PCM-impregnated fabrics, and highly reflective aluminum foil was used. As shown in Fig. 2.14, the PCM roof also used 10-cm air channels to exhaust excess heat during peak irradiance (subventing). Two reflective membranes were placed above the roof sheathing boards with the aluminized surfaces facing each other across the 10-cm air gap. Loading of PCM was about 0.39 kg/m^2 of the surface area. Two types of PCM were used with two melting temperatures, which were around 26 and 32 °C. The total storage capacity of the PCM heat sink was about 54 kJ/m^2. The field test results show that when comparing the heat flow penetrating the conventional shingle roof to the similar metal roof assembly using cool-roof pigments, reflective insulation, and subventing air channels, the summertime peak heat flow crossing the roof deck was reduced by about 70 %. Installation of the PCM heat sink on the metal roof generated an additional 20 % reduction in the peak-hour heat flow, bringing the total load reduction to over 90 % (see Fig. 2.15).

A similar configuration of a roof containing metal roof panels with photovoltaic (PV) laminates and PCM heat sink was field tested during 2009–2010 in East Tennessee, USA, climatic conditions (Kośny et al. 2012b). It was a new solar roof

Fig. 2.14 Construction details of a roof assembly containing, reflective insulation, and subventing air channels, followed with two PCM insulation systems produced by Outlast, USA (PCM-enhanced foam and PCM-impregnated fabric, see *left photograph*)

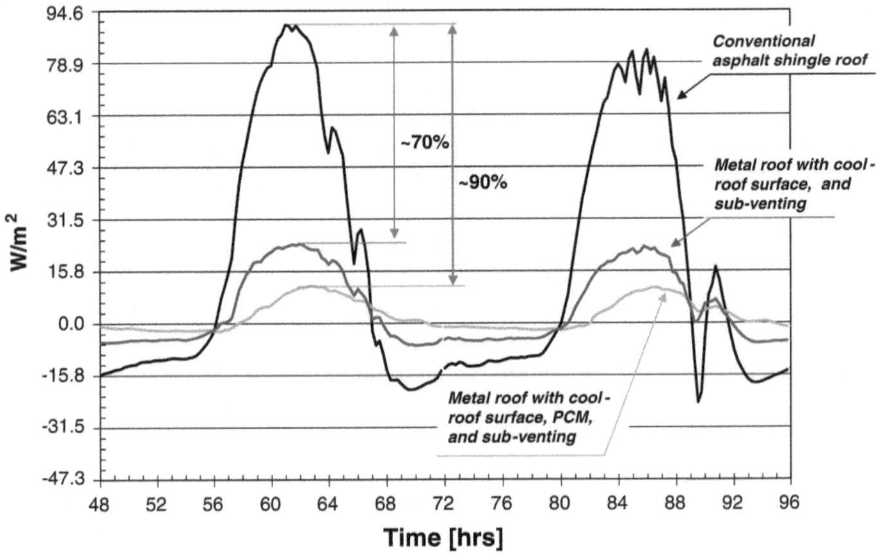

Fig. 2.15 Heat transfer penetrating the roofs of a direct nailed asphalt shingle, standing seam metal roof (containing, cool-roof pigments, reflective insulation, and subventing air channels), and a similar metal roof containing PCM

technology utilizing metal roof panels with integrated amorphous silicon PV laminates, a ventilated air cavity, dense fiberglass over-the-deck insulation with reflective surface, and arrays of bio-based PCM cells (see Fig. 2.16). The thermal performance of the experimental roof was compared to a control attic with a conventional asphalt shingle roof. The test data demonstrated that, during winter, without the phase-change contribution, the PV-PCM attic had a 30 % reduction in roof-generated heating loads compared to a conventional shingle attic. Conversely,

Fig. 2.16 Construction of the experimental PCM-PV roof—Oak Ridge, TN, USA

during the cooling season, the attic-generated cooling loads from the PV-PCM attic were about 55 % lower than the shingle attic. In addition, about 90 % reductions in peak daytime roof heat fluxes were observed with the PV-PCM roof.

Heat transmission across the concrete roof structure containing PCM was also analyzed for Indian climatic conditions (Sathyamangalam, Tamilnadu State) by Ravikumar and Srinivasan (2012, 2014). A transient numerical procedure was developed to analyze thermal performance of the PCM roof. A numerical model was validated against the available experimental data. The analysis shows that the melting and solidification of PCM, over the day–night cycle and over summer–winter cycle during the year, compensate for all the external climatic excitations and keep the roof bottom surface temperature almost constant and close to room air temperature. On a yearly basis, there was about a 56 % reduction in heat transmission observed into the room with a PCM roof in comparison with the conventional Indian concrete roof design.

A development of microencapsulated PCM which can be installed as a part of the attic insulation system was a critical step in development of modern building envelopes. Subsequent introduction of the insulation blends with microencapsulated PCM was another noteworthy technology improvement during the first decade of twenty-first century. PCM-enhanced cellulose was one of the first dynamic insulation products successfully developed for building applications (Kośny et al. 2007). In about the same time, PCM blended with blown fiberglass (Kośny et al. 2010a) and plastic foams (Mehling and Cabeza 2008; Kośny 2008) were introduced. The major advantage of PCM-enhanced insulations is their capability of significantly lessening and shifting peak-hour thermal loads generated by building envelopes.

Dynamic hot-box experiments were performed on a residential attic module containing PCM-enhanced attic floor insulation (see Fig. 2.17). The attic module was tested under periodic temperature changes in the Large Scale Climate Simulator (LSCS) at the Oak Ridge National Laboratory, USA (see Kośny 2008). Two concentrations of microencapsulated PCM were tested (5 and 20 % by weight). The main focus of the attic tests included discharging time of the PCM-enhanced

Fig. 2.17 Experimental attic module containing microencapsulated PCM blended with cellulose insulation, ORNL, USA, testing facility

insulation (time necessary to refreeze PCM). PCM charging is not a problem in attics because of the intensive fluctuations of the attic air temperature during sunny days (a rapid increase in temperature caused by the sun). However, the attic cooling process is significantly slower. In a well-designed PCM application, 100 % of the PCM material should be able to fully discharge its energy before daytime operation the next day. During the dynamic LSCS tests, the attic test module was subjected to periodic changes of temperature. The following temperature schedule was used: 18 °C for about 16 h, rapid temperature ramp to 50 °C and kept for about 4 h, followed by natural cooling back to 18 °C. One of the noteworthy findings was that only layers of insulation located higher than 10 cm from the attic floor were involved in the phase-change process. It took about 6–8 h to fully discharge the energy stored in these layers. It is interesting that even in attic insulation containing only 5 % PCM by weight, notable evidences of phase change were observed (Kośny 2008).

During the summer of 2008, a full-scale experimental attic was constructed and instrumented in order to field test blown fiber glass insulation combined with microencapsulated PCM in the ORNL test facility. Bio-based microencapsulated PCM, manufactured by Microtek Labs., USA, was utilized in this experiment. The main goal of this experiment was to investigate at what level and how often PCM was going through the phase-change process. As shown on Fig. 2.18, a full-scale residential attic was filled with about 26-cm blown fiber glass insulation with the approximate density of 29 kg/m^3. Next, on top of this insulation, four 1.3-cm-thick layers of the PCM-adhesive blend were installed with 1.3-cm-thick layers of blown fiber glass installed in-between. The total thickness of added PCM-fiberglass multilayer sandwich was approximately 10 cm (Kośny et al. 2010a). Results of DSC testing and numerical analysis indicated that in order to make PCM fully melt, attic air temperature should be—during the peak of the day—higher than 32 °C. It was also found that during the night, attic air temperature should be below 20 °C to allow PCM to refreeze. Melting temperature of microencapsulated PCM used was at about at 29 °C. A temperature difference between melting and freezing peaks

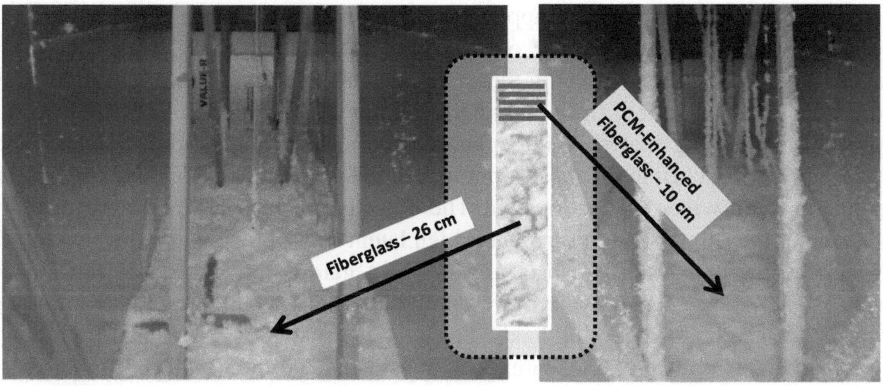

Fig. 2.18 Field test attic module containing PCM-enhanced fiberglass insulation

(PCM subcooling effect) was about 6 °C with freezing temperature close to 23 °C. The phase-change enthalpy was about 170 kJ/kg. Detailed temperature profiles across the roof, attic space, and within the attic insulation were collected for two summer seasons of 2008 and 2009. It was found that for East Tennessee, USA, climate, the second week of May was a beginning week for PCM to have at least two full phase changes a week. This process ended during the first week of October. In May and September, calculated number of active days for PCM was close to 50 % of total number of days. During the months of June and August, during about 75 % of days, full phase transition processes took place. In July, due to increased night temperature, a number of days when PCM was fully active went down to below 50 %. Theoretical analysis showed that, in order to improve PCM effectiveness during July, it is possible to use PCM of higher melting point (Kośny 2008; Kośny and Kossecka 2013). However in that case, a number of active days can be reduced for May and September. This work demonstrated that numerical analysis is usually necessary to optimize the PCM roof design for a specific climate.

A series of thermal simulations of the 25-cm-thick attic floor insulation showed significant reductions to the total thermal loads, generated by the PCM attic. In general, PCM-enhanced attic insulation assemblies may reduce peak-hour loads by 25–44 % when compared to non-PCM systems of equivalent *R*-value—subject to the PCM load and PCM location (see Kośny et al. 2014).

2.9 PCM-Enhanced Windows and Window Attachment Products

As presented in earlier sections, an application of PCM is an interesting solution for improving opaque building envelopes. However, PCM can be also an attractive alternative for fenestration products as they generate a significant part of building thermal loads. In conventional applications, thermal performance of fenestration is most-often improved by the use of absorbing gases filling the gap between glass sheets, or by an application of thermally insulated glass windows. When only translucency is required, the other thermal improvement possibilities may incorporate novel filling materials into glass panes, such as silica aerogel or a semi-transparent PCM. Both of these options require serious consideration of optical properties and window functionality. The objective of using PCM in the window glazing or the shutters is to utilize their high latent heat of fusion to reduce window-generated thermal loads by absorbing the heat gain before it reaches the indoor space. The phase-change cycle may stabilize the indoor building temperature and reduce the heating and cooling loads. From the thermal perspective, PCM windows work like the optically transparent or translucent Trombe walls. They usually consist of a single- or multilayer glazing panel made of conventional glass, integrated with a layer of a transparent or translucent PCM product—Fig. 2.19a. In dynamic windows, PCM changes its phase from solid state to liquid when heated, thus absorbing available solar energy in the endothermic process. At the same time, PCM's translucency is usually changing as well. When the ambient temperature drops during the night, the melted PCM of required optical properties solidifies into a solid-state material while giving off the earlier absorbed heat in the exothermic process. A suitable phase-change temperature range, depending on climatic conditions and desired comfort temperatures, as well as the ability to absorb and release large amounts of heat energy are needed for proper work of PCM windows. It is essential that both phase-change transition ability, as well as PCM optical characteristics do not deteriorate in time, which is a crucial condition when selecting PCM for fenestration application.

Translucent Trombe walls are constructed in a way that they can transmit light and illuminate the building interior. In conventional solar fenestration systems, the

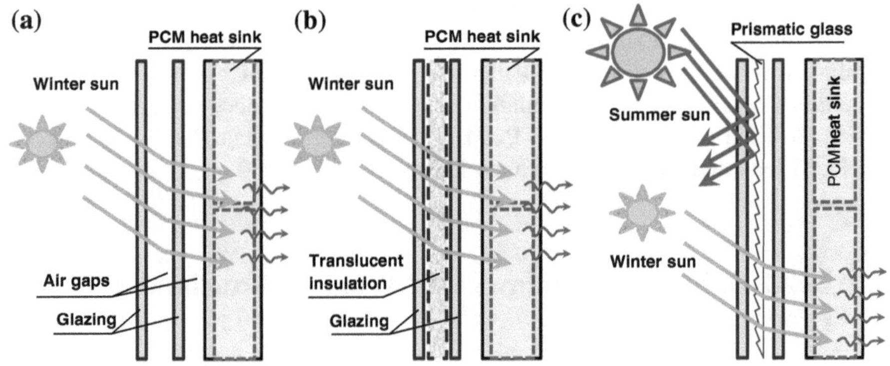

Fig. 2.19 Different configuration options of semi-transparent PCM solar fenestration: **a** Semi-transparent Trombe Wall containing PCM heat sink. **b** Translucent Trombe Wall with PCM heat sink and translucent insulation. **c** Solar fenestration system using PCM heat sink and selective prismatic glass

temperature gradient within the assembly results in winter time significant heat losses from the heated interior to the cold exterior. As shown in Fig. 2.19b, heat losses through the glazing can be reduced using a transparent insulation (vacuum) or translucent insulating fillers (i.e., aerogels) placed between window panes. The transparent or translucent thermal insulation transmits solar radiation, which is then absorbed by a layer of PCM, which is usually partly transparent when melted. In order to further improve thermal performance, translucent/transparent storage walls may use selective glass. During the summer, the outer prismatic glass may be used to reflect the high-angle rays of the sun—Fig. 2.19c.

Various paraffins are typical examples of PCM in building products, but a low thermal conductivity (Farid et al. 2004), significant volume change during phase transition (Hasnain 1998) and poor optical properties limit their fenestration applications. Askew (1978) used a collector panel made of a thin slab of paraffin placed behind the double glazing of a building and found that thermal efficiency was comparable to a conventional flat plate solar thermal collector. In a following research performed by Nayak (1987), a Transwall solar system was used as a transparent modular wall that provides both heating and illumination of the dwelling space.

Further design possibilities emerge from the variable translucency of some PCM-based products. They provide dynamic thermal characteristics and a source of natural lighting to the building. The energy state of these assemblies is visualized as transparent or translucent when PCM is melted and milky when PCM is frozen (Mehling and Cabeza 2008). Manz et al. (1997) studied a solar facade composed of transparent insulation material and translucent PCM used both for solar heat storage and daylighting. The PCM was hexahydrated calcium chloride ($CaCl_2 \cdot 6H_2O$) with 5 % of additives. The numerical model was developed for analysis of the radiative heat transfer inside the PCM-enhanced solar window. Experimental data were gathered over a period of 5 months. The authors concluded that overall system performance could be improved by changing of the PCM melting temperature from 26.5 to about 21 °C. Another semi-transparent solar window system containing PCM has been introduced by the INGLAS company form Friedrichshafen, Germany.[4] This technology combines design principles of passive solar walls with fenestration function and a semi-transparent heat reservoir. As a result, this solar window efficiently transfers solar light and absorbs the heat developed in the process. The absorbed heat is stored by the heat sink utilizing organic PCM. According to manufacture, large amounts of solar energy can be stored during daytime and released into the building at night, when PCM cools down and solidifies.

In similar research performed in Germany, an application of semi-transparent PCM components from Dorken has been jointly investigated by the glass company Glaswerke Arnold and research institute ZAE Bayern. A complete system is made of two glass sheets on the outside and a macro-encapsulated PCM on the inside (Mehling and Cabeza 2008). In this technology, a variable transparency is utilized

[4]http://www.inglas.eu/glass/company.html.

Fig. 2.20 PCM-enhanced window component with variable optical properties, Dorken, Germany (http://www.cosella-dorken.com/bvf-ca-en/projects/pcm/kempen.php)

for an effective diffuse illumination of the interior space. PCM optical properties are changing to some degree between the solid and liquid states (see Fig. 2.20).

Parametric analysis and performance comparisons with other advanced fenestration products are important factors, in designing of low energy buildings, which needs to be undertaken with advanced thermal modeling on a system scale, with use of whole building energy simulations. As an example, Ismail et al. (2008) numerically analyzed performance of the PCM-filled window and the window filled with the infrared absorbing gas. The use of absorbing gases and filling the gaps between glazing sheets appears to be an alternative solution for thermally insulated glass windows. In this work, a thermal efficiency comparison between glass window, filled with an absorbing gas and the other one containing a PCM heat sink was analyzed for a hot cooling-dominated climate.

A concept of solar windows using PCM-enhanced window attachment products is an attractive alternative for described above fenestration systems with PCM heat sinks. However, this technology is still awaiting successful implementation. Additionally, many semi-transparent and translucent solar walls have several disadvantages. One is reduced optical transparency. The other one is limited thermal storage capacity. These systems need to have sufficient thermal mass resulting in significant thickness coming from added PCM components. This is where PCM-enhanced window attachments (blinds, shutters, or curtains) offer a unique solution: a thin, movable layer of PCM which can be easily used "as needed" without compromising window optical properties (see Alawadhi 2012). A fenestration system using movable PCM heat sink needs to be installed on the south-side windows. When PCM window attachments are in use (during the sunny day), the window heat sink is heated by incoming solar radiation, melting the PCM. At night that heat is utilized to warm the interior space.

The window shutters and blinds are often made of extruded plastic or foam-filled aluminum profiles. The same extruded profiles can be used as PCM containers. PCM can fill existing cavities or replace foam as a filling material there. PCM-enhanced window attachments can be installed on the interior side of south-facing windows. In cooling scenario, the PCM blinds are used in the same way as conventional blinds, yet absorb excess solar radiation. The PCM is regenerated at night through radiation to the exterior or/and by cool air (from outside ventilation flaps or tilted-open windows). According to Alawadhi, the numerical result indicates that heat gain through the windows can be reduced as high as 20–30 % depends of climatic conditions, window type, and PCM configuration (Alawadhi 2012). A similar window attachment systems using PCM have been installed in four offices[5] (two in Kassel City Hall and two in the EnBW building in Karlsruhe, Germany). The solar protection system consists of vertical blinds; there is an approximately 1-cm-thick layer of PCM in each of the slats.

In another concept, window shutters containing PCM are placed outside of window areas. During daytime, they remain opened to the outside. However, the shutter surface is still exposed to solar radiation and heat is absorbed causing PCM to melt. At night, the shutters are closed, insulating the windows and heat from the PCM radiates into the rooms. Buddhi et al. (2003) studied the thermal performance of a test cell (1 m × 1 m × 1 m) with and without PCM. Lauric acid with melting point of 49 °C was used as a PCM latent heat storage material. It was found that the heat storing capacity of the cell due to the presence of PCM increases up to 4 °C for 4–5 h, which was used during nighttime. In another research project focused on PCM-enhanced window attachments, Ismail and Henriquez (2001) numerically and experimentally studied thermally effective windows with moving PCM curtains.

References

Abhat A (1983) Low temperature latent heat thermal energy storage. Solar Energy J 30(4):313

Alawadhi EM (2012) Using phase change materials in window shutter to reduce the solar heat gain. Energy Build 47:421–429

Antonopoulos KA (1992) Analytical and numerical heat transfer in cooling panels. Int J Heat Mass Transf 35(11):2777–2782

Antonopoulos KA, Democritou F (1993) Periodic steady-state heat transfer in cooling panels. Int J Heat Fluid Flow 14(1):94–100

Antonopoulos KA, Tzivanidis C (1997) Numerical solution of unsteady three-dimensional heat transfer during space cooling using ceiling-embedded piping. Energy Int J 22:59–67

Antonopoulos A, Vrachopoulos M, Tzivanidis C (1997) Experimental and theoretical studies of space cooling using ceiling embedded piping. Appl Therm Eng 17(4):351–361

Askew GL (1978) Solar heating utilization a Paraffin's phase change material. In: Proceedings of the 2nd national passive solar conference. University of Pennsylvania, Philadelphia, 16–18 Mar 1978

Athienitis AK, Liu C, Hawes D, Banu D, Feldman D (1997) Investigation of the thermal performance of a passive solar test-room with wall latent heat storage. Build Environ 32 (5):405–410

Balcomb JD (1979) Heat Storage effectiveness of concrete masonry units. Los Alamos Scientific Laboratory report no. LA–UR-82-966

Balcomb JD, McFarland RD (1978) Passive testing in Los Alamos. Los Alamos National Lab. Report LA-UR-7B-1178

Balcomb JD, Kosiewicz CE, Lazarus GS, McFarland RD, Wray WO (1983) Passive solar design handbook, vol 3. Los Alamos National Lab. ISBN 0-89553-106-2

[5]http://pcm-demo.info/objective.htm.

Barbour JP, Hittle DC (2006) Modeling phase change materials with conduction transfer functions for passive solar applications. J Solar Energy Eng-Trans ASME 128(1):58–68

Benard C, Body Y, Zanoli A (1985) Experimental comparison of latent and sensible heat thermal walls. Sol Energy 34(6):475–487

Benson DK, Webb JD, Burrows RW, McFadden JDO, Christensen C (1985) Materials research for passive solar systems: solid state phase-change materials. Solar Energy Research Institute, US Department of Energy Contract EG-77-C-01-4042

Bentz DP, Turpin R (2007) Potential applications of phase change materials in concrete technology. Cem Concr Compos 29:527–532

Bourdeau L (1982) Utilisation d'un Materiau a` Changement de Phase Dans un Mur Trombe sans Thermocirculation. Revue Phys Appl 17:633–642

Bourdeau L, Jaffrin A (1979) Actual performance of a latent heat diode wall. In: Proceedings of Izmir international symposium II on solar energy fundamentals and applications, Izmir

Bourdeau L, Jaffrin A, Moisan A (1980) Captage et Stockage d'E´nergie Solaire dans l'Habitat par le Moyen de Mur Diode a` Chaleur Latente. Revue Phys Appl 15:559–568

Bromley AKR, McKay EM (1994) Incorporating phase change materials into the building fabric. In: Proceedings of CIBSE conference 1994, p 101

Buddhi D, Sharma SD (1999) Measurements of transmittance of solar radiation through stearic acid: latent heat storage material. Energy Convers Manag 40:1979–1984

Buddhi D, Mishra HS, Sharma A (2003) Thermal performance studies of a test cell having a PCM window in south direction. Presentation at the IEA, Annex 17 Meeting, Indore, India

Cabeza LF, Castellon C, Nogues M, Medrano M, Leppers R, Zubillaga O (2007) Use of microencapsulated PCM in concrete walls for energy savings. Energy Build 39:113–119

Castellón C, Medrano M, Roca J, Nogués A, Castell M, Cabeza LF (2007) Use of microencapsulated phase change materials in building applications. In: Proceedings of X conference—thermal performance of the exterior envelopes of buildings, Clearwater

Chandra S, Kumar S, Kaushik S, Kaul S (1985) Thermal performance of a non-air-conditioned building with PCM thermal storage wall. Energy Convers manag 25:15–20

Childs K, Stovall T (2012) Potential energy savings due to phase change material in a building wall assembly: an examination of two climates. Report—ORNL/TM-2012/6 March 2012, Oak Ridge National Laboratory, USA

Collier RK, Grimmer DP (1979) The experimental evaluation of phase change material building walls using small passive test boxes. In: Proceedings of 3rd national passive solar conference. San Jose, Jan 11–13 1979

Conroy CL, Mumma SA (2001) Ceiling radiant cooling panels as a viable distributed parallel sensible cooling technology integrated with dedicated outdoor air systems. ASHRAE Trans 107(1):578–585

Darkwa J, Zhou T (2011) Enhanced laminated composite phase change material for energy storage. Energy Convers Manag 52:810–815 (Department of Energy project report. Award No. DE-FC26-00NT40999)

Drake JB (1978) PCMSOL—simulation code for a passive solar structure incorporating phase change materials. U.S. Department of Energy, Oak Ridge National Laboratory. Report ORNL —6359

Entrop AG, Brouwers HJH, Reinders AHME, Muthing F (2009) Experimental research on the use of phase change materials to come to passive solar energy concepts. In: 3rd CIB international conferences on smart and sustainable built environments, Delft, 15–19 June 2009

Entrop AG, Brouwers HJH, Reinders AHME (2011) Experimental research on the use of micro-encapsulated phase change materials to store solar energy in concrete floors and to save energy in Dutch houses. Sol Energy 85:1007–1020. ISSN 0038-092X

Evers AC, Medina MA, Fang Y (2010) Evaluation of the thermal performance of frame walls enhanced with paraffin and hydrated salt phase change materials using a dynamic wall simulator. Build Environ 45(8):1762–1768

Farid MM, Khudhair AM, Razack SAK, Al-Hallaj S (2004) A review on phase change energy storage: materials and applications. Energy Convers Manag 45:1597–1615

Feldman D, Shapiro MM (1989) Fatty acids and their mixtures as phase-change materials for thermal energy storage. Sol Energy Mater 18:201–216

Feldman D, Banu D, Haves D, Ghanbari E (1991) Obtaining an energy storing building material by direct incorporation of an organic phase change material in gypsum wallboard. Sol Energy Mater 22:231–242

Feustel HE (1995) Simplified numerical description of latent storage characteristics for phase change wallboard. In: Indoor Environmental Program, Energy and Environment Division. Lawrence Berkeley Laboratory, University of California, Berkeley

Feustel HE, Stetiu C (1997) Thermal performance of phase change wallboard for residential cooling application. Indoor environment program energy and environment division Lawrence Berkeley national laboratory, University of California Berkeley, CA 94720, USA. Report—LBL-38320 UC-1600

Frysinger GA, Sliwkowski J (1987) Phase change material storage assisted heating systems. University of Delaware, USA. PH-87-5, No. 4, p 516

Ghoneim AA, Klein SA (1991) Phase change materials analysis. Sol Energy 47(3):239

Ghoneim AA, Klein SA, Duffie JA (1991) Analysis of collector-storage building walls using phase change materials. Sol Energy 47(1):237–242

Gutherz JM, Schiler ME (1991) A passive solar heating system for the perimeter zone of office buildings. Energy Sources 13:39–54

Hadjieva M, Stoykov R, Filipova T (2000) Composite salt-hydrate concrete system for building energy storage. Renew Energy 19:11–15

Hasnain SM (1998) Review on sustainable thermal energy storage technologies, part 1: heat storage materials and techniques. Energy Convers Manag 397(11):1127–1138

Haussmann T, Schossig P, Grossmann L (2009) Experiences with LowEx PCM chilled ceilings in demonstration buildings. In: Proceedings of 8th IIR conference on phase change materials and slurries for refrigeration and air conditioning: PCM 2009, Karlsruhe, 3–5 June 2009

Hawes DW, Feldman D (1993) Latent heat storage in building materials. Energy Build 20:77–86

Hawes DW, Banu D, Feldman D (1990) Latent heat storage in concrete. Sol Energy Mater 21:61–80

Hawes DW, Banu D, Feldman D (1992) The stability of phase change materials in concrete. Sol Energy Mater Sol Cells 27(2):103–118

Hawlader MNA, Uddin MS, Khin MM (2003) Microencapsulated PCM thermal-energy storage system. Appl Energy 74:195–202

Hittle DC (2002) Phase change materials in floor tiles for thermal energy storage. U.S. Dept. of Energy project report. Award no. DE-FC26-00NT40999

Ismail KAR, Castro JNC (1997) PCM thermal insulation in buildings. Int J Energy Res 21:1281–1296

Ismail KAR, Henríquez JR (2001) Thermally effective windows with moving phase change material curtains. Appl Therm Eng 21(18):1909–1923

Ismail KAR, Salinas CT, Henriquez JR (2008) Comparison between PCM filled glass windows and absorbing gas filled windows. Energy Build 40(5):710–719

Jie J, Hua Y, Wei H, Gang P, Jianping L, Bin J (2007) Modeling of a novel Trombe wall with PV cells. Build Environ 42(3):1544–1552

Jin X, Zhang X (2011) Thermal analysis of a double layer phase change material floor. Appl Therm Eng 31(2011):1576–1581

Kalz D, Pfafferott J, Schossig P, Herkel S (2007) Thermoaktive Bauteilsysteme mit Integrierten Phasenwechselmaterialine – eine Simulationsstudie. Bauphysik 29(1):27–32

Khudhair AM, Farid M (2004) A review on energy conservation in building applications with thermal storage by latent heat using phase change materials. Energy Convers Manag 45 (2):263–275

Kim S, Drzal L (2009) High latent heat storage and high thermal conductive phase change materials using exfoliated graphite nanoplatelets. Sol Energy Mater Sol Cells 93:136–142

Kissock KJ, Limas S (2006) Diurnal load reduction through phase-change building components. ASHRAE Trans 112(1):509–517

Kissock KJ, Hannig JM, Thomas IW (1998) Testing and simulation of phase change wallboard for thermal storage in buildings. In: Morehouse JM, Hogan RE (eds) Proceedings of 1998 international solar energy conference, Albuquerque, American Society of Mechanical Engineers, New York

Knowles T (1983) Proportioning composites for efficient thermal storage walls. Sol Energy 31 (3):319–326

Kochendörfer C (1996) Standard testing of cooling panels and their use in system planning. ASHRAE Trans 102(1):651–658

Kondo T, Ibamoto T (2006) Research on thermal storage using rock wool phase-change material ceiling board. In: ASHRAE winter conference, Chicago ASHRAE Transactions, vol 112, p 1

Kondo T, Ibamoto T, Yuuji T (2000) Research on thermal storage of PCM wall-board. In: 5th Workshop of the IEA ECES IA Annex 10, Tsu, Japan

Kośny J (2008) 2006/07 field testing of cellulose fiber insulation enhanced with phase change material. Oak Ridge National Laboratory report. ORNL/TM-2007/186, Oak Ridge

Kośny J, Starakiewicz A (1987) Ocena Efektow Stosowania w Budownictwie Niewentylowanych Przegrod Trombe'go. PZITB Lodz, Konferencja Naukowo – Techniczna, Fizyka Budowli w Teorii i Praktyce, Lodz, 11–13 Listopad 1987

Kośny J, Kossecka E (2013) Understanding a potential for application of phase-change materials (PCMs) in building envelopes. ASHRAE Trans 119(1):1 (Part 1.DA-13-001)

Kośny J, Yarbrough DW, Wilkes KE, Leuthold D, Syed AM (2006) PCM-enhanced cellulose insulation thermal mass in lightweight natural fibers. In: Proceedings of ECOSTOCK—IEA international conference on thermal energy storage, Stockton College of New Jersey, 31 May–02 June 2006

Kośny J, Yarbrough DW, Miller WA, Childs P, Syed AM (2007) Thermal performance of PCM-enhanced building envelope systems. In: Proceedings of X conference. Thermal performance of the exterior envelopes of buildings, Clearwater

Kośny J, Fallahi A, Shukla N, Kossecka E, Ahbari R (2014) Thermal load mitigation and passive cooling in residential attics containing PCM-enhanced insulations. Sol Energy 108:164–177

Kośny J, Yarbrough DW, Riazzi T, Leuthold D, Smith JB, Bianchi M (2009) Development and testing of ignition resistant microencapsulated phase change material. In: Proceedings of effstock 2009. The 11th international conference on thermal energy storage, Stockholm

Kośny J, Yarbrough D, Kossecka E, Miller W, Bianchi M, Smith J, Lee E (2010a) Theoretical and experimental thermal performance analysis of building shell components containing blown fiber glass insulation enhanced with phase change material (PCM). In: Proceedings of DOE, ASHRAE, ORNL conference—thermal envelopes XI. Thermal performance of the exterior envelopes of buildings, Clearwater

Kośny J, Miller W, Zaltash A (2010b) Dynamic thermally-disconnected building envelopes—a new paradigm for walls and roofs in low energy buildings. In: Proceedings of DOE, ASHRAE, ORNL conference—thermal envelopes XI. Thermal performance of the exterior envelopes of buildings, Clearwater

Kośny J, Kossecka E, Brzezinski B, Tleoubaev A, Yarbrough D (2012a) Dynamic thermal performance analysis of fiber insulations containing bio-based phase change materials (PCMs). Energy Build 52:122–131

Kośny J, Biswas K, Miller WA, Kriner S (2012b) Field thermal performance of naturally ventilated solar roof with PCM heat sink. Solar Energy 86(9):2504–2514

Kuznik F, Virgone J (2009) Experimental investigation of wallboard containing phase change material: data for validation of numerical modeling. Energy Build 41(5):561–570

Lai CM, Chiang CM (2006) How phase change materials affect thermal performance: hollow bricks. Build Res Inf 34:118–130

Lane GA (1983) Latent heat storage: background and scientific principles, vol 1. CRC Press, Boca Raton p 15

Lee AS (2005) Phase change material in floor tiles for thermal energy storage. PhD thesis, Colorado State University, 2005; publication number: AAI3173039. ISBN: 9780542098499 (Source: Dissertation Abstracts International, vol 66-04, Section: B, page: 2274; 145 p)

Lee T, Hawes DW, Banu D, Feldman D (2000) Control aspects of latent heat storage and recovery in concrete. Solar Energy Mater Solar Cells 62:217–237

Li J, Xue P, Ding W, Han J, Sun G (2009) Micro-encapsulated paraffin/high-density polyethylene/wood flour composite as form stable phase change material for thermal energy storage. Sol Energy Mater Sol Cells 93:1761–1767

Mancini NA (1980) Use of paraffins for thermal storage. In: Den Ouden C (ed) Thermal storage of solar energy. Martinus Nijhoff Publishers, Leiden, pp 99–109

Manz H, Egolf PW, Suter P, Goetzberger A (1997) TIM–PCM external wall system for solar space heating and daylighting. Solar Energy 61(6):369–379

Mehling H, Cabeza LF (2008) Heat and cold storage with PCM. Springer, Berlin. ISBN 978-3-540-68556-2

Mehling H, Krippner R, Hauer A (2002) Research project on PCM in wood lightweight concrete IEA, ECES IA Annex 17, Slovenia, 3–5 Apr 2002

Miller W, Kośny J (2008) Next generation roofs and attics for residential homes. In: Proceedings of ACEEE summer study on energy efficiency in buildings, Pacific Grove

Nayak JK (1987) Transwall versus Trombe wall: relative performance studies. Energy Convers Manag 27(4):389–393

Neeper DA (2000) Thermal dynamics of wallboard with latent heat storage. Sol Energy 68:393–403

Peippo K, Kauranen P, Lund PD (1991) A multi component PCM wall optimized for passive solar heating. Energy Build 17:259–270

Petrie TW, Childs KW, Childs PW, Christian JE, Shramo DJ (1997) Thermal behavior of mixtures of perlite and phase change material in a simulated climate, In: Graves RS, Zarr RR (eds) Insulation materials: testing and applications, vol 3, ASTM STP 1320. ASTM—American Society for Testing and Materials, West Conshohocken, pp 180–194

Ravikumar M, Srinivasan PSS (2012) Analysis of heat transfer across building roof with phase change material. J Comput Inf Syst 8(4):1497–1505

Ravikumar M, Srinivasan PSS (2014) Heat transfer analysis in PCM-filled RCC roof for thermal management. J Mech Sci Technol 28(3):1073–1078

Romanowska A, Jedraszek K, Szulc J, Klemm P (1991) Badanie Efektu Cieplnego w Materialach Ceramicznych Nasyconych Materialem Fazowo-Zmiennym. In: proceedings of 3rd conference building physics in theory and practice, Politechnika Lodzka, Lodz

Salyer IO, Sircar AK (1989) Development of PCM wallboard for heating and cooling of residential buildings. Thermal Energy Storage Research Activities Review, U.S. Department of Energy, New Orleans, pp 97–123, 15–17 Mar 1989

Salyer IO, Sircar AK (1990) Phase change materials for heating and cooling of residential buildings and other applications. In: Proceedings of the 25th intersociety society energy conversion engineering, pp 236–243

Salyer IO, Sircar AK, Kumar A (1995) Advanced phase change materials technology: evaluation in lightweight solite hollow-core building blocks. In: Proceedings of the 30th intersociety energy conversion engineering conference, Orlando, pp 217–224

Sari A, Karaipekli A (2007) Thermal conductivity and latent heat thermal energy storage characteristics of paraffin/expanded graphite composite as phase change material. Appl Therm Eng 27(8–9):1271–1277

Sari A, Sari H, Onal A (2004) Thermal properties and thermal reliability of eutectic mixtures of some fatty acids as latent heat storage materials. Energy Convers Manag 45:365–376

Schossig P, Henning HM, Raicu A, Haussmann T (2003) Mikroverkapselte Phasenwechselmaterialien in Baustoffen. In: Proceedings of 13th symposium Thermische Solarenergie OTTI Technologie-Kolleg, Staffelstein, pp 489–494

Schossig P, Henning H, Gschwander S, Haussmann T (2005) Microencapsulated phase change materials integrated into construction materials. Sol Energy Mater Sol Cells 89:297–306

Shapiro M (1989) Development of the enthalpy storage materials, mixture of methyl stearate and methyl palmitate. Subcontract Report to Florida Solar Energy Center, USA

Shapiro MD, Feldman D, Hawes D, Banu D (1987) P.C.M. thermal storage in drywall using organic phase-change material. Passive Solar J 4(4):419–438

Sharma SK (2013) Zero energy building envelope components: a review. Int J Eng Res Appl 3 (2):662–675. ISSN: 2248-9622

Sharma SD, Sagara K (2005) Latent heat storage materials and systems: a review. Int J Green Energy 2:1–56. ISSN: 0197-1522

Sharma AK, Bansal NK, Sodha MS, Gupta V (1989) Vary-therm wall for cooling/heating of buildings in composite climate. Int J Energy Res 13(6):733–739

Shilei L, Neng Z, Guohui F (2006) Impact of phase change wall room on indoor thermal environment in winter. Energy Build 38:18–24

Stamatiadou M, Antonopoulos K, Founti M (2009) Energy savings using floor tiles with phase change materials in a S. European country. In: Proceedings of 11th international conference on thermal energy storage; Effstock 2009. Thermal energy storage for energy efficiency and sustainability, Stockholm

Stetiu C, Feustel HE (1996) Phase change wallboard as an alternative to compressor cooling in californian residences. In: Proceedings of 96 ACEEE, summer study for energy efficient buildings, Asilomar

Stovall TK, Tomlinson JJ (1995) What are the potential benefits of including latent storage in common wallboard? J Sol Energ Eng Trans ASME 117:318–325

Stritih U, Novak P (1996) Solar heat storage wall for building ventilation. Renew Energy 8(1–4): 268–271

Swet CJ (1980) Phase change storage in passive solar architecture. In: Proceedings of the 5th national passive solar conference, Amherst, pp 282–286

Tardieu A, Behzadi S, Chen JJ, Farid M (2011) Computer simulation and experimental measurements for an experimental PCM-impregnated office building. In: Proceedings of the building simulation 2011: 12th conference of international building performance simulation association, vol 1, pp 56–63

Telkes M (1947) Solar house heating—a problem of heat storage. J. Heat Ventilating 44:68–75

Telkes M (1978) Trombe wall with phase change storage material. In: Proceedings of the 2nd national passive solar conference, Philadelphia

Telkes M (1980) Thermal energy storage in salt hydrates. Solar Energy Mater 2:381–393

Tomlinson JJ, Heberle D (1990) Analysis of wallboard containing a phase change material. In: Proceedings of the 25th intersociety energy conversion engineering conference, vol 4. Reno

Tomlinson J, Jotshi C, Goswami D (1992) Solar thermal energy storage in phase change materials. In: Proceedings of solar '92: the american solar energy society annual conference, Cocoa Beach, pp 174–79

Turnpenny JR, Etheridge DW, Reay DA (2000) Novel ventilation cooling system for reducing air conditioning in buildings. Part I: testing and theoretical modelling. Appl Therm Eng 20 (11):1019–1037

Tyagi VV, Buddhi D (2007) PCM thermal storage in buildings: a state of art. Renew Sustain Energy Rev 11(6):1146–1166

van Haaren M (2012) Application of PCM in concrete; improvement of the indoor comfort and reducing energy demand. MSc thesis, Technische Universiteit Eindhoven, The Netherlands

Zalba B, Marin JM, Cabeza LF, Mehling H (2004) Free-cooling of buildings with phase change materials. Int J Refrig 27:839–849

Zalewski L, Chantant M, Lassue S, Duthoit B (1997) Experimental thermal study of a solar wall of composite type. Energy Build 25(1):7–18

Zalewski L, Lassue S, Duthoit B, Butez M (2002) Study of solar walls, validating a simulation model. Build Environ 37(1):109–121

Zamalloa A, Embil MJ, Zuñiga J, Zubillaga O, Cano F, Flores I (2006) PCM containing indoor plaster for thermal comfort and energy saving in buildings. In: Proceedings of ECOSTOCK. IEA international conference on thermal energy storage, Stockton College of New Jersey, 31 May–02 June 2006

Zhang D, Li Z, Zhou J, Wu K (2004) Development of thermal energy storage concrete. Cem Concr Res 34:927–934

Zhang M, Medina MA, King J (2005) Development of a thermally enhanced frame wall with phase-change materials for on-peak air conditioning demand reduction and energy savings in residential buildings. Int J Energy Res 29(9):795–809

Zhou G, Zhang Y, Wang X, Lin K, Xiao W (2007) An assessment of mixed type PCM-gypsum and shape-stabilized PCM plates in a building for passive solar heating. Sol Energy 81:1351–1360

Zhou T, Darkwa J, Kokogiannakis G (2014) Numerical and experimental analysis into the thermal performance of a novel phase change material gypsum board. In: Proceedings of 12th international energy conversion engineering conference, Cleveland. ISBN: 978-1-62410-304-9

Zhuang C-L, Deng A-Z, Chen Y, Li S-B, Zhang H-Y, Fan G-Z (2010) Validation of veracity on simulating the indoor temperature in PCM light weight building by energyplus. In: LSMS/ICSEE'10 Proceedings of the 2010 international conference on Life system modeling and intelligent computing, and 2010 international conference on Intelligent computing for sustainable energy and environment: Part I. Springer, Berlin, pp 486–496

Chapter 3
Overview of Basic Solid–Liquid PCMs Used in Building Envelopes—Packaging Methods, Encapsulation, and Thermal Enhancement

3.1 Introduction

Modern developments in areas of building technologies and new construction methods have led to reduced build time and cost for residential and commercial buildings. However, they have also resulted in lightweight construction materials with very little thermal mass. Wood- or steel-framed buildings can be five times less thermally massive than their stone or brick counterparts (Campbell and Sailor 2011). This lack of thermal mass combined with dynamic thermal excitations (climate-and internal load-based) results with an inability to mitigate dynamic thermal loads, leading to large diurnal indoor air temperature fluctuations, followed with increased space-conditioning energy consumption. Several demonstration projects discussed in previous chapters showed that adding PCM thermal mass will yield an increase of the building thermal storage capability and thus help in improving overall building energy performance. In this chapter, an overview has been made of the thermal energy storage materials with solid–liquid phase change. The following four major aspects have been the focus of this overview: the type of materials, heat transfer characteristics followed by methods of thermal conductivity enhancements, and durability. In addition, PCM encapsulation and packaging methods are discussed here.

PCM's heat storage capability is very attractive alternative for conventional building thermal mass, such as solid wood, stone, brick, gypsum, and concrete, because of their high storage density within small temperature swing. Approximate temperature range for PCMs used in building envelopes is usually between −5 °C (building envelope surface temperature during winter) and 70 °C (peak roof surface temperature during summer). For more traditional internal building applications, such as wall, floor, or ceiling finish, PCM's phase-transition temperature should be within the human comfort range of 20–30 °C.

Flame resistance and long-term durability of PCM-enhanced building envelopes are important design factors. Paraffinic PCMs are very chemically stable and

J. Kośny, *PCM-Enhanced Building Components*, Engineering Materials and Processes, DOI 10.1007/978-3-319-14286-9_3

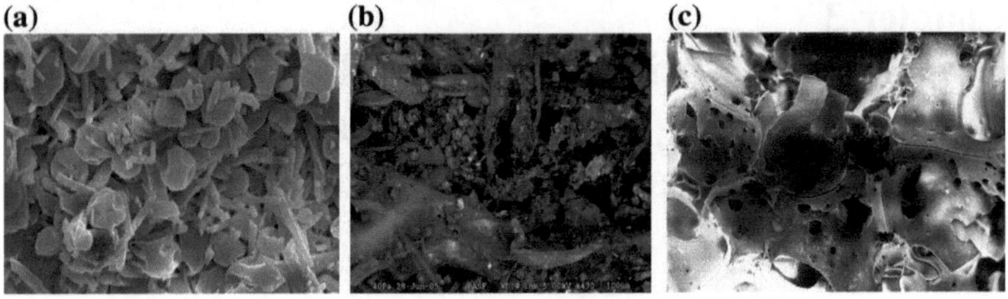

Fig. 3.1 Electron microscope images of PCM-enhanced building materials—author's archive. **a** PCM-enhanced gypsum composite. **b** Cellulose-PCM blend. **c** Polyurethane foam–PCM blend

durable materials. For decades, they have been adding thermal mass to the walls, ceilings, or floors. In early applications (mostly in North America and Europe), these PCMs were used in a passive way in buildings without space conditioning. Figure 3.1 shows electron microscope images of microencapsulated paraffinic PCMs mixed with different building envelope products. Unfortunately, in North America, some of paraffinic PCM applications failed during the required US flammability tests (ASTM E119—12a—"Standard Test Methods for Fire Tests of Building Construction and Materials,"[1] the Corner Fire Test- NFPA 265,[2] and "Standard Test Method for Room Fire Test of Wall and Ceiling Materials and Assemblies"—ASTM E2257—13a[3]). That is why, in more current approaches, designers either use different types of less flammable organic PCMs, or place PCMs in different locations, taking advantage of larger temperature fluctuations in residential attics, roofs, and at the exterior wall surfaces. Climate-generated energy fluctuations, which can represent a significant part of the building cooling and heating loads, are largely absorbed by the PCM-enhanced building material and insulations and are later transferred to the environment without affecting the interior building energy balance (Kośny et al. 2008, 2012a). In buildings with space conditioning, this simple change in thermal mass configuration means real energy savings. This new placement method for PCM significantly improves fire-safety, since PCM is not directly exposed to the interior of the building. However, it is good to remember that this change also dictates the different temperature ranges for selected PCMs.

A large number of solid–liquid PCMs are commercially available today for almost any required temperature range. Zalba et al. (2003) classifies 118 organic substances, inorganic substances, organic eutectics, inorganic eutectics, fatty acids, and non-eutectic mixtures with potential for use as PCM. Lane (1978, 1980, 1983, 1989) identifies 210 organic, inorganic, and eutectic materials that undergo phase change without component separation, suitable for building temperature regulation.

[1] http://www.astm.org/Standards/E119.htm.

[2] http://www.nfpa.org/codes-and-standards/document-information-pages?mode=code&code=265.

[3] http://www.astm.org/Standards/E2257.htm.

Water/ice, with its heat of phase transition of about 330 kJ/kg, is the best known PCM abandon in human environment. PCM materials used in building envelopes need to have relatively high energy of fusion, with phase-transition process at a certain temperature range specific for building enclosure components and type of application. They should be capable of storing and releasing relatively large amounts of energy on daily bases for a typical life span of a building envelope (in most cases at least 25 years). In addition, for PCMs used in building envelopes, the amount of latent heat should be significantly larger compared to the sensible heat gain/loss available in typical building materials. Most of PCMs discussed here have latent heat between 100 and 250 kJ/kg. PCMs used in building envelopes are at least 2–10 times lighter from structural materials and should store up to 15 times more heat per unit volume than conventional construction materials such as masonry, concrete, or stone, for temperature changes of $\sim$10 K. This allows an application of lightweight construction methods and still taking advantage of thermal mass benefits. Energy is absorbed or released when the material changes from solid to liquid and vice versa; thus, PCM-enhanced building envelopes are often classified as lightweight latent heat storage components. For all internal surface applications, it is very important that a PCM temperature should remain relatively constant (or should vary within a very small temperature range) during the phase-change process. This is useful for keeping the conditioned internal space of the building at a desired uniform temperature.

There is also a group of solid–solid PCMs which are under investigation right now (Gu et al. 2010; Sarı et al. 2011). They can be potentially useful for building envelopes, because some of them change phase in temperatures close to 25–30 °C and do not encompass problems associated with handling of PCM in liquid phase, i.e., containment, potential leakage, etc. Also, unlike many inorganic solid/liquid PCMs, they do not require nucleation aid additives to prevent supercooling. Solid–solid PCMs change their crystalline structure from one lattice configuration to another at a fixed, single temperature range, and their phase transition involves significant amount of latent heats comparable to the most effective solid–liquid PCMs. Because it is a solid–solid phase transition, there is no visible change in the appearance of the material (insignificant expansion–contraction). In the future, some of solid–solid PCMs with lower phase-change temperature may have a chance to be used in building envelops.

3.2 PCM Classification

In general, PCMs latent heat storage can be achieved through a solid–solid, solid–liquid, solid–gas, and liquid–gas phase changes. However, phase change used in most building envelope applications is a solid–liquid process. As stated earlier, there is also considered a limited number of solid–solid PCM applications, which due to their very slow phase-transition processes, maybe helpful in a future building dynamic load management.

3.2.1 Introduction—Phase-Change Processes

From the thermodynamic standpoint, the PCM absorbs or releases thermal energy or heat because of the entropy change. This portion of heat is associated with the unit mass of the material, and it is defined as the latent heat of the material. The latent heat is absorbed by a material during the melting, or it is released during the freezing process. In the case of melting, added thermal energy is used to break the bonds between the molecules. When heat is entering a substance, this energy is used to break the molecular connections. The phase changing substance is, at this time, a mixture of solid and liquid molecules. In solids, the molecules, ions, or atoms are held relatively close to each other by strong intermolecular forces. Although the particles move due to their kinetic energies, their relative movement is quite small. This gives a solid the property of having a definite shape. It can be observed that during a phase transition, both liquid and solid molecules of PCM are moving with the same rate (vibration). This means that molecules have the same average speed and thus the same average kinetic energy (since they have the same mass). It is followed with the same Kelvin temperature. When a solid is heated, the particles gain kinetic energy. When the particles gain enough energy to overcome the attractions that hold them in a solid structure, the substance is starting to melt. Finally, after some time, PCM's molecules rearrange themselves, causing an entropy change and the material is starting to melt. Melting process can be also called fusion.

Freezing is an exothermic process (releases thermal energy). Freezing or solidification occurs when a substance changes from liquid to solid. In case of freezing, energy is taken from PCM. This energy exchange does not cause a change in kinetic energy. While in the liquid state, the particles in a substance possess enough kinetic energy to move around in close proximity to each other, because of existence of intermolecular forces. When temperature drops down, the particles lose their kinetic energy and bond together. Gradually, the particles settle into fixed positions, causing the substance to take shape and become a solid. Sometimes, if a liquid PCM is rapidly chilled down, the temperature drops faster from the speed of the solidification. This results in a liquid PCM that can be colder from its freezing point (Skripov 1974; Kurz and Fisher 1992; Kashchiev 2000). This process is called supercooling. It is good to remember that supercooled liquid PCMs, because they are very unstable, will crystallize rapidly upon any slightest disturbance, such as stirring the liquid, vibration, or small amounts of foreign substances. There are numerous inorganic PCMs (i.e., hydrated salts), called congruent, which melt and solidify cleanly with the liquid phase being exactly the same chemical composition as the solid phase. In semi-congruent PCMs, a salt hydrate and an aqueous solution are formed during the fusion process. This reaction needs to be overturned before freezing to avoid volume reduction of the active PCM, which compromises the overall system thermal performance. Usually, semi-congruent melting PCMs can be modified with additives to render them congruent. Finally, anhydrous salt and an aqueous solution are formed in incongruent PCMs during the melting. This process is often called PCM separation. Unless such PCM is modified, the anhydrous salts settle down due to gravity, which is resulting in a gradual loss in performance of the PCM. The most suitable methods of preventing this process are either through thickening the salt solution or through PCM gelling. Thickening means an addition of a material to the salt hydrate that increases the viscosity and hereby holds the salt hydrate together (Zalba et al. 2003; Farid et al. 2004; Mehling and Cabeza 2008). Gelling means adding a cross-linked material (e.g., polymer) to the salt to create a three-dimensional network that holds the salt hydrate together. So, when anhydrous salt crystals are formed, PCM separation does not occur. Simply, anhydrous salt crystals remain dispersed throughout the PCM, making it easier for them to redissolve in the aqueous solution (Farid et al. 2004).

Numerous distinctions between organic and inorganic PCMs can be investigated from various perspectives including operating temperature range, phase-transition enthalpy, chemical stability, long-term durability, and flammability [see Lingayat and Suple (2013)]. Figure 3.2 shows relationship between PCM's enthalpy and temperature for different types of PCMs. These characteristics are especially important for their application in building envelopes, since PCM has to be incorporated into the structural components, finish materials, or thermal insulation. A very common situation is when for a specific application, there is no a single PCM product which possesses all the required properties and performance characteristics. In such circumstances, overall design of the PCM system has to be modified to make up for the missing physical properties. As discussed earlier, supercooling effect can be suppressed by introducing nucleation aid additives. PCM semi-congruent melting can be inhibited by using additives to render them congruent. Finally, performance of potentially incongruent PCMs can be improved by thickening the salt solution (Farid et al. 2004). Other key factors important for building envelope applications are compatibility with different packaging or encapsulation methods and PCM flammability.

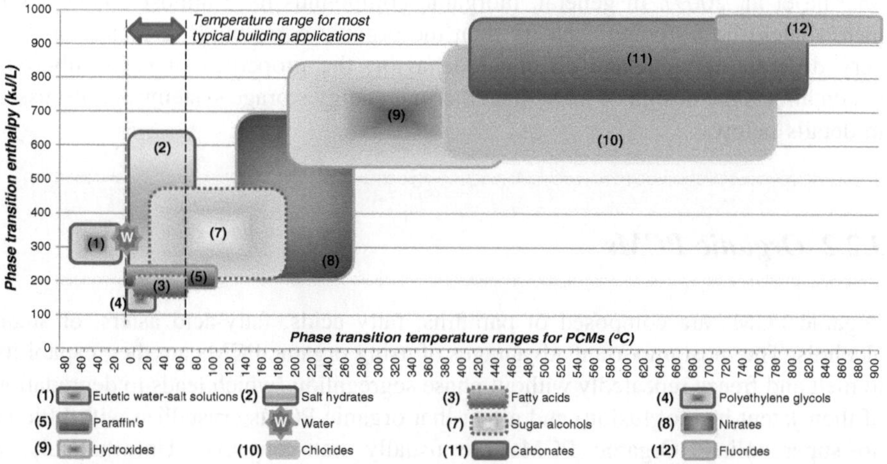

Fig. 3.2 Relationship between PCM's enthalpy and temperature for different types of PCMs

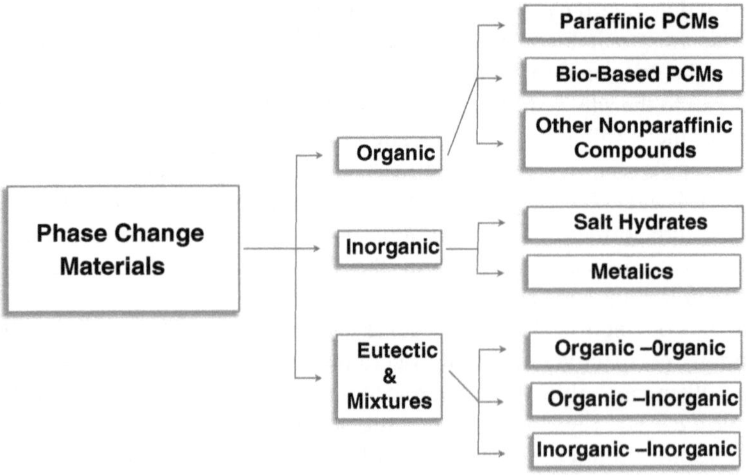

Fig. 3.3 Classification of PCMs

As shown in Fig. 3.3, multiple types of PCMs exist. The main categorization of PCMs is the differentiation between inorganic PCMs and organic PCMs. The commonly used phase-change substances for technical applications are as follows: paraffins (organic), fatty acids (organic), and salt hydrates (inorganic)—IEA Annex 39 (2005). Additionally, numerous combinations of eutectic PCMs can be used for this purpose (see Fig. 3.3). They are compositions of two or more PCMs, each of which melt and freeze congruently forming a mixture of the component crystals during crystallization.

As mentioned above, the main division between PCMs is made by the fact that they can either be organic or inorganic(Farid et al. 2004; Mehling and Cabeza 2008; Sharma et al. 2009). In general, inorganic compounds have almost double volumetric latent heat storage capacity than the organic compounds. Because of their very different thermal and chemical behavior, the properties of each subgroup which affects the design of latent heat thermal energy storage systems are discussed in details below.

3.2.2 Organic PCMs

Organic PCMs are composed of paraffins, fatty acids, fatty-acid asters, or sugar alcohols. The most important advantages of using organic PCMs are their capability to melt and freeze repeatedly without phase segregation (which leads to degradation of their latent heat of fusion) and a fact that organic PCMs crystallize with little or no supercooling. Organic PCMs are usually non-corrosive. Unfortunately, a relatively low ignition resistance of some organic PCMs is an important drawback, in case of building envelope applications.

Historically, paraffinic PCMs have been the most commonly used in building envelope applications because of ease of microencapsulation using well-established and relatively simple process. This fact enables blending with cement and gypsum composites and PCM enhancement of thermal insulations. Most of paraffins available today on the market are petroleum-based straight chains of n-alkane hydrocarbons (CH_3–(CH_2)–CH_3). There are also non-petroleum paraffins, produced out of food industry waste products. Melting temperature and phase-transition enthalpy increase for paraffins with the length of the carbon chain. When the number of carbon atoms in the paraffin molecule is between 13 and 28, the melting temperature falls within a range of approximately −5 to 60 °C—Mehling, Cabeza (2008). This temperature range covers great majority of building envelope applications. When comparing to inorganic compounds, major advantages of paraffinic PCMs are their superior chemical stability, non-corrosiveness, and non-toxicity. In addition, paraffins show negligible subcooling effect (Boh and Sumiga 2008). Major disadvantages are their high cost and flammability.

Fatty acids are represented by the chemical formula $CH_3(CH_2)2nCOOH$, for example, capric acid, lauric acid, palmitic acid, etc. Fatty acids have heat storage densities very similar to paraffins, and like paraffins, their melting temperatures increase with the length of the molecule. Although chemically stable upon cycling, they tend to react with the environment due to their acidic nature.

Sugar alcohols are a hydrogenated form of a carbohydrate such as D-sorbitol, and xylitol. They generally have higher latent heat and density than paraffins and fatty acids. However, they melt at temperatures between 90 and 200 °C, making them unsuitable for building envelope applications.

3.2.2.1 Paraffinic PCMs

An origin of paraffinic PCMs is coming from the earliest candle production. For centuries, candles have been made of many different bio-based materials. In ancient Egypt and during early Roman times, the fat rendered from butchered animals was used to make a type of candle. This material is known as tallow. China relied on bees for their wax, while Japan used tree nut extract to produce wax. In early India, wax was produced through boiling of the cinnamon tree fruits. In the eighteenth century, most wax came from the whales' oil. In the second half of nineteenth century, commercial production of paraffinic wax was initiated in Europe and North America. The feedstock for paraffin is a mixture of oil and wax, called slack wax. It is a by-product from the refining of lubricating oil.

Paraffin is the name given to a mineral wax and oil and is also used as a generic name of a particular series of hydrocarbons. In 1830, Carl Von Reichenbach distilled paraffin out of tar of beech wood. At this time, Reichenbach called his new developed mixture of hydrocarbon oils a "eupione." The production of paraffin wax from coal started in 1845, in Glasgow, UK.[4] In 1850, James Young patented the production procedure of paraffin oil by distillation of coal. Following these developments, it was discovered in the USA that paraffin could be extracted from crude oil as well. During 1890s, this technology spread around the world and became one of the basic petroleum by-products. Today, production of paraffin waxes is based on slack wax, which is a mixture of oil and waxes. Most often, they are by-products of the lubrication oil refinery processes. Paraffin wax consists of a mixture of mostly straight chain n-alkanes CH_3–(CH_2)–CH_3.

In organic chemistry, paraffin is the common name of the alkane hydrocarbons with the general formula C_nH_{2n+2}. Paraffin waxes have typically a clearly defined crystal structures composed of macroscopic crystals, which are called the macro-crystalline. In case of low-carbon-number hydrocarbons, these individual crystals can be directly observed with use of the magnifying glass or with use of the microscope. Macro-crystalline crystals are usually hexagonal or needle shaped (orthorhombic) (Becker 1997). Paraffin's potential for thermal storage applications is coming from the fact that the crystallization of the (CH_3) chain releases a large amount of latent heat during phase-change processes. Most of paraffinic PCMs used to enhance building envelopes are not 100 % pure. Usually, latent heat of phase transition varies for these compounds between 150 and 270 kJ/kg and depends on chemical purity (Sharma et al. 2009).

Poor heat conduction of paraffinic PCMs is an often reported problem associated with PCM applications in building products facing interior space (wall, floor, and ceiling surfaces). This group of PCM applications requires fast heat accumulation and PCM discharging. This can be either achieved through direct increase of thermal conductivity of the PCM-enhanced product, or indirectly through increase of the surface heat transfer coefficient in place of PCM location (Stovall and Tomlinson 1995). Ceiling panels containing PCM or PCM-enhanced wall boards are good examples of PCM applications facing interior of the building, where relatively slow heat transport to and from the PCM effects system's thermal performance. During last decade, several authors investigated the improvement of heat conduction by the addition of highly conductive fillers into the paraffin or by using high conductive matrix to carry paraffin. Most typical heat conduction enhancing agents are as follows:

1. Metal particles (Kozumi 2004; Molefi et al. 2010)
2. Graphite powder, expanded graphite, and exfoliated graphite (Kim and Drzal 2009; Xia et al. 2010; Cheng et al. 2010a, b)
3. Carbon fibers and nano-fibers (Fukai et al. 2002; Wang et al. 2010; Cui et al. 2011)

Table 3.1 summarizes basic physical properties of paraffins having carbon numbers between C(14) and C(33). As shown in Table 3.1, melting temperatures of n-alkanes increase with the increasing number of carbon atoms. It can be observed

[4]http://europecandles.org/pages/uk/an-everlasting-history/raw-materials-and-candles-production-processes.php?lang=FR.

Table 3.1 Basic thermal properties of n-paraffins

Name	Carbon number	Chemical symbol	Melting point[a]	Heat of fusion[b]	Molecular weight[a]
Units			°C	kJ/kg	g/mole
n-Tetradecane	C(14)	$C_{14}H_{30}$	6	228	198.39
n-Pentadecane	C(15)	$C_{15}H_{32}$	10	205	212.42
n-Hexadecane	C(16)	$C_{16}H_{34}$	18	237	226.44
n-Heptadecane	C(17)	$C_{17}H_{36}$	22	213	240.47
n-Octadecane	C(18)	$C_{18}H_{38}$	28	244	254.5
n-Nonadecane	C(19)	$C_{19}H_{40}$	32	222	268.53
n-Eicosane	C(20)	$C_{20}H_{42}$	37	246	282.55
n-Henicosane	C(21)	$C_{21}H_{44}$	40.5	200	296.58
n-Docosane	C(22)	$C_{22}H_{46}$	44.5	249	310.61
n-Tricosane	C(23)	$C_{23}H_{48}$	47.5	232	324.63
n-Tetracosane	C(24)	$C_{24}H_{50}$	52	255	338.66
n-Pentacosane	C(25)	$C_{25}H_{52}$	54	238	352.69
n-Hexacosane	C(26)	$C_{26}H_{54}$	56.5	256	366.71
n-Heptacosane	C(27)	$C_{27}H_{56}$	59	236	380.74
n-Octacosane	C(28)	$C_{28}H_{58}$	64.5	253	394.77
n-Nonacosane	C(29)	$C_{29}H_{60}$	65	240	408.79
n-Triacontane	C(30)	$C_{30}H_{62}$	66	251	422.82
n-Hentriacontane	C(31)	$C_{31}H_{64}$	67	242	436.85
n-Dotriacontane	C(32)	$C_{32}H_{66}$	69	170	450.87
n-Triatriacontane	C(33)	$C_{33}H_{68}$	71	268	464.9

[a]Becker (1997)
[b]Sharma et al. (2009)

that these paraffins have melting points between 6 and 71 °C, which is almost exactly within expected temperature range for the PCM-enhanced building envelopes. Paraffin wax is non-corrosive, chemically stable, and non-reactive below 500 °C. Pure paraffin waxes of carbon number above C(16) are colorless, odorless, and tasteless solids. Paraffinic PCMs usually have relatively sharp enthalpy profiles with well notable melting and freezing points (KCPC 2003). They show very little, or no supercooling and little volume changes during phase change (Farid et al. 2004). Paraffin can react with halogens under UV, light, or heat. When heated at high temperatures in the absence of air, paraffin wax can also crack and break up into smaller, lighter paraffin molecules. The density of paraffin wax generally increases with increasing molecular mass, but remains below the density of water (1.0 g/cm^3). At 15 °C, the density of paraffins is about 0.773 g/cm^3 for carbon numbers C(14) to C(17) and 0.798 g/cm^3 for C(17) to C(20).[5] Paraffin waxes are also non-polar; therefore, they are insoluble in water but are soluble in other

[5]http://www.chemicalland21.com/petrochemical/n-PARAFFINS.htm.

nonpolar petroleum-based solvents, such as benzene, ether, and certain esters. Solubility of paraffin wax decreases as a number of carbon atoms increases (Becker 1997).

Paraffin does burn readily if ignited. The US Environmental Protection Agency (EPA) claims that smoke and fumes generated by paraffin during its burning can be toxic (EPA 2001). It has been found that during burning of paraffins, styrene, naphthalene, toluene (aromatic hydrocarbons that are extremely toxic), and formaldehydes are released. Burning paraffin wax can cause irritation of the respiratory system. According to EPA, paraffin burning products exceeds the EPA's standards and poses an increased risk of cancer. That is why, when paraffinic PCMs are used in buildings, the flammability of the PCM-enhanced products may be a concern. The flash point of paraffin wax is about 120 °C for carbon numbers C(14) to C(17) and 160 °C for C(17) to C(20).[6] It is the lowest temperature at which paraffin can vaporize to form an ignitable mixture in air. It is measured according to ASTM D92 standard (ASTM 1992) and it requires an ignition source.

Various flame retardants may be used to improve flame resistance of PCM building components. The flame retardant may contain boric acid, borates, ammonium polyphosphates, sodium carbonate, sodium silicate, aluminum hydroxide, magnesium hydroxide, antimony trioxide, various hydrates, tetrakis (hydroxymethyl) phosphonium salts, halocarbons, including chlorendic acid derivates, halogenated phosphorus compounds including tri-o-cresyl phosphate, tris (2,3-dibromopropyl) phosphate (TRIS), bis (2,3-dibromopropyl) phosphate, tris (1-aziridinyl)-phosphine oxide (TEPA), and others.[7] The flame retardant may be applied as a solution, dispersion, a suspension, or a colloid that forms a coating on the microencapsulated paraffinic product to provide flame resistant characteristics.

Summarizing, paraffinic PCMs have numerous favorable characteristics useful in building applications including chemical stability, long-term durability, congruent melting, and good nucleating properties. However, they also show some undesirable properties such as:

1. Low thermal conductivity
2. Relatively high flammability, and
3. Paraffin burning products are toxic

Most of these undesirable effects can be partly eliminated by slightly modifying the wax through addition of highly conductive additives, or fire retardants.

In addition, a great majority of paraffinic PCMs used today in building products are produced out of crude oil. That is why, they are relatively expensive and their price strongly depends on fluctuations of the world's oil market. That is why, in the USA, an effort was undertaken to develop a similar PCM technology based on non-petroleum source. In 2010, Syntroleum Corporation of Tulsa, OK, and USA received the US Department of Energy grant (D.E-EE0003924) for the

[6]http://www.chemicalland21.com/petrochemical/n-PARAFFINS.htm.

[7]http://www.faqs.org/patents/app/20100087115#b.

development of low cost, bio-based phase-change material for building envelope applications.[8] A new production facility was built in Geismar, LA, USA, to convert fats and oils into hexadecane and octadecane at the biodiesel price range. This novel paraffinic PCM technology is manufactured with utilization of low-cost food industry waste products.

3.2.2.2 Non-paraffinic Bio-based PCMs

In the past, the PCM market has been dominated by paraffin products and salt hydrates. Paraffins have recently become more popular than the salt hydrates. However, the use of bio-based compounds has started to make strong inroads in the world's PCM market. The ability of bio-PCMs to compete in the building market depends on their price/performance characteristics relative to paraffinic PCMs. Despite of many advantages of paraffinic PCMs, their relatively high cost, along with high flammability, is proving to be major barriers to their widespread acceptance in buildings. Finally, since paraffins are derived from crude oil, their prices are susceptible to the seasonal changes and geopolitical scenarios, which makes economical planning extremely difficult for manufactures of paraffin-based PCMs. All these factors underline a need to shift the major focus in research and development away from paraffinic PCMs to some other types of non-petroleum-based equivalents. The following two alternative groups of chemical compounds have a great potential to substitute paraffin in the future and compete with the existing energy-efficient building materials and technologies. They are bio-based PCMs and inorganic PCMs. Fire safety and environmental benefits of using bio-based PCMs are significantly greater than those of paraffinic hydrocarbons.

The non-paraffin organic PCMs constitute a highly diversified group of chemical compounds. Each of these materials has its own physical characteristics unlike paraffins, which all have very similar properties. This is the largest category of potential materials for future sustainable heat storage products. Bio-based PCMs are obtained from animal fat and vegetation such as beef tallow, lard, palm, coconut, and soybean. They are hydrogenated hydrocarbons, which are chemically stable and can last for decades. They are non-toxic and can be cycled thousands of cycles without experiencing any material degradation. That is why they are a sustainable alternative to paraffinic PCMs. A family of non-paraffin organic PCMs includes numerous materials with highly varied properties. Abhat (1978) and Buddhi and Sawhney (1994) have conducted an extensive survey of organic materials and identified a number of esters, fatty acids, alcohols, and glycols suitable for heat storage.

---

[8]Low-cost phase-change material (PCM) for building envelopes. http://www.recovery.gov/Transparency/RecipientReportedData/pages/RecipientProjectSummary508.aspx?AwardIdSur=117469.

Fatty acids are given by the formula $CH_3(CH_2)2nCOOH$. They perform solidification with very little supercooling and have similar heat of fusion as paraffins, which are both positive qualities. However, they are also relatively expensive if not produced from waste products and show a tendency toward mild corrosiveness (Sharma et al. 2009). Fatty acids, possess some superior properties over inorganic PCMs such as for example melting congruency. They also show (i) good chemical stability, (ii) they are non-toxic, and (iii) have suitable melting temperature range for building applications. For most of fatty acids with melting point appropriate for building envelope applications, densities in solid and liquid phases are ranging between 800 and 1100 kg/m^3. According to Rozanna et al. (2005), in the liquid phases, these materials have surface tensions in the order of 20–30 dyne cm^{-1} and are therefore high enough to be retained (when melted) in the structure of the material carrier such as concrete, gypsum, and wood composite. The raw fatty acids are derived from sustainable vegetable and animal sources. This assures an environment-friendly non-pollutant source of supply.

Fatty-acid esters come from combination of fatty acids with alcohols. Their densities are similar to paraffins. When the alcohol component is glycerol, the fatty-acid esters produced can be monoglycerides, diglycerides, or triglycerides, which are components of vegetable fats and oils. Compared to paraffinic PCMs, fatty-acid esters are significantly less flammable. They are made from underused feedstocks such as soybean oils, coconut oils, palm oils, and beef tallow. Because they are fully hydrogenated, the fatty-acid esters PCMs remain stable for years with ability to undergo many thousands of phase-change cycles with no risk of oxidation. These bio-based PCMs are capable of absorbing, storing, and releasing large amounts of latent heat with no or little supercooling, similar to paraffin-based PCMs. In addition, fatty-acid esters are equally easy to microencapsulate. Table 3.2 shows basic phase-change properties of selected organic acids and esters, which can be considered for building envelope applications—due to their melting temperatures ranging between 6 and 70 °C.

Polyethylene glycols, or short PEGs, are polymers with heat of fusion between 100 and 200 kJ/kg (see Table 3.3). The chemical formula for PEGs is $C_{2n}H_{4n} + 2O_{n+1}$. They are produced from ethylene glycol. Polyethylene glycols are available in a molecular weight range from about 200 to 35,000. PEGs with molecular weight below 6000 g/mol are suitable for building envelope applications. Melting temperature for PEGs with an average molecular weight between 200 and 400 g/mol is between 17 and 22 °C. These PEGs are in a liquid stage at room temperature. If higher melting temperature (ranging between upper twenties and forties) is desired, PEGs with molecular weight between 400 and 4000 g/mol can be used. The melting temperature of all PEGs with a molecular weight exceeding 4000 g/mol is around 58–65 °C.

Sugar alcohols are getting interest today as potential PCMs. Sugar alcohols are a hydrogenated form of a carbohydrate such as D-sorbitol and xylitol. They generally have higher latent heat and density than paraffins and fatty acids. According to Kakiuchi et al. (1998), sugar alcohols do not present health hazard; erythritol and xylitol are used as sweeteners. Unfortunately, they have melting temperatures over 90 °C, which limits their building applications.

Table 3.2 Comparison of melting points and heat of fusion of selected bio-based fatty acids and fatty-acid esters

Name	Molecular formula	Melting point	Heat of fusion	References
Units		°C	kJ/kg	
Isopropyl palmiate	$C_{19}H_{38}O_2$	10	186	Haves et al. (1993)
Isopropyl stearate	$C_{21}H_{42}O_2$	14–18	142	Lutton (1967)
Caprylic acid	$C_8H_{16}O_2$	16.5	148.5	Schossig et al. (2003)
Butil stearate	$C_{22}H_{44}O_2$	19	200	Lutton (1967)
Dimethyl sabacate	$C_{12}H_{22}O_4$	21	120–135	Feldman et al. (1986)
Vinyl stearate	$C_{20}H_{38}O_2$	27–29	122	Lutton (1967)
Capric acid	$C_{10}H_{20}O_2$	32	152	Lutton (1967)
Lauric acid	$C_{12}H_{24}O_2$	42–44	178	Lutton (1967)
Myristic acid	$C_{14}H_{28}O_2$	49–58	187–204	Abhat (1983), Lane (1980), Sari and Kaygyusz (2001)
Palmitic acid	$C_{16}H_{32}O_2$	61–64	185–203	Dicer and Rosen (2002), Lane (1980), Sari and Kaygyusz (2002)
Stearic acid	$C_{18}H_{36}O_2$	69–70	203	Dincer and Rosen (2002), Lane (1980), Lutton (1967)

In the USA, the US Department of Agriculture and the National Science Foundation has sponsored research to investigate the potential vegetable-derived compounds becoming significant in the PCM market.[9,10] During this work, the researchers investigated over 200 different fat- and vegetable oil-based PCMs with latent heats between 150 and 220 J/g. Table 3.3 gives fusion energies and melting temperature for selected organic non-paraffinic PCMs with phase-transition temperature range making them suitable for building envelope applications.

The fire safety and environmental benefits of using bio-based PCMs are significantly greater than those of paraffinic hydrocarbons (Suppes et al. 2002a).

Waste disposal of organic compounds used in buildings is another important fact to consider. Paraffins are long chain alkanes with no "active" element on their molecule for bio agents (such as microbes and bacteria) to begin their breakdown. While this is very attractive from long-term durability perspective, it presents environment disposal problems. In comparison, most of bio-based PCMs are naturally degradable. However, when properly encapsulated, they remain chemically stable. During accelerated aging tests, vegetable-based PCMs have demonstrated chemical stability for equivalence of 30 years of service (Suppes et al. 2002a). Accordingly, bio-based PCMs will most likely degrade in half a year, when discarded in landfills. Many vegetable-based and animal fat-based PCMs can be derived locally using common agricultural crops and food industry waste products.

[9]US Department of Agriculture-SBIR Phase 1 grant #2002-00401 and Phase 2grant # 2003-04094.

[10]National Science Foundation, USA—SBIR, Award Number—0750470.

Table 3.3 Fusion energies and melting temperature for selected organic non-paraffinic PCMs

Name	Melting point	Heat of fusion	References
Units	°C	kJ/kg	
Formic acid	7.8	247	Lingayat and Suple (2013)
Polyglycol E400	8	99.6	Lane (1980), Mehling and Cabeza (2008)
Dimethyl-sulfoxide	16.5	86.7	Farid et al. (2004)
Glycerin	17.9	198.7	Lingayat and Suple (2013)
Octadecyl 3-mencaptopropylate	21	143	Kenisarin (1993), Cassedy (2000)
Polyglycol E600	22	127.2	Dicer and Rosen (2002), Lane (1980)
I-Dodecanol	26	200	Hawes et al. (1993)
Octadecyl Thioglycate	26	90	Kenisarin (1993), Cassedy (2000)
Methyl Palmitate	29	205	Lingayat and Suple (2013)
I-Tetradecanol	38	205	Hawes et al. (1993)
Camphenilone	39	205	Sharma and Sagara (2005), Lingayat and Suple (2013)
Docasyl Bromide	40	201	Lingayat and Suple(2013)
Caprylone	40	259	Sharma and Sagera (2005), Lingayat and Suple(2013)
Phenol	41	120	Lingayat and Suple (2013)
Methyl-12-hydroxy stearate	42–43	120–126	Feldman et al. (1986)
Cyanamide	44	209	Lingayat and Suple (2013)
Hydrocinnamic acid	48	118	Mehling and Cabeza (2008), Lingayat and Suple (2013)
Camphene	50	238	Lingayat and Suple (2013)
Nitro Naphthalene	56.7	103	Sharma and Segera (2005) Lingayat and Suple (2013)
Beeswax	61.8	177	Ramnanan-Singh (2012), Lingayat and Suple (2013)
Glycolic acid	63	109	Mehling and Cabeza (2008), Lingayat and Suple (2013)
Polyglycol E6000	66	190	Dicer and Rosen (2002), Lane (1980)
Acrylic acid	68	115	Lingayat and Suple (2013)

This leads to a more stable price and regionally based feedstocks. In addition, according to USDA, bio-based PCMs are considered to be nearly carbon neutral (Suppes et al. 2002b).

3.2.3 Inorganic PCMs

Numerous studies demonstrated that hydrated salts have a phase-change latent heat value of around mid 200 s kJ/kg and reasonably thermal conductivity of about 0.5 W/mK. Starting from 1948, when the first passive solar house utilizing PCM was constructed by Maria Telkes,[11] most of early PCM buildings used inorganic phase changing compounds (Telkes 1952, 1978). These materials were often installed in solar walls containing massive metal containers. Inorganic compounds include salt hydrates, salts, metals, and alloys. They cover a wide range of phase-transition temperatures. Only salt hydrates, salt mixtures, and metallics are used as PCMs today. Inorganic PCMs demonstrate relatively high heat storage density because they have high latent heat of fusion per unit volume and their conductivity may be at least twice as high as that of organic materials. Compared to organic PCMs, inorganic materials usually have similar enthalpies per mass, but higher ones per volume due to their high density. Another advantage is that they are readily available at a low cost. They are also not considered as a safety hazard, since inorganic compounds are non-flammable and only slightly toxic (Sharma et al. 2009).

Certain disadvantages were identified by using inorganic PCMs, though. The major behavioral characteristics of salt hydrates impeding the scale of their heat storage applications are that of incongruent or semi-congruent melting and poor nucleation characteristics. In case of a large number of inorganic PCMs, the phase-transition process that occurs at the melting and freezing temperatures involves the hydration and dehydration. Sometimes the inadequate amount of water released during the melting process is not capable to dissolve the crystalline structure of the PCM. At the same time, a significant difference in densities between the solid and liquid material, combined with gravity forces, results in the less hydrated salt that desegregates and is settled at the bottom of the container. A similar situation is observed during the solidification process, where poor nucleating properties of hydrated salts lead to supercooling and thus a reduction in the overall latent heat storage performance (Hawes et al. 1993). If such inorganic compound is not properly tailored using special additives, these problems may become irreversible tampering the long-term system thermal effectiveness (Farid et al. 2004; IEA 2005). Another main problem effecting building applications is material compatibility with metals, yielding a potential for corrosion in case of installations utilizing metal enclosures or containments (IEA 2005; Mehling and Cabeza 2008). There is still a significant number of other inorganic compounds which are expected to have desired latent heat storage properties and which need to be investigated (IEA 2005).

3.2.3.1 Salt Hydrates

Salt hydrates are the most important group of inorganic PCMs, which have been extensively studied for their use in latent heat thermal energy storage systems. Salt hydrates are inorganic salts containing one or multiple water molecules, such that the resulting crystalline solid has a chemical formula of $AB \cdot nH_2O$. Some of the best known examples are sodium sulfate decahydrate ($Na_2SO_4 \cdot 10H_2O$), lithium nitrate trihydrate ($LiNO_3 \cdot 3H_2O$), and potassium fluoride tetrahydrate ($KF \cdot 4H_2O$).

---

[11] http://web.mit.edu/invent/iow/telkes.html.

They may be also considered as alloys of inorganic salts and water, forming typical crystalline solids. Salt hydrates are non-toxic, non-flammable, and moderately corrosive. They possess higher thermal conductivity compared to organic PCMs, and they demonstrate volume changes on melting. The solid–liquid phase-change process of salt hydrates is actually a dehydration of hydrated salt, although this reaction resembles melting or freezing thermodynamically. A salt hydrate usually melts either to a salt hydrate with fewer moles of water, or to its anhydrous form (Telkes 1978).

A large number of salt hydrates have melting temperatures in a range suitable for building applications. Table 3.4 shows basic phase-change properties of selected inorganic PCMs of known phase-transition characteristics. In general, for inorganic salt hydrates, densities of solid and liquid phases are ranging between 1400 and 2000 kg/m^3 (for compounds with melting point suitable for building envelope applications). In building applications, where building materials are usually exposed to the hygrothermal processes, it is good to remember that salt hydrates are not stable in water environments and vapors and that they need to be protected from the negative environmental impacts. Macro-packaging, in form of plastic pouches or containers, is possible since salt hydrates are compatible with plastics. This fact makes PCM packaging process relatively easy and inexpensive. The only challenge is long-term durability of such products (tightness).

As stated earlier, the major performance issue, which affects a wide spread application of salt hydrates in buildings, is their incongruent melting. Incongruent melting in salt hydrates is caused by the fact that the amount of available water during crystallization is not always sufficient to dissolve all the solid phase salt. It leads to an irreversible phase separation during the PCM cycling. During the melting process, the lower hydrate (or anhydrous salt) settles down due to density difference. Most often, phase separation leads to formation of three stratified zones. At the bottom, there is a solid layer formed by salt precipitation from the solution. Above this, there is an active salt hydrate zone. At the top, there is a dilute solution of the salt hydrate. Unfortunately, this hydrate is later unavailable for recombination with water during the crystallization process. Furthermore, during the following melting/freezing processes, additional salt may break out from the solution, followed with its accumulation on the bottom. This phenomenon results in an irreversible melting–freezing of the salt hydrate leading to lower phase-change efficiency with each phase-transition cycle. Incongruent melting decreases the overall heat storage capacity since phase separation results in the top and bottom layers providing only sensible heat to the total quantity of heat of an energy storage system. Control of phase separation can be accomplished by chemical modification of the salt hydrate through an addition of suitable additives. Sometimes, an extra water principle is used to avoid incongruent melting. The method involves using extra water to dissolve the entire anhydrous salt during melting.

The other improvement method is thickening of PCM. Thickening of the material to gel form was suggested by Telkes (1952). Different thickening materials have been tried to minimize the phase separation in solid and liquid phases and also

Table 3.4 Basic phase-transition properties for selected inorganic PCMs

Name	Chemical formula	Melting point	Heat of fusion	References
Units		°C	kJ/kg	
Dipotassium phosphate hexahydrate	$K_2HPO_4\ 6H_2O$	14	109	Sharma et al. (2009)
Potassium fluoride tetrahydrate	$KF \cdot 4H_2O$	18.5	231	Abhat (1983), Heckenkamp and Baumann (1997), Zalba et al. (2003), Khundhair and Farid (2004)
Ferric bromide hexahydrate	$FeBr_3\ 6H_2O$	21	105	Sharma et al. (2009)
Manganese nitrate hexahydrate	$Mn(NO_3)_2 \cdot 6H_2O$	25.8	125.9	Nagano et al. (2000)
Calcium chloride hexahydrate	$CaCl_2 \cdot 6H_2O$	29–30	171, 190.8	Abhat (1983), Heckenkamp and Baumann (1997), Khundhair and Farid (2004)
Lithium nitrate trihydrate	$Li\ NO_3 \cdot 3H_2O$	30	296	Heckenkamp and Baumann (1997)
Calcium bromide hexahydrate	$CaBr_2 \cdot 6H_2O$	30.2	115.5	Lane (1980), Dincer and Rosen (2002)
Sodium sulfate decahydrate	$Na_2SO_4 \cdot 10H_2O$	32.4	254	Telkes (1975), Dincer and Rosen (2002)
Yavapaiite dodecahydrate	$KFe(SO_4)_2\ 12H_2O$	33	173	Sharma et al. (2009)
Lithium bromite dihydrate	$LiBr_2\ 2H_2O$	34	124	Sharma et al. (2009)
Sodium carbonate decahydrate	$Na_2CO_3 \cdot 10H_2O$	34.2	146.9	Telkes (1975), Abhat (1983)
Sodium orthophosphate dodecahydrate	$Na_2HPO_4 \cdot 12H_2O$	35.5–44	265	Belton and Ajami (1973), Telkes (1975)
Zinc nitrate hexahydrate	$Zn(NO_3)_2 \cdot 6H_2O$	36.2	246.5	Lane (1980), Dincer and Rosen (2002)
Ferric chloride hexahydrate	$FeCl_3\ 6H_2O$	37	223	Sharma et al. (2009)
Manganese nitrate tetrahydrate	$Mn(NO_3)_2\ 4H_2O$	37.1	115	Sharma et al. (2009)
Cobalt sulfate heptahydrate	$CoSO_4\ 7H_2O$	40.7	170	Sharma et al. (2009)
Potassium fluoride dihydrate	KF_2H_2O	42	162	Sharma et al. (2009)
Magnesium iodide octahydrate	$MgI_2\ 8H_2O$	42	133	Sharma et al. (2009)
Calcium iodide hexahydrate	$CaI_2\ 6H_2O$	42	162	Sharma et al. (2009)
	$K_2HPO_4\ 7H_2O$	45	145	Sharma et al. (2009)

(continued)

Table 3.4 (continued)

Name	Chemical formula	Melting point	Heat of fusion	References
Units		°C	kJ/kg	
Potassium orthophosphate heptahydrate				
Zinc nitrate tetrahydrate	$Zn(NO_3)_2$ $4H_2O$	45	110	Sharma et al. (2009)
Magnesium nitrate tetrahydrate	$Mg(NO_3)$ $4H_2O$	47	142	Sharma et al. (2009)
Calcium nitrate tetrahydrate	$Ca(NO_3)$ $4H_2O$	47	153	Sharma et al. (2009)
Ferric nitrate nano-hydrate	$Fe(NO_3)_3$ $9H_2O$	47	155	Sharma et al. (2009)
Sodium silicate tetrahydrate	Na_2SiO_3 $4H_2O$	48	168	Sharma et al. (2009)
Potassium orthophosphate trihydrate	K_2HPO_4 $3H_2O$	48	99	Sharma et al. (2009)
Sodium thiosulfate pentahydrate	$Na_2S2O_3 \cdot 5H_2O$	48–55	187, 209	Telkes (1975), Heckenkamp and Baumann (1997)
Magnesium sulfate heptahydrate	$MgSO_4$ $7H_2O$	48.5	202	Sharma et al. (2009)
Calcium nitrate trihydrate	$Ca(NO_3)_2$ $3H_2O$	51	104	Sharma et al. (2009)
Zinc nitrate bihydrate	$Zn(NO_3)_2$ $2H_2O$	55	68	Sharma et al. (2009)
Ferric chloride bihydrate	$FeCl_3$ $2H_2O$	56	90	Sharma et al. (2009)
Nickle nitrate hexahydrate	$Ni(NO_3)_2$ $6H_2O$	57	169	Sharma et al. (2009)
Sodium acetate trihydrate	$Na(CH_3COO) \cdot 3H_2O$	58	226, 264	Heckenkamp and Baumann (1997)
Magnesium chloride tetrahydrate	$MnCl_2$ $4H_2O$	58	151, 178	Sharma et al. (2009)
Ferric nitrate hexahydrate	$Fe(NO_3)_2$ $6H_2O$	60.5	126	Sharma et al. (2009)
Sodium aluminum sulfate decahydrate	$NaAl(SO_4)_2$ $10H_2O$	61	181	Sharma et al. (2009)
Sodium phosphate dodecahydrate	Na_3PO_4 $12H_2O$	65	190	Sharma et al. (2009)
Tetrasodium pyrophosphate decahydrate	$Na_2P_2O_7 \cdot 10H_2O$	70	184	Heckenkamp and Baumann (1997)
Lithium acetate dihydrate	$LiCH_3COO$ $2H_2O$	70	150	Sharma et al. (2009)

to prevent nucleating agents from settling down due to the gravity (Lane 1983; Ryu et al. 1992; Cabeza et al. 2003; Gok et al. 2009). In early PCM research with Glauber salt, a bentonite clay has been used as a thickening agent (Farid et al. 2004). In addition, a bentonite clay, due to the lower thermal conductivity of the mixture, provides an important feature for some building envelope applications, which is a reduction of overall heat transfer rate. PCMs modified in this way, if applied as a part of the building fabric, can generate a peak hour load shifting.[12]

Biswas (1977) found that for Glauber's salt ($Na_2SO_4 \cdot 10H_2O$), incongruent melting formed a heavy anhydrous salt (anhydrous sodium sulfate—Na_2SO_4) and a saturated solution of in the water. Initial attempts to solve this problem were by using a composition of 68.2 wt% sodium sulfate decahydrate ($Na_2SO_4 \cdot 10H_2O$) and 31.8 wt% H_2O. This salt modification enables more stable cycling; however, it also resulted in a reduction in the heat storage density. It also made the system to operate at a large temperature swing. Another similar research performed by McMullin and Ben-Abdallah (1987) demonstrated that at levels of excess water greater than 10 %, a separation of anhydrous salt and following this process, compound stratification was increasing significantly.

In later years, super-absorbent polymers (SAPs) synthesized from the acrylic acid have been tried as thickening agents. It was found that SAPs were effectively preventing phase separation of the high hydrate inorganic salts ($Na_2SO_4 \cdot 10H_2O$, $Na_2HPO_4 \cdot 12H_2O$, $Na_2CO_3 \cdot 10H_2O$) (Ryu et al. 1992). Inclusion of 3–5 wt% SAP was found to be an effective thickener for most of these salt hydrates. For the low hydrate inorganic salts ($CH_3COONa \cdot 3H_2O$, $Na_2S_2O_a \cdot 5H_2O$), adding small amounts of 2–4 wt% of carboxymethyl cellulose (CMC)[13] stabilized the salt hydrate (Lane 1980, 1983). In research study performed by Gok et al. (2009), Glauber's salt (Na_2SO_4 $10H_2O$) was successfully stabilized with different concentrations of polyacrylamide and gelatin gels, which were used to prevent incongruent melting.

Incongruent melting of salt hydrates is not their only problem. Their relatively high latent heat storage density is often compromised by a large supercooling effect. Most salt hydrates have poor nucleating properties resulting in supercooling of the liquid before crystallization process begins. This means that in order to achieve a reasonable rate of PCM crystallization, the salt hydrate solution has to be supercooled, since the rate of nucleation is generally very low at a temperature level close to the melting point. As a result, the phase-transition heat, instead of being discharged at fusion temperature, is discharged at much lower temperature. A supercooled liquid is thermodynamically not stable; the solid crystals can only grow due to the surface tension, once a solid nucleation aid additive is available (Skripov 1974; Kurz and Fisher 1992; Kashchiev 2000). Different foreign substances, introduced to the salt hydrates, can reduce their surface tension and improve their ability to crystallize. This process is sometimes called " PCM seeding." Small

[12]http://www.google.com/patents/US3986969.

[13]http://www.google.com/patents/US4585572.

amounts of added foreign materials are called "seeds," "nucleators," or "nucleation agents."

A large number of experimental works performed during last decades has demonstrated that the nucleation problem posed by PCM supercooling in salt hydrates can be adequately solved by adding suitable nucleation agents. One of the first inorganic PCMs for which nucleators were investigated was Glauber's salt—$Na_2SO_4 \cdot 10H_2O$—Telkes (1952, 1978), Biswas (1977). Glauber's salt has a phase-change temperature between 32 and 35 °C. Maria Telkes experimented with some success using sodium tetraborate, to minimize supercooling effect. Experimental results showed that supercooling in Glaber's salt could be reduced to less than 2 °C. The sodium tetraborate has a high density, and therefore, an additional thickening agent had to be added to the Glauber's salt to prevent settling. However, the above attempt was not fully successful. It was observed that upon continued testing, supercooling was still present, causing the material to remain in liquid form below its freezing point without solidifying.

In many commercially available PCM products, the performance target is to reduce supercooling of the used hydrated salt to less than 5 °C. At the same time, it is preferred that the chemical additives do not act as nucleating agents for the lower hydrated salts. In this way, any lower hydrate which might form could supercool to a temperature at which the first or original hydrate begins to solidify and segregation could be avoided. An effective nucleation agent has to provide sufficient characteristics to uphold surface tension and help in initiation of local crystallization (Günther et al. 2007). However, it is good to remember that most often each PCM compound requires a specific combination of performance enhancing additives.

Today, a development of a novel commercial inorganic PCM has most often to involve a development of the proprietary nucleation enhancement strategy. Furthermore, nucleation additives that are effective for one specific salt hydrate may not be necessarily effective when a different PCM is used. The nucleating additives are preferably added in an amount from 0.005 to 2 wt%. Very often, preferred nucleating additives include $Ba(OH)_2$, BaO, $BaCO_3$, $BaSO_4$, $Sr(OH)_2$, SrO, $SrCO_3$, $Mg(OH)_2$, MgO, $MgCO_3$, $NaOH$, Na_3AlF_6, CaO, $Ca(OH)_2$, $CaCO_3$, and CaC_2O_4.[14] Ryu et al. (1992) experimented with several other nucleators to reduce the effect of supercooling. Powders of carbon (1.5–6.7 μm), copper (1.5–2.5 μm), and titanium oxide (2–200 μm) were found to reduce supercooling of thickened $Na_2HPO_4 \cdot 12H_2O$. For thickened $CH_3COONa \cdot 3H_2O$, supercooling was reduced from 20 to 2–3 °C by adding 2 wt% potassium sulfate. Numerous compositions of PCMs were developed in the range of 30–60 °C to prevent both supercooling and phase separation (Abhat 1978, 1983; Sharma and Sagara 2005; Gok et al. 2006). This included Glauber's salt/SAP/sodium tetraborate at 94/3/3 wt%, with a melting temperature of 35 °C.

[14]http://www.freepatentsonline.com/EP0807150.html.

3.2.3.2 Metallics

This category includes low-temperature melting metals and metal eutectics. In most of metallics, melting temperatures are pretty high over 3000 °C. For example, carbon melts at 3527 °C. However, there is also a group of materials with lower melting temperatures, which may be useful for building applications. Mercury is a liquid at room temperature, and gallium melts at slightly below body temperature. In metallic substances, atoms are held together by metallic bonds, which are much stronger than intermolecular forces in other types of chemical compounds. In addition, in metals, atoms are spaced much tighter than the molecules in molecular solids. However, the fluid nature of the valence electrons in metals allows the nuclei and core electrons to slip past one another without breaking the metallic bonds. Since the atoms can move without disrupting bonds, metals have a wide range of melting points. Because of their high weight, metallics have not been seriously considered for PCM building applications, yet. However, when low volume or limited thickness of building envelope applications is required, metallics may be likely considered because of the high heat of fusion per unit volume and their high thermal conductivity. Unfortunately, thermal performance of metallics, suitable for building applications, is often compromised by their large supercooling effect (see Ge and Liu 2013). Some of the additional features of metallics are low specific heat and relatively low vapor pressure.

Metallic alloys have been already employed in several high-performance military systems and electronic applications (Antohe et al. 1996; Ge and Liu 2013). They are also considered for high-temperature solar applications[15] (Kotzé et al. 2012). The use of metallics poses a number of unusual engineering problems related with their unique physical characteristics. A major difference between the metallics and other PCMs is their high thermal conductivity. A list of phase-change properties for selected metallics and their eutectics with potential for building envelope application is given in Table 3.5.

3.2.4 Eutectics and PCM Mixtures

PCMs can be used in form of an individual chemical compound or as a eutectic mixture of several components (multi-component blend which acts as a single component). Eutectic compositions are mixtures of two or more chemical components, which solidify simultaneously out of the liquid at a minimum freezing point. On melting, all of these mixture constituents liquefy simultaneously. A large number of eutectics of inorganic and organic compounds have been reported Hasnain (1998), George (1989). Eutectics usually perform better than straight

[15]http://www1.eere.energy.gov/solar/sunshot/pdfs/csp_review_meeting_042413_chen2.pdf.

Table 3.5 Melting point and fusion heat for selected metallics

Name	Melting point	Heat of fusion
Units	°C	kJ/kg
Gallium–gallium antimony eutectic	29.3	n.a.
Gallium	30	80.3
Cerrelow eutectic	58	90.9
Bi–Cd–In eutectic	61	25
Cerrbend eutectic	70	32.6
Bi–Pb–In eutectic	70	29

Reference Sharma et al. (2009)

chemical compounds with respect to potential segregation during the phase transition (Lane 1989). Each component of the eutectic composition melts and freezes congruently forming a mixture of crystals during the solidification process (George 1989), again with unlikely separation. Therefore, most often, none of the mixture components can sink down due to a different density.

Eutectics have been relatively well known since their early applications dated back to the late eighteenth century. However, the separation and the life expectancy of these mixtures were poorly explored at that time, and therefore, their wide spread usage as PCMs is still limited.

Eutectic PCMs can be broadly grouped into two categories: "organic eutectic PCMs" and "salt-based products." However, there is also a third group of "organic–inorganic eutectic mixtures." The fundamental performance requirements for eutectic PCMs can be classified in the following three categories: (i) stable solution (no separation or degrading in long term), (ii) minimum supercooling, and finally (iii) close freezing and melting temperatures for all components of the eutectic composition.

Mixtures of fatty acids and fatty-acid esters are promising PCMs for building envelope applications. Their major advantages over paraffins and salt hydrates are relatively high phase-change enthalpy, long-term durability, and capability for easy impregnation of porous materials. Gypsum and concrete composites are superior PCM carrier candidates when it comes to building products using organic eutectic mixtures. Florida Solar Energy Center, USA, investigated impregnation of gypsum building products with eutectic mixtures of methyl esters, methyl palmitate, methyl stearate, capric acid, and lauric acid (Feldman et al. 1986; Shapiro 1987). Shilei et al. (2007) investigated thermal performance of wall board impregnated with eutectic mixture of capric and lauric acids. Karaipekli and Sari (2007) used eutectic mixture of capric and palmitic acid to impregnate building gypsum board. Dimaano and Watanabe (2002a, b) analyzed thermal performance eutectic blends of capric–lauric acid with pentadecane as a component of dynamic building envelopes. Table 3.6 shows phase-change thermal properties for a selection of best known fatty-acid blends and other eutectics, which can be used in building envelopes. Densities of fatty-acid eutectics are close to or just below a density of water.

Table 3.6 Melting point and fusion heat for selected eutectic mixtures of fatty acids and other organic eutectics

Name	Proportions	Melting point	Heat of ion	References
Units	wt%	°C	kJ/kg	
Fatty-acid eutectics				
Caprylic acid-1-dodacanol	70–30	6.5	171	Zou et al. (2011)
Capric–lauric acid plus pentadecane	90–10	13.3	142.2	Dimanno and Watanabe (2002a)
Capric–lauric acid	65–35	18	148	Dimanno and Watanabe (2002b)
Lauric–capric acid	55–45	21	143	Hawes et al. (1993)
Capric–myristic acid	73.5–26.5	21.4	152	Karaipekli and Sari (2008a, b)
Capric–palmitic acid	76.5–23.5	21.8	171.2	Karaipekli and Sari (2008a, b)
Capric–myristic acid	66–54	24	147	Lane (1980)
Capric–stearic acid	82–17	24.7	178.6	Karaipekli et al. (2009)
Lauric–palmitic acid	77–23	33	150.6	Nikolic et al. (2003)
Lauric–myristic acid	66–34	34.2	168.8	Sari (2005)
Lauric–palmitic acid	69–31	35.2	166.3	Sari (2005)
Lauric–stearic acid	75.5–24.5	37	183.7	Sari (2004)
Myristic–palmitic acid	58–42	42.6	169.7	Sari (2004)
Myristic–stearic acid	64–36	44.1	182.4	Sari (2005)
Palmitic–stearic acid	64.2–35.8	52.3	181.7	Sari (2004)
Naphthalene–benzoic acid	67.1–32.9	67	123	Hawes et al. (1993)
Other organic eutectics				
Triethylolethane + water + urea	45–55	13.4	160	Sharma et al. (2009)
Butyl stearate–butyl palmitate	49–48 + 3 addditives	17	138	Feldman et al. (1986)
Methyl stearate–cetyl stearate	91–9	22.2	180	Nicolic et al. (2003)
Methyl stearate–methyl palmitate	86–14	23.9	220	Nicolic et al. (2003)
Methyl stearate–cetyl palmitate	91–9	28.2	189	Nicolic et al. (2003)
Triethylolethane + urea	62.5–37.5	29.8	218	Sharma et al. (2009)
Acetamide–urea	63.5–37.5	53	na	Abhat (1983)

Elimination of component separation and reduction of subcooling are two major reasons for mixing of inorganic compounds into eutectic PCM solutions. Sometimes, the following performance enhancing additives need to be added as well: (i) nucleating agents, which encourage crystal formation and therefore minimize super cooling; (ii) freeze depressants to achieve a lower phase-change temperature; and (iii) gelling/thickening agents in order to overcome separation and long-term degrading problems. As shown in Table 3.7, several inorganic eutectic compositions show a melting temperature within a range required for building envelope applications and relatively high heat storage density. For this group of eutectics, densities are ranging most often between 1400 and 1600 kg/m^3.

3.3 Long-Term Durability of PCMs in Building Envelopes

Durability of building envelope materials can be defined in many different ways. In contrast to the simple dictionary definition, durability as applied to buildings must offer not only mere material survival, but also capability of performing desired functions. The Canadian Standards Association provides one of the most recognized definitions of building durability in North America. According to their "Guideline on Durability in Buildings" (CSA S478-95, Rev. 1997),[16] the broad definition of durability of the building system is its ability to perform for a desired or designed time period without significant deterioration or loss of functionality and without unforeseen cost for maintenance or repair.

Long-term durability of building materials is usually considered as one of the basic performance requirements. It is measured as the length of the service time. However, the useful life of a building component is always related to the particular combination of environmental factors to which it is subjected. So, both long-term durability and service life span must always be related to the particular conditions involved (Sereda 1969). Long-term durability criteria for basic PCMs products are pretty well defined today. However, specific long-term performance characteristics for PCM-enhanced building envelopes have not been fully understood yet, and corresponding performance regulations are still waiting for development. In 2004, in Europe, several German companies founded the Gütegemeinschaft PCM e.V. (quality control association) and commissioned the Bayerische Zentrum für Angewandte Energieforschung e.V. (ZAE Bayern—Bavarian Centre for Applied Energy Research) and the Fraunhofer Institute for Solar Energy Systems (ISE) to develop appropriate procedures for PCM quality control. At that time, the main objective was to guarantee the quality of both PCMs as well as PCM-enhanced products. After completion of the initial work, the RAL[17] Quality Mark was introduced in April 2007. The main quality criteria defined in RAL are phase-transition heat as a function of temperature, the cyclic reproducibility of the heat storage process, and the thermal conductivity of the PCM.

When applied to building envelopes containing PCMs, long-term durability can be defined in a similar manner, however, with several necessary significant distinctions. In building envelopes, PCMs represent a direct part of the building structure, in contrary to PCMs used in forms of heat storage containers, heat

---

[16]http://webstore.ansi.org/RecordDetail.aspx?sku=CSA%20S478-1995%20(R2007)&source=msn&adgroup=csa.

[17]http://195.20.235.12/pdf/RAL-GZ_896.pdf.

Table 3.7 Basic phase-change characteristics of inorganic eutectics

Chemical formula	Proportions	Melting point	Heat of fusion	Reference
Units	%	°C	kJ/kg	
Na_2SO_4–NaCl–NH_4Cl–H_2O	32–14–12–42	11	na	Abhat (1983), Sharma and Sagara (2005)
$C_5H_5C_5H_5$–$(C_6H_5)_2O$	26.5–73.5	12	97.9	Abhat (1983), Sharma and Sagara (2005)
Na_2SO_4–NaCl–H_2O	37–17–46	18	na	Abhat (1983)
$CaCl_2$ $6H_2O$–NH4NO_3–NH_4Br	96–2–2	20	141	Kimura and Kai (1984)
$CaCl_2$ $6H_2O$–KNO_3–KBr	96–2–2	23	138	Kimura and Kai (1984)
$CaCl_2$ $6H_2O$–$Ca(NO_3)_2$ $4H_2O$–$Mg(NO_3)_2$ $6H_2O$	93–5–2	24	125	Kimura and Kai (1984)
$CaCl_2$ $6H_2O$–$MgCl_2$ $6H_2O$	66.6–33.3	25	127	Heckenkamp and Baumann (1997)
$CaCl_2$–NaCl–KCl–H_2O	48–4.3–0.4–47.3	26.8	188	Heckenkamp and Baumann (1997)
$Ca(NO)_3 \cdot 4H_2O$–Mg $(NO)_3 \cdot 6H_2O$	47–53	30	136	Abhat (1983), Zalba et al. (2003)
$Na(CH_3COO)$ $3H_2O$–$CO(NH_2)_2$	60–40	31.5	226	Wada et al. (1983)
$CaCl_2$ $6H_2O$ $Ca(NO_3)_2$ $4H_2O$ Mg $(NO_3)_2$ $6H_2O$	85–15	49	170	Kimura and Kai (1984)
$NaCH_3COO$ $3H_2O$ NaBr $2H_2O$	90–10	51	175	Kimura and Kai (1984)
$Mg(NO_3)_2$ $6H_2O$–NH_4NO_3	61.5–38.5	52	125.5	Lane (1980)
$Mg(NO_3)$ $6H_2O$–$MgCl_2$ $6H_2O$	58.7–41.3	58	132.2	Lane (1980), Heckenkamp and Baumann (1997)
$Mg(NO_3)_2$ $6H_2O$–$Al(NO_3)_2$ $9H_2O$	53–47	61	148	Abhat (1983)

exchangers, hydronic tubing, etc. The major difference is that PCM building systems cannot be freely replaced or repaired in case of their failure. They have to stay fully functional as long as building is in operation. For example, recent surveys indicate that the average building life in North America is about 80 years, in Switzerland within 70–90 years, and in Britain even over 130 years (Bowyer 2005). The National Standard of China stipulates that the building life is 100 years and 50–100 years for the main construction of high-rise buildings and common buildings, respectively.[18] Detail reports in this field are not available for China yet, however, according to Zhisheng et al. (2006), the building service life, especial for residential buildings in large cities, may often be less than 30 years. Since different configurations of PCM-enhanced building envelops can be located in different parts of the

[18]Uniform credibility standard of building construction. (GB50068-2001) MOC-2001: Beijing, China.

building, these technologies have to be individually designed for each specific location and application. That means, a nominal life span for the PCM-enhanced building envelope materials has to be at least as long as expected service time of the building component, which hosts the PCM system.

PCMs, when applied to building envelopes, cannot compromise their desired attributes, performance characteristics, and overall functionality. Although some building shell components may perform according to a single required task, the modern building envelope must usually fulfill many functions. First and foremost, the building envelope acts as a prime construction component of a building. It must resist all environmental loads such as wind, rain, snow, dynamic and static service, and seismic loads, effectively transferring these loads to the building's structural components. Building shell serves also as an environmental barrier to resist the intrusion of moisture in many forms, including rain, snow, hail, ice, and vapor. Environmental protection function also includes building air sealing, which plays a very important role both in building's long-term hygrothermal and energy performance. Finally, as one of the most significant contributors to a building's energy efficiency, building exterior shell must control heat transfer between interior of the building and external environment. The basic performance requirements for building envelopes include noise and interior lighting control, as well. Finally, the building envelope must also provide a satisfactory level of fire resistance to protect against ignition in case of contact with a fire source, enable evacuation of the building, and to reduce the spread of fire to adjacent buildings.

The life expectancies of building envelope components depend on used material configuration, on the quality of installation, the level of maintenance, climatic conditions, and the intensity of use. Some components may remain functional but become obsolete due to changing code requirements, architectural, styles and preferences, or improvements in newer products, while others may have a short life expectancy due to intensive use. As stated above, live expectancy of the PCM-enhanced building envelopes should be at least as long it is considered now for analogical conventional systems without PCM. Based on the National Home Builders Association (NHAB) data, Table 3.8 shows estimated life expectancies for building envelope components of North American residential buildings (NAHB 2007).

Unfortunately, very little of service life data are available today for modern building envelopes containing PCM, and what data are available appear to contain many limitations and contradictions coming from the fact that both general building lifetime as well as life span data for different building envelopes varies from country to country. Evaluation of PCM's stability and potential of failure over a long period of application are important challenges for the successful application of suitable PCMs in building envelopes. Theoretically, most of PCM-enhanced building envelopes should undergo at least one melt–freeze cycle per day. However, it is good to remember that many floor and ceiling PCM applications integrated with hydronic heating/cooling systems can easily reach even 20 cycles per day. In addition, an annual number of phase-change cycles is usually geographic location specific, being a function of exact climatic conditions. Table 3.9 presents example minimum cycling requirements for most typical PCM applications in

Table 3.8 Estimated life expectancies for building envelope components of North American residential buildings

Material name	Comment	Live expectancy (years)
Engineered wood systems	Floor and roof trusses and laminated strand lumber are expected to last at least 30 years	30
Wood framing and plywood boards	Framing and structural systems have extended longevities: poured concrete systems, timber frame houses, and structural insulated panels will all last a lifetime. Wall panels and roof and floor trusses will similarly last a lifetime	Lifetime of the building depends on country
Plywood		30
Oriented strand boards (OSB) and particle boards		25–30
Gypsum boards		30
Concrete and masonry walls	Masonry is one of the most durable components of a building. Brick veneers and brick walls have an average life span over 100 years	100
Building foundations	Poured concrete foundations and concrete block foundations—lifetime of the building	Lifetime of the building depends on country
Hardwood floors		100 or more
Marble, slate, or granite floors		100
Vinyl floors		50
Linoleum floors		25
Carpet floors		8–10
Roofing	The lifetime of a roof depends on climatic conditions, building design, material quality, and maintenance	
Slate, metal, copper, and clay/concrete tile roofs		50
Asphalt shingles		20
Fiber cement shingles		25
Wood windows		30
Aluminum windows		15–20

Source NAHB (2007)

building envelopes, considering that PCM is used for cooling application in southern US climate. Conservatively, 10 cycles per day were considered for dynamic building envelope applications integrated with mechanical systems. Building life span was assumed as 80 years (data available for North American houses—see above). It can be seen that an expected lifetime number of cycles can

Table 3.9 Minimum cycling requirements estimated for most typical PCM applications in building envelopes

PCM application name	Expected number of PCM cycles per day	Expected number of days when PCM is cycling per year	Life expectancy of building envelope component (years)	Minimum number of PCM cycles during the system lifetime
PCM floor tiles, or PCM-enhanced concrete slab—passive use	1	300	80	24,000
PCM floor tiles, or PCM-enhanced concrete slab—used in conjunction with floor hydronic system	10	300	80	240,000
PCM-enhanced gypsum board or plaster facing building interior—installed on the lightweight wall	1	300	30	9000
PCM-enhanced composite board, or array of PCM pouches installed behind the gypsum board or interior plaster	1	300	30	9000
PCM-enhanced concrete wall	1	240	80	19,200
PCM-enhanced thermal insulation installed in the wall cavity, wall core	1	240	80	19,200
PCM-enhanced composite board, or array of PCM pouches installed on the exterior side of the wall	1	240	30	7200
PCM ceiling tiles—passive use	1	300	30	9000
PCM ceiling tiles used in conjunction with dynamic ceiling system	1–10	300	30	9000–90,000
PCM-enhanced attic floor insulation[a]	1	240	80	19,200
Composite PCM board or array of PCM pouches installed on the roof deck level	1	240	50	12,000
PCM-enhanced flat roof membrane	1	240	15–20	3600–4800

*There is an easy access to PCM-enhanced attic insulation, so cycling requirement for this application can be lower considering that insulation can be replaced during its lifetime

be between 3.600 for flat roof membrane applications and 240.000 for floor and ceiling technologies integrated with mechanical cooling systems. This is a 66-fold difference, which needs to be considered during selection of the PCM technology for specific application. In most common PCM applications in gypsum boards and plasters, an expected lifetime number of PCM cycles may reach 9000.

However, long-term durability performance requirements for PCM-enhanced building envelopes can not only be focused on PCM thermal characteristics. Depends on their locations within the building, they need to also follow specific design and performance requirements for specific building envelope components. This includes structural, thermal, hygrothemal, and fire safety requirements. These regulations and corresponding testing methods do not exist yet for PCM-enhanced building envelopes. It is expected that following a current growth in the area of PCM building envelope applications, new engineering performance analysis methods and tools will be developed soon for these systems. They should enable a development of necessary testing methods and following code requirements.

3.4 PCM Packaging and Encapsulation Methods

In building envelopes, PCMs store heat or cold during their phase-transition cycles. It is important to keep in mind that any potential volume loss of the acting PCM may directly affect an overall performance of the entire envelope system. In addition, since PCMs are often directly embodied into building materials, it is obvious that they should not damage their carriers. Unfortunately, a large number of the solid/liquid PCMs need to be encapsulated because their liquid state is very mobile, which may either cause esthetical surface discolorations (PCM leaking to the surface), or compromise PCMs long-term performance due to a drop of acting PCM volume. Some inorganic PCMs may also cause corrosion of building materials. Incorporation of PCMs into building envelopes poses several design challenges, critical ones being the integration method with building structure as well as the location of PCMs within the volume of the building envelope component. Such location is essential for building envelope performance in terms of peak load shifting and control of heat transfer rates, which is essential for energy conservation and dynamic load management (Kossecka and Kośny 2010; Childs and Stovall 2012; Kośny and Kossecka 2013a). For many additional reasons, such as thermal performance, structural requirements, building code regulations, esthetics, fire safety, and long-term durability, PCM integration method needs to be carefully analyzed in order to provide maximum overall performance at a minimum cost and with a minimum chance for system failures.

The following PCM containment methods can be considered: (i) bulk material packaging using glass, metal, or plastic containers (semi-transparent windows, partitions, ceilings with metal PCM containers, etc.…), (ii) foil packaging (PCM panels or arrays of PCM pouches), (iii) direct PCM integration with the carrier material (impregnated or shape-stabilized PCM products), (iv) macro-encapsulation

Fig. 3.4 Different methods of PCM encapsulation which can be used in building envelopes. **a** Transparent, modular window containers with melted inorganic PCM—Dörken GmbH & Co. KG, Germany. **b** Aluminum and plastic foil packages containing PCM—PCM Products Ltd, UK. **c** Macroencapsulation of PCM in form of spheres—PCM Products Ltd, UK. **d** Addition of the microencapsulated PCM pellets to cellulose insulation during experimental production—author's archive

(spheres, tubing, etc.…), and finally (v) microencapsulation. Figure 3.4 presents a selection of encapsulated PCM products.

Packaging or encapsulation serves as a barrier between PCM and the surrounding environment. It prevents liquid PCM from leaking out of the carrier material. Encapsulant also protects PCM from reacting with the outside environment and sometimes provides mechanical stiffness to the PCM product. Several additional functions of PCM packaging or encapsulation can be considered as well, for example, (i) improved heat transfer characteristic of the whole assembly (packaging material or encapsulant may contain heat conduction enhancing additives), (ii) encapsulant may contain fire retarder in order to improve ignition and flame resistances, (iii) packaging material can be a part of the building's moisture and air sealing membrane, and (iv) reflective surface of the PCM package may serve as a radiant barrier. Today, a number of different packaging and encapsulation techniques are offered depending on properties of the PCM and the operational requirements. They are containers, boards containing PCM, bags, arrays of pouches, spheres, microcapsules, etc. It is good to remember that most of PCMs are not very costly by themselves, but their packing or encapsulation processes, necessary to achieve reliable and consistent long-term performances, make a final product cost significantly higher.

The largest packaged PCM products which are used in building envelopes are PCM containers. Historically, transparent glass or plastic modules with PCM have been considered for solar window applications (Fig. 3.4a). These systems require extensive optical design and heat transfer optimization enabling sufficient solar energy extraction. Likewise, opaque containers can be used for harvesting heat from roofs and attics. Flat metal containers with PCM are often used for heat storage in dynamic ceiling systems as well. Main concerns with bulk heat storage systems may be their long-term tightness, potential corrosion of the container material, and, in case of window applications, long-term optical properties of the system. On the other hand, bulk PCM containers in building envelopes can significantly reduce the cost of PCM application through elimination of the expensive encapsulation processes.

A major advantage of using PCMs encapsulation is that, in general, it helps to overcome some problems associated with phase segregation (Hadjieva et al. 2000; Mehling and Cabeza 2008; Pasupathy et al. 2008). Encapsulation is absolutely critical for all water-based PCMs (i.e., salt hydrates, eutectics). If these materials are exposed to the surrounding atmosphere, their water content changes, either by evaporation or, in some cases, by absorption. Because most of salt hydrates only perform well as PCMs for the specific water to salts ratios, any change in this ratio results in PCM desegregation and ultimately loss of phase-transition performance. The other prospective source of durability problems is corrosion potential of some PCMs. This issue has been thoroughly studied during last decade. The corrosion caused by some salt hydrates and salt eutectics was experimentally analyzed by Porosino (1988), Ettouney et al. (1990), and Groll et al. (1990). Cabeza et al. (2001a, b) investigated corrosion resistance of aluminum, brass, copper, steel, and stainless steel in contact with a large selection of molten salt hydrates (zinc nitrate hexahydrate, sodium hydrogen phosphate dodecahydrate, calcium chloride hexahydrate, sodium carbonate, potassium hydrogen carbonate, potassium chloride, water, sodium acetate trihydrate, and sodium thiosulfate pentahydrate). In addition, Sari and Kaygusuz (2003) studied in long term a corrosion potential of fatty acids (stearic, palmitic, myristic, and lauric acid) against stainless steel, carbon steel, aluminum, and copper. No major signs of corrosion caused by fatty acids were found in all tested metals.

Finally, PCM containers if exposed to high temperatures may start to leak. During summer, roof surfaces and attic air temperatures may easily reach 70 °C during mid of the day. As shown in Fig. 3.5, PCM leaking through building skin materials may cause esthetic surface discolorations.

Metallic or plastic foil packages are relatively low-cost options for encapsulation of inorganic and organic PCMs (Figs. 3.4b and 3.5). Foil packaging is a well-

Fig. 3.5 Leaky inorganic PCM–foil pouches leaving esthetic discoloration marks on the metal roofing panels—*left* discoloration of the bitumen roofing membrane caused by the inorganic PCM sipping out from the leaky containers—*right*. *Source* Author's archive

established technology widely used by food and healthcare industries. It has been lately adopted by PCM companies. Plastic foil containers, bags, or arrays of pouches are the cost effective way of PCM encapsulation, which have been tried for last two decades. Because of cost advantage over microencapsulated products, they may soon become the dominant trend when it comes to PCM packaging. Very often, these systems are installed within the core of building envelopes as continues impermeable layers of foil containing arrays of PCM pouches, or just as membranes with directly enclosed PCMs. It is important to remember that such configurations may significantly modify hygrothermal performance of the hosting building envelope. That is why detailed hygrothermal analysis is recommended for foil PCM assemblies of that type. Potential disadvantages associated with this technology are possible plastic corrosion by some bio-based PCMs and, in case of some PCM applications, a need for compensation for PCM's volume change during phase transition. Another limitation is coming from the fact that not all package sealing methods (high-temperature welding) may be resistive enough to high temperatures, as observed in, for example, roof and attic applications (see Fig. 3.5). It is necessary that these foil packages remain tight after desired number of cycles with temperature peak reaching its summer maximum. Aluminum foil can be used for packaging of PCM when reflective surface characteristics are desired, or PCM may cause corrosion of the plastic foil. An application of aluminum foil also improves heat transfer between PCM and surrounding environment. Most popular applications of these technologies are in walls and ceilings—Murugananthama et al. (2010) as well as in attics and roofs (Kośny et al. 2012b).

Another direct encapsulation method, which is gaining popularity today, enables fabrication of a novel PCM material known as a shape-stabilized (SS) PCM. In this material, PCM can retain the shape of the solid structure during phase transition. SS-PCM is a composite of PCM with another material. For example, SS-paraffin composite can consist of paraffin incorporated on a microscopic level into a supporting porous structure created by high-density polyethylene (HDPE) (Xu et al. 2005). SS-PCMs offer a large number of benefits comparing to more traditional encapsulation methods. Most important are as follows: (i) apparent heat capacity comparable with other PCM products, (ii) suitable overall thermal conductivity, (iii) good compatibility with building envelope materials, (iv) carrier material can serve as a structural components of building envelopes, and finally (v) no more need for an encasement as PCM does not tend to flow out of the porous structure during melting. Inaba and Tu observed no leakage of paraffin through high-density polyethylene (HDPE) network for paraffin loading levels as high as 75 % by weight (Inaba and Tu 1997). In another project, Yinping prepared SS-PCM with 80 % by weight loading of paraffin and found that no containment was required, i.e., samples can be cut into pieces without any drainage(Yinping et al. 2006). Keep in mind that even with all the benefits mentioned above, shape-stabilized application usually adds up notable to the cost of the final PCM product

Macro-encapsulation utilizes different three-dimensional geometries to prevent PCM from contacting with outside environment. Most often, macro-encapsulated PCMs are manufactured in form of small rectangular containers, tubes, and spherical products (Fig. 3.4c). The skin of the container provides self-supporting structure isolating PCM from the surrounding environment. It has been found by Wei et al. (2005) that from the heat transfer point of view, the highest thermal performance can be observed for spherical PCM capsules, followed by cylindrical, plate type, and tubular encapsulations. Figure 3.6 shows different phases of paraffinic PCM melting inside a 85-mm spherical container. Because the sizes of macro-encapsulated PCM products usually vary between 1 and 10 cm, it is a designing challenge to place them within building envelopes. Tubular encapsulations are probably the easiest for building envelope applications (Zhang et al. 2005).

Barba and Spiga (2003) have investigated PCM phase-change processes in different sizes of encasements, and they have demonstrated that the PCM solidification time is increasing proportionally to the square of the capsule radius. These findings were just a confirmation of a long-known awarness that a size of the PCM encasement plays an important role. In years, a size of encapsulated particles was made significantly smaller and PCM microencapsulation was widely adopted by PCM industry worldwide. As a result, a notable increase of heat transfer intensity was reported in case of building products using microencapsulated PCMs.

Outside of inorganic PCMs, the second group of heat storage compounds, which really benefits from encapsulation consists of paraffins and some organic PCMs. For decades, it had been a practical challenge to find a safe way of adding organic PCMs to building materials. At the beginning, PCMs were added to porous building materials using immersion method (Salyer and Sircar 1989; Shapiro 1987; Neeper 1989; Tomlinson et al. 1992). Unfortunately, it was common at that time that in PCM-enhanced building products (gypsum boards, or concrete components), liquid PCM was leaking from the surface leaving unesthetic surface discoloration in the best case. That problem was successfully solved by the development of PCM

Fig. 3.6 Phases of paraffin melting in 85-mm-diameter spherical glass container at 70 °C. *Source* Khot et al. (2011). **a** 0 min, **b** 20 min, **c** 40 min, **d** 50 min, **e** 60 min, **f** 70 min

microencapsulation processes. In 1953, a first major patent claim was filed by Green et al.[19] for a new type of a copy paper which worked with oil-filled microcapsules. This invention initiated the large industrial use of microcapsules having composite or polymeric shells with sizes varying between 3 and 100 µm in diameter. Since microcapsules are very robust, they allow multiple material applications performed by different industries. Since 1950s, this technology has been widely employed, for example, in production of pharmaceuticals, cosmetics, PCMs, and in printing.

PCM microencapsulation is a technique by which solid or liquid active material with phase-change capability is enclosed within a second material. The major purpose

[19]US Patent 2,730,456, "Manifold record material", NCR Company, filed June 30, 1953.

of this process is shielding the PCM from the surrounding environment.[20,21,22] Thus, the active ingredient constitutes a core of the small micro-pellet, and the surrounding material forms the protective shell. In microencapsulation, PCM loading is usually between of 70 and 85 % (Zalba et al. 2003; Mehling and Cabeza 2008; Kośny et al. 2013). Microencapsulation can be either achieved by means of physical or chemical processes. Phase changing particles enclosed in such manner can be introduced to almost any carrier material used in buildings (i.e., cement or gypsum composites, thermal insulations, building membranes and foils, coatings, particle boards) (Riza 2007). Water-insoluble paraffinic hydrocarbons are widely considered as the easiest materials to encapsulate today. The best known microencapsulation processes are as follows: complex coacervation with gelatine, interfacial polycondensation, polycondensation of melamine (or urea) formaldehyde resins, and the radical polymerization of suitable monomers.[23] Amino resins and melamine–formaldehyde are probably the most common today among a multitude of possible shell materials for paraffinic PCM microcapsules. In addition, microencapsulation helped in improvement of the product ignition resistance, through possible incorporation of fire retardants into the skin of microcapsules (Kośny et al. 2009). However, PCM microencapsulation exhibits the highest production cost among most of PCM containment methods, which limits the economic viability of such technology (Zhao and Zhang 2011; Kośny et al. 2013b).

Microencapsulation of salt hydrates becomes even more challenging due to the fact that salt hydrates are hydrophilic in nature and contain well-defined water fraction. To be able to be used in building applications, the microcapsule shell needs to act as a water–vapor barrier and be made of hydrophobic materials to prevent the salt hydrate from evaporating. Hessbrugge and Vaidya (1997) developed a novel technique for the encapsulation of water-soluble salts. The method employs a simple process, where additives such as emulsifiers and acid acceptors, that are needed to stabilize the aqueous–organic interface and to remove the excess acid formed due to hydrolysis, are not required. Further development of his technology may result in the future in a less expensive microencapsulation process since the post-treatment required to remove additives and their byproducts is no longer necessary. In addition, there is less likelihood of contamination of encapsulated product due to the absence of trace amounts of the reaction components and their by-products.

A similar attempt of microencapsulation of inorganic PCMs was performed by the Bavarian Center for Applied Energy Research (ZAE Bayern), in Würzburg, Germany. Brandt et al. (2006) encapsulated powdered Na_2CO_3 hydrate particles with an inorganic SiO_x layer via a solgel process.

Recently, there have been several successful efforts to microencapsulate salt hydrates. Fraunhofer Institute for Silicate Research (ISC) in Würzburg, Germany,

---

[20]http://www.zae.uni-wuerzburg.de/files/11-jahns_mikroverkapselung.pdf.

[21]http://microteklabs.com/technical_overview.html.

[22]http://www.functionalpolymers.basf.com/portal/basf/ien/dt.jsp?setCursor=1_290868.

[23]http://microteklabs.com/u-s-patents.html.

has developed a method where a salt hydrate melt is encapsulated by an inorganic–organic hybrid polymer known as ORMOCER®.[24,25] The process involves injecting salt hydrate melt drops into the ORMOCER® solution that solidifies locally as it comes in contact with the surface of the drop, thus forming a hard encapsulating layer around the salt hydrate drop. The encapsulated drops are removed from the coating solution via mechanical means and washed thoroughly before allowing them to dry in air under ambient condition. Salt hydrate drops subsequently crystallize as they cool down. Capsules produced in such manner have diameters ranging from a few μm to a few mm. Although the final microencapsulated product still has some water permeability and stability issues, nonetheless it is an encouraging step.

3.5 Enhancement Methods for Improved Thermal Conductivity of PCMs

One of the major advantages of using PCM-enhanced building envelopes is that these systems have capability to amend thermal response of the building components. In some designs of low-energy buildings, it is beneficial when dynamic thermal excitations, generated by the exterior environment, can be transferred to the interior space with significant delay time (Kośny and Kossecka 2013). In these situations, high thermal conductivity of PCMs is not necessary because the major design objectives for these thermal systems are peak load reduction and time shift of thermal response. Based on earlier investigations by Kossecka and Kośny (1998, 2002), Halford and Boehm (2007), Kośny et al. (2008), and Farid and Mecaial (2008), such performance requirements can be easily satisfied by different configurations of thermally resistive and thermally massive components sandwiched together or thick layers of PCM-enhanced thermal insulations. In some other building applications, where PCM-enhanced products are expected to absorb and release heat in a fast manner, high thermal conductivity is one of the key characteristics of interest. Very good examples of such PCM products are all PCM-enhanced boards or plasters facing interior of the building. Typically, these products are using either microencapsulated or shape-stabilized organic PCMs (paraffins or fatty acids) of relatively low thermal conductivities. There is significant performance improvement potential in building applications requiring fast absorption and release of latent heat. Most of these applications use today low conductive organic PCMs. In addition, because of thermal comfort prerequisites for building interior space, these PCM-enhanced systems usually have to operate within small temperature differences. Furthermore, these materials have relatively high latent heat compared to their sensible heat storage. That is why a key challenge which needs to be addressed in

[24]http://www.messib.eu/about_project/messib_innovative_elements/phase_change_materials/microencapsulation_of_salt_hydrates.php.

[25]http://www.freepatentsonline.com/y2011/0017944.html.

such PCM applications is enhancement of their thermal conductivity. Since building envelope acts as a major building's thermal shield, the above approach is conflicting with high R-value requirements so typical for low-energy buildings. While salt hydrates exhibit thermal conductivities that rarely surpass 0.6 W/m K, most of organic PCMs have thermal conductivities two to four times lower Naumann and Emons (1989), Zalba et al. (2003), IEA (2005), Sharma et al. (2009). So potentially, the easiest solution would be to switch in building envelopes from using organic PCMs to inorganic PCMs' applications. This would bring improved thermal conductivity and higher phase-change enthalpy. In addition, an application of naturally non-flammable microencapsulated inorganic PCMs would also solve fire safety problems allied to many building technologies utilizing paraffinic PCMs today. However, microencapsulation of these materials needs to reach a mature development stage, when long-term durability will be not an issue any more.

Various heat conduction enhancement methods have been already considered for PCMs. Theoretically, there are two options for solving this problem. On one hand, the nominal distances for heat transfer by conduction in PCM's enclosures needs to be reduced. This can be achieved by stuffing the PCM into relatively small capsules or by using microcapsulation. On the other hand, the thermal conductivity can be enhanced by the addition of highly conductive materials to the PCM. Most commonly, graphite, carbon fibers, or metal in form of shavings, or powders can be added to the PCM to form a composite. Alternatively, PCM may be incorporated into the highly conducting matrix such as expanded graphite, metallic, or polymer matrix. Ideally, this matrix network should hold the PCM inside the pores even when it is in a liquid phase. In case of microencapsulated PCMs, it is also possible to add conduction aid additives to the microcapsule shell material.

Practically, numerous means of enhancing the PCM heat transfer characteristics have been already developed and tested. An example of early work in this area is blending of PCMs with highly conductive substances. Different material configuration had been tested starting from 1980s. More recent work performed by Mettawee and Assassa (2007), focused on organic PCMs, demonstrated that in extreme scale, an introduction of 50 % by weight of aluminum powder to paraffinic PCM can improve the overall thermal conductivity by a factor of 20. However, it is good to remember that such high percentage of foreign non-PCM additive, introduced to the mixture, will significantly reduce the overall latent heat storage capacity, which will later impact dynamic system performance. The most often analyzed method for thermal conduction enhancement is an addition of highly conducting metal particles and powders to the PCM. Siegel (1977), Hafner and Schwarze (1999) analyzed thermal performance of PCM dispersed with high conducting particles. Promising results on improving the conductivity of PCMs have been also reported by Koizumi (2004), who analyzed an impact of copper plates on thermal performance of n-octadecane, and Wang et al. (2010) who tested an addition of aluminum powder to the PCM. Analogously, Fukai et al. (2000) experimented with carbon fiber as a heat transfer improving material for organic PCMs. Different forms of graphite are also often considered as highly conductive additives. Composite, paraffin–graphite blends have been analyzed by Py et al. (2001) and Marín et al. (2005). This research demonstrated that moderately compressed exfoliated graphite leads to highly porous graphite matrices which can be used as a highly conductive carrier for PCM applications. These graphite/PCM composites offer excellent heat conduction and thermal storage characteristics. They are usually fabricated by uniaxial compression of a mixture of solid PCM powders and expanded natural graphite. An impact of graphite powder and expanded graphite as heat conduction enhancement factors was investigated, for different organic PCMs, by Zhang and Fang (2006), Pincemin et al. (2008), Kim and Drzal (2009), Cheng et al. (2010a, b), Chen et al. (2013). In addition, Yin et al. (2008) tested absorption of PCM by the expanded graphite and its impact on thermal conductivity of the final product. Other developments in this area include (i) incorporation the PCM into the highly conducting metal matrix (Kamimoto et al. 1986; Hodendom and Bart 1992; Tong et al. 1996; Hafner and Schwarzer 1999), (ii) a usage of exfoliated graphite for improving the thermal conductivity of the shape-stabilized PCM composite (Sari 2004), or (iii) an application of metal screens/spheres placed inside the PCM (Ettouney et al. 2004). A unique heat storage system, based on n-octadecane retained by capillary forces in a porous silica carrier, was presented by Erk and Dudukovic (1996). This configuration reduces overall cost of application, provides higher energy density, and gives higher rates of energy storage and heat release. Finally, Leoni and Amon (1997) employed aluminum foam to enhance the heat transfer process in latent heat storage system.

In PCM applications described above, thermal conductivity improvements can vary from 20 to up to 1000 times and depends on the specific material configurations and blending methods (Ning-Wei et al. 2009). For example:

1. The introduction of 10 % by weight of expanded graphite to the composite made of capric–myristic acids and expanded perlite, increased the overall thermal conductivity by 58 % as reported by Karaipekli and Sari (2008a);
2. After the introduction of 2 % of expanded graphite into the mixture of capric–myristic acid and vermiculite, 85 % thermal conductivity improvement was observed by Karaipekli and Sari (2008b);
3. An addition of 10 % of expanded graphite into stearic acid, increased overall thermal conductivity by 266 % as reported by Karaipekli et al. (2007).

References

Abhat A (1978) Investigation of physical and chemical properties of phase change materials for space heating/cooling applications. In: Proceedings of the international solar energy congress, New Delhi, India

Abhat A (1983) Low temperature latent heat thermal energy storage: heat storage materials. Sol Energy 30(1983):313–332

Antohe BV, Lage JL, Price DC, Weber JL (1996) Thermal management of high frequency electronic systems with mechanically compressed microporous cold plates. Thermal management of commercial and military electronics. In: Proceedings of ASME national heat transfer conference, pp 179–186

ASTM (1992) ASTM D92-12b Standard test method for flash and fire points by cleveland open cup tester. ASTM International, USA

Barba A, Spiga M (2003) Discharge mode for encapsulated PCMs in storage tanks. Sol Energy 74:141–148

Becker JR (1997) Crude oil waxes, emulsions and asphaltenes. Penn Well Publishing Company, Tulsa

Belton G, Ajami F (1973) Thermochemistry of salt hydrates. National Science Foundation Report No. NSF/RANN/SE/GI27976/TR/73/4, Philadelphia, USA

Biswas DR (1977) Thermal energy storage using sodium sulphate decahydrate and water. Sol Energy 19:99–100

Boh B, Sumiga B (2008) Microencapsulation technology and its applications in building construction materials. Mater Geoenviron 55(3):329–344. http://www.rmz-mg.com/letniki/rmz55/RMZ55_0329-0344.pdf

Bowyer J (2005) Phase I research report. Life cycle environmental performance building material in the context of residential construction. Consortium for Research on Renewable Industrial Materials, Madison, Wisconsin, USA. http://www.corrim.org/pubs/reports/2005/Phase1/index.asp

Brandt R, Meier C, Reichenauer G, Ebert HP (2006) Small and wide angle X-ray scattering investigation of SiO*x* micro-encapsulated Na_2CO_3 hydrates. Bavarian Center for Applied Energy Research (ZAE Bayern), Am Hubland, 97074 Würzburg, Germany

Buddhi D, Sawhney RL (1994) Proc: thermal energy storage and energy conversion. School of Energy and Environmental Studies, Devi Ahilya University, Indore, India

Cabeza LF, Illa J, Roca J, Badia F, Mehling H, Hiebler S, Ziegler F (2001a) Immersion corrosion tests on metal-salt hydrate pairs used for latent heat storage in the 32 to 36 °C temperature range. Mater Corros 52:140–146

Cabeza LF, Illa J, Roca J, Badia F, Mehling H, Hiebler S, Ziegler F (2001b) Middle term immersion corrosion tests on metal-salt hydrate pairs used for latent heat storage in the 32 to 36 °C temperature range. Mater Corros 52:748–754

Cabeza LF, Svensson G, Hiebler S, Mehling H (2003) Thermal performance of sodium acetate trihydrate thickened with different materials as phase change energy storage material. Appl Therm Eng 23:1697–1704

Campbell KR, Sailor DJ (2011) Phase change materials as thermal storage for high performance homes. In: Proceedings of the 2011 ASME international mechanical engineering congress and exposition, IMECE 2011, Denver, Colorado, USA, 11–17 Nov 2011

Cassedy ES (2000) Prospects for sustainable energy. Cambridge University Press, Cambridge

Chen YS, Shiha YF, Yub YH (2013) Stearyl alcohol modified expanded graphite/paraffin phase change composite. Int J Appl Sci Eng 11(2):227–233

Cheng W, Zhang R, Xie K, Liu N, Wang J (2010a) Heat conduction enhanced shape-stabilized paraffin/HDPE composite PCMs by graphite addition: preparation and thermal properties. Sol Energy Mater Sol Cells 94(10):1636–1642

Cheng W, Zhang R, Xie K, Liu N, Wang J (2010b) Heat conduction enhanced shape-stabilized paraffin/HDPE composite PCMs by graphite addition: preparation and thermal properties. Sol Energy Mater Sol Cells 94:1636–1642

Childs K, Stovall T (2012) Potential energy savings due to phase change material in a building wall assembly: an examination of two climates. Report—ORNL/TM-2012/6 Mar 2012, Oak Ridge National Laboratory, USA

CSA S478-95 (1997) Durability guidelines for building wall envelopes: real property program and the program on energy research and development (PERD). Energy efficiency-buildings Task 1.4. Public works and government services, Canada, Mar 1997

Cui Y, Liu C, Hu S, Yu X (2011) The experimental exploration of carbon nanofiber and carbon nanotube additives on thermal behaviour of phase change materials. Sol Energy Mater Sol Cells 95(2011):1208–1212

Dimaano MNR, Watanabe T (2002a) The capric–lauric acid and pentadecane combination as phase change material for cooling applications. Appl Therm Eng 22(2002):365–377

Dimaano MNR, Watanabe T (2002b) The capric and lauric acid mixture with chemical additives as latent heat storage materials for cooling application. Energy 27(2002):869–888

Dincer I, Rosen MA (2002) Thermal energy storage. Systems and applications. Wiley, Chichester

EPA (2001) Candles and Incense as potential sources of indoor air pollution: market analysis and literature review US environmental protection agency (EPA), Office of research and development, Prepared by national risk management research laboratory EPA-600/R-01-001, Jan 2001

Erk HF, Dudukovic MP (1996) Phase-change heat regenerators: modeling and experimental studies. AIChE 42(3):791–808

Ettouney HM, Alatiqi I, Al-Sahali M, Groll M, Brost O, Heine D (1990) Corrosion of steels in contact with salt eutectics as latent heat storage materials: influence of water and other impurities. Heat Recover Syst CHP 10:567–572

Ettouney HM, Alatiqi I, Al-Sahali M, Al-Ali SA (2004) Heat transfer enhancement by metal screens and metal spheres in phase change energy storage systems. Renew Energy 29:841–860

Farid M, Mecaial N (2008) Use of phase change materials for thermal comfort and electrical energy peak load shifting. In: Proceedings of U21 international conference on energy technologies and policy, Birmingham, UK, 8–10 Sept 2008

Farid MM, Khudhair AM, Razack SAK, Al-Hallaj S (2004) A review on phase change energy storage: materials and applications. Energy Convers Manage 45(9–10):1597–1615

Feldman D, Shapiro MM, Banu D (1986) Organic phase change materials for thermal energy storage. Sol Energy Mater 13:1–10

Fukai J, Kanou M, Kodama Y, Miyatake O (2000) Thermal conductivity enhancement of energy storage media using carbon fibers. Energy Convers Manage 41:1543–1556

Fukai J, Hamada Y, Morozumi Y, Miyatake O (2002) Effect of carbon-fibre brushes on conductive heat transfer in phase change materials. Int J Heat Mass Transfer 45(2002):4781–4792

Ge H, Liu J (2013) Keeping smartphones cool with gallium phase change material. J Heat Transfer 135(5):054503

George A (1989) Phase change thermal storage materials. In: Guyer C (ed) Handbook of thermal design. McGraw Hill Book Co, US

Gok O, Yilmaz MO, Yilmaz S, Metim O, Paksoy HO (2006) Stabilization of Glauber's salt for latant heat storage. In: Proceedings of ecostock 2006—the 10th international conference on thermal energy storage, Stockton College, NJ

Gok O, Yilmaz MO, Yilmaz S, Turgut B, Mazman M, Beyhan B, Sahan N, Paksoy HO, Evliya H (2009) Phase change material for domestic appliances. In: Proceedings of Effstock 2009—the 11th international conference on thermal energy storage. Stockholm, Sweden, 14–17 June 2009

Groll M, Brost O, Heine D (1990) Corrosion of steels in contact with salt eutectics as latent heat storage materials: influence of water and other impurities. Heat Recover Syst CHP 10:567–572

Gu X, Xi P, Cheng B, Niu S (2010) Synthesis and characterization of a novel solid–solid phase change luminescence material. Polym Int 59(6):772–777

Günther E, Mehling H, Werner M (2007) Melting and nucleation temperatures of three salt hydrate phase change materials under static pressures up to 800 MPa. J Phys D: Appl Phys 40(15)

Hadjieva M, Stoykov R, Filipova T (2000) Composite salt-hydrate concrete system for building energy storage. Renew Eng 19:111–115

Hafner B, Schwarzer K (1999) Improvement of the heat transfer in a phase-change-material storage. In: Proceedings of the IEA, ECES IA Annex 10, 5th workshop, Benedikbauern, Germany

Halford CK, Boehm RF (2007) Modeling of phase change material peak load shifting. Energy Buildings 39(2007):298–305

Hasnain SM (1998) Review on sustainable thermal energy storage technologies. Part 1: heat storage materials and techniques. Energy Convers Manage 39:1127–1138

Hawes DW, Feldman D, Banu D (1993) Latent heat storage in building materials. Energy Buildings 20(1993):77–86

Heckenkamp J, Baumann H (1997) Latentwärmespeicher. Sonderdruck aus Nachrichten 11 (1997):1075–1081

Hessbrugge BJ, Vaidya AM (1997) Preparation and characterization of salt hydrates encapsulated in polyamide membranes. J Membr Sci 128(2):175–182

Hoogendoorn CJ, Bart CCJ (1992) Performance and modeling of latent heat stores. Sol Energy 48:53–58

IEA (2005) Inventory of phase change materials (PCM). A report of IEA solar heating and cooling programme—Task 32. Advanced storage concepts for solar and low energy buildings. Report C2 from Subtask C. International Energy Agency Solar Heating and Cooling Program. http://www.ieashc.org/publications/category.aspx?CategoryID=64

Inaba H, Tu P (1997) Evaluation of thermophysical characteristics on shape stabilized paraffin as a solid–liquid phase change material. Heat Mass Transf 32:307–312

Kakiuchi H, Yamazaki M, Yabe M, Chihara S, Terunuma Y, SakataY (1998) A study of erythritol as phase change material. In: Proceedings of the 2nd workshop IEA ECES Annex 10. Phase change materials and chemical reactions for thermal energy storage. Sofia, Bulgaria, 11–13 April 1998. www.fskab.com/annex10

Kamimoto M, Abe Y, Kanari K, Swata S, Tani T, Ozawa T (1986) Heat transfer in latent heat thermal storage units using pentarythritol slurry thermal energy storage. World Congress of Chemical Engineering, Tokyo, Japan

Karaipekli A, Sari A (2007) Capric acid and palmitic acid eutectic mixture applied in building wallboard for latent heat energy storage. J Sci Ind Res 66(6):470–476

Karaipekli A, Sari A (2008a) Capric–myristic acid/expanded perlite composite as form-stable phase change material for latent heat thermal energy storage. Renew Eng 33(2008):2599–2605

Karaipekli A, Sari A (2008b) Capric–myristic acid/vermiculite composite as form-stable phase change material for thermal energy storage. Sol Energy 2008 (Article in Press)

Karaipekli A, Sari A, Kaygusuz K (2007) Thermal conductivity improvement of stearic acid using expanded graphite and carbon fiber for energy storage applications. Renew Eng 32:2201–2210

Karaipekli A, Sari A, Kaygusuz K (2009) Thermal properties and thermal reliability of capric Acid/Stearic acid mixture for latent heat thermal energy storage. Energy Sources Part A, 31:199–207

Kashchiev D (2000) Nucleation—basic theory with applications. Butterworth Heinemann, Oxford. ISBN: 0-7506-4682-9

KCPC—The Key Centre for Polymer Colloids (2003) Properties of alkanes, alkenes. The Australian Research Council's Research Centres Program. http://discovery.kcpc.usyd.edu.au/9.2.1/9.2.1_Properties.html

Kenisarin MM (1993) Short-term storage of solar energy. low temperature phase-change materials. Appl Sol Energy 29(2):48–65

Khot SA, Sane NK, Gawali BS (2011) Experimental investigation of phase change phenomena of paraffin wax inside a capsule. Int J Eng Trends Technol 2(2)

Khundhair AM, Farid MM (2004) A review on energy conservation in building applications with thermal storage by latent heat using phase change materials. Energy Convers Manage 45 (2004):263–275

Kim S, Drzal LT (2009) High latent heat storage and high thermal conductive phase change materials using exfoliated graphite nanoplatelets. Sol Energy Mater Sol Cells 3(2009):136–142

Kimura H, Kai J (1984) Phase change stability of $CaCl_2$ $6H_2O$. Sol Energy 33(1):49–55

Koizumi H (2004) Time and spatial heat transfer performance around an isothermally heated sphere placed in a uniform, downwardly directed flow (in relation to the enhancement of latent heat storage rate in a spherical capsule). Appl Therm Eng 24(17):2583–2600

Kośny J, Kossecka E (2013) Understanding a potential for application of phase-change materials (PCMs) in building envelopes. ASHRAE 2013 winter conference, Dallas, TX, 26–30 Jan 2013, ASHRAE Transactions vol 119, Part 1. DA-13-001

Kośny J, Yarbrough DW, Miller WA (2008) Use of PCM-enhanced insulations in the building envelope. J Build Enclosure Des

Kośny J, Yarbrough DW, Riazzi T, Leuthold D, Smith JB, Bianchi M (2009) Development and testing of ignition resistant microencapsulated phase change material. In: Proceedings of

Effstock 2009—the 11th international conference on thermal energy storage, Stockholm, Sweden, 14–17 June 2009
Kośny J, Kossecka E, Brzezinski B, Tleoubaev A, Yarbrough D (2012a) Dynamic thermal performance analysis of fiber insulations containing bio-based phase change materials (PCMs). Energy Build 52:122–131
Kośny J, Biswas K, Miller WA, Kriner S (2012b) Field thermal performance of naturally ventilated solar roof with PCM heat sink. Sol Energy 86(9):2504–2514
Kośny J, Shukla N, Fallahi A (2013) Cost analysis of simple phase change material-enhanced building envelopes in southern U.S. climates. U.S. Department of Energy-Building Technologies Program, Report. http://www.nrel.gov/docs/fy13osti/55553.pdf
Kossecka E, Kośny J (1998) Effect of insulation and mass distribution in exterior walls on the dynamic thermal performance of a whole buildings. In: Proceedings of DOE, ASHRAE, ORNL conference—thermal envelopes VII, Clear Water, Florida
Kossecka E, Kośny J (2002) Influence of insulation configuration on heating and cooling loads in continuously used building. Energy Build 2002(34):321–331
Kossecka E, Kośny J (2010) Thermal balance of a wall with PCM-enhanced thermal insulation. In: Proceedings of the 1st central European symposium on building physics, Cracow, Poland
Kotzé JP, von Backström TW, Erens PJ (2012) High temperature thermal energy storage utilizing metallic phase change materials and metallic heat transfer fluids. J Sol Energy Eng 135(3): 035001
Kurz W, Fisher D (1992) Fundamentals of solidification, 3rd edn. Trans Tech Publications, Switzerland. ISBN:0-87849-522-3
Lane GA (1978) Macro-encapsulation of PCM. Report no. oro/5117-8, Dow Chemical Company, Midland, Michigan, USA
Lane GA (1980) Low temperature heat storage with phase change materials. Int J Ambient Energy 1(1980):155–168
Lane GA (1983) Solar heat storage: latent heat materials, vol 1. CRC Press Inc, USA
Lane GA (1989) Hand book of thermal design, Chap. 1. In: Guyer C (ed) Phase change thermal storage materials. McGraw Hill Book Co, New York
Leoni N, Amon CH (1997) Transient thermal design of wearable computers with embedded electronics using phase change materials. ASME HTD 437(5):241–250
Lingayat AB, Suple YS (2013) Review on phase change material as thermal energy storage medium: materials, application. Int J Eng Res Appl (IJERA) 3(4):916–921 Jul-Aug2013. ISSN:2248–9622
Lutton ES (1967) In: Markley KS (ed) Fatty acids: their chemistry, properties, production and uses, 2nd edn, Part 4. Interscience, New York
Marín JM, Zalba B, Cabeza LF, Mehling H (2005) Improvement of a thermal energy storage using plates with paraffin-graphite composite. Int J Heat Mass Transfer 48(2005):2561–2570
McMullin M, Ben-Abdallah N (1987) An investigation of low-temperature phase-change materials for short-term energy storage in greenhouses. Can Agri Eng. http://www.csbe-scgab.ca/docs/journal/30/30_1_65_ocr.pdf
Mehling H, Cabeza LF (2008) Heat and cold storage with PCM. Springer, Berlin. ISBN 978-3-540-68556-2
Mettawee E, Assassa G (2007) Thermal conductivity enhancement in a latent heat storage system. Sol Energy 81(2007):839–845
Molefi JA, Luyt AS, Krupa I (2010) Investigation of thermally conducting phase change materials based on polyethylene/wax blends filled with copper particles. J Appl Polym Sci 116 (2010):1766–1774
Muruganantham K, Phelan P, Horwath P, Ludlam D, McDonald T (2010) Experimental investigation of a bio-based phase-change material to improve building energy performance. In: Proceedings of ASME 2010 4th international conference on energy sustainability, ES2010, Phoenix, Arizona, USA, 17–22 May 2010. http://proceedings.asmedigitalcollection.asme.org/volume.aspx?volumeid=15077

Nagano K, Mochida T, Iwata K, Hiroyoshi H, Domanski R (2000) Thermal performance of $Mn(NO_3)_2$. In: Proceedings of the 5th Workshop IEA ECES IA Annex 10, $6H_2O$ as a new PCM for cooling system, Tsu, Japan

NAHB (2007) Study of life expectancy of home components, Feb 2007—report. In: Jackson J (ed) National Association of Home Builders/Bank of America Home Equity

Naumann R, Emons HH (1989) Results of thermal analysis for investigation of salt hydrates as latent heat storage materials. J Therm Anal 35:1009–1031

Neeper DA (1989) Potential benefits of distributed PCM thermal storage. In: Coleman MJ (ed) Proceedings of 14th National passive solar conference. Denver, Colorado, USA: American solar energy society, 1989, pp 283–288

Nikolic R, Marinovic-Cincovic M, Gadzuric S, Zsigrail J (2003) New materials for solar thermal storage solid/liquid transitions in fatty acid esters. Sol Energy Mater Sol Cells 79:285–292

Ning-Wei CJ, Martin V, Setterwall F (2009) A review of thermal energy storage systems with salt hydrate phase change materials for comfort cooling. In: Proceedings of the 11th international conference on energy storage. International Energy Agency conference—Effstock 2009: thermal energy storage for efficiency and 2009: thermal energy storage for efficiency and sustainability, Stockholm, Sweden, 14–17 June 2009

Pasupathy A, Velraj R, Seeniraj RV (2008) Phase change material-based building architecture for thermal management in residential and commercial establishments. Renew Sust Energy Rev 12 (2008):39–64

Pincemin SS, Olives R, Py X, Christ M (2008) Highly conductive composites made of phase change materials and graphite for thermal storage. Sol Energy Mater Sol Cells 92 (2008):603–613

Porisino FG (1988) Salt hydrates used for latent heat storage—corrosion of metals and reliability of thermal performance. Sol Energy 41:193–197

Py X, Olives R, Mauran S (2001) Paraffin/porous-graphite-matrix composite as a high and constant power thermal storage material. Int J Heat Mass Transfer 44(2001):2727–2737

Ramnanan-Singh R (2012) Formulation and thermophysical analysis of a beeswax microemulsion and the experimental calculation of its heat transfer coefficient. Master of Science Thesis, Department of Mechanical Engineering, CUNY Energy Institute, NY, USA

Riza M (2007) Thermal characteristics of eutectic mixture of capric-lauric acids as phase change material (PCM) in gypsum board. Jurnal Teknologi Proses, Media Publikasi Karya Ilmiah Teknik Kimia 6(1):31–38. ISSN:1412-7814

Rozanna D, Salmiah A, Chuah TG, Medyan R, Choong T, Saari M (2005) A study on thermal characteristics of phase change material (PCM) in gypsum board for building application. J Oil Palm Res 17:41–46

Ryu HW, Woo SW, Shin BC, Kim SD (1992) Prevention of supercooling and stabilization of inorganic salt hydrates as latent heat storage materials. Sol Energy Mater Sol Cells 27:161–172

Salyer I, Sircar A (1989) Development of PCM wallboard for heating and cooling of residential buildings. Thermal energy storage research activities review, U.S. Department of Energy, New Orleans, LA, 15–17 March 1989, pp 97–123

Sari A (2004) Form-stable paraffin/high density polyethylene composites as solid-liquid phase change material for thermal energy storage: preparation and thermal properties. Energy Convers Manage 45:2033–2042

Sari A (2005) Eutectic mixtures of some fatty acids for low temperature solar heating applications: thermal properties and thermal reliability. Appl Therm Eng 25:2100–2107

Sari A, Kaygusuz K (2001) Thermal performance of mystiric acid as a phase change material for energy storage application. Renew Energy 24(2001):303–317

Sari A, Kaygusuz K (2002) Thermal performance of palmitic acid as a phase change energy storage material. Energy Convers Manage 43(2002):863–876

Sari A, Kaygusuz K (2003) Some fatty acids used for latent heat storage: thermal stability and corrosion of metals with respect to thermal cycling. Renew Energy 28:939–948

Sarı A, Alkan C, Biçer A, Karaipekli A (2011) Synthesis and thermal energy storage characteristics of polystyrene-graft-palmitic acid copolymers as solid–solid phase change materials. Sol Energy Mater Sol Cells 95(12):3195–3201

Schossig P, Henning HM, Haussman T, Raicu A (2003) Encapsulated phase-change materials integrated into construction materials. In: Proceedings of the 9th international conference on thermal energy storage—Futurestock'2003, Warsaw, Poland

Sereda PJ (1969) Performance of building materials. Canadian Building Digest 115, July 1969, 4 p

Shapiro M, Feldman D, Hawes D, Banu D (1987) P.C.M. thermal storage in drywall using organic phase-change material. Passive Sol J 4(4):419–438

Sharma SD, Sagara K (2005) Latent heat storage materials and systems: a review. Int J Green Energy 2: 1–56. ISSN:0197-1522

Sharma A, Tyagi VV, Chen CR, Buddhi D (2009) Review on thermal energy storage with phase change materials and applications. Renew Sust Energy Rev 13:318–345

Shilei L, Guohui F, Nenga Z, Li D (2007) Experimental study and evaluation of latent heat storage in phase change materials wallboards. Energy Build 39(2007):1088–1091

Siegel R (1977) Solidification of low conductivity material containing dispersed high conductivity particles. Int J Heat Mass Transfer 20:1087–1089

Skripov VP (1974) Metastable liquids. Halsted Press, New York, USA. ISBN:0-470-79546-8

Stovall TK, Tomlinson JJ (1995) What are the potential benefits of including latent storage in common wallboard? J Sol Energy Eng 117

Suppes GJ, Lopes S, Goff M (2002a) Phase change materials from fats and oils. In: Proceedings of bioenergy'2002 conference, paper #2062, published by Pacific Regional Biomass Energy Program, Boise, Idaho, 1–5 Sept 2002

Suppes GJ, Goff MJ, Lopes S (2002b) Use of renewable phase change materials to reduce carbon dioxide emissions. In: Proceedings of the 6th annual green chemistry and engineering conference, An American Chemical Society Publication, Washington, D.C. USA

Telkes M (1952) Nucleation of super saturated inorganic salt solution. Ind Eng Chem 44:1308

Telkes M (1975) Thermal storage for solar heating and cooling. In: Proceedings of the workshop —solar energy storage subsystems for the heating and cooling of buildings. Charlottesville, Virginia, USA

Telkes M (1978) Remarks on thermal energy storage using sodium sulfate decahydrate and water. Sol Energy 20(1):107

Tomlinson J, Jotshi C, Goswami D (1992) Solar thermal energy storage in phase change materials. In: Proceedings of solar '92: proceedings of the American Solar Energy Society annual conference, Cocoa Beach, FL, 15–18 June 1992, pp 174–179

Tong X, Khan J, Amin MR (1996) Enhancement of heat transfer by inserting a metal matrix into a phase change material. Num Heat Trans Part A Appl 30(2):125–141

Wada T, Kimura F, Yamamoto R (1983) Studies on salt hydrate for latent heat storage. II. Eutectic mixture of pseudo-binary system CH_3CO_2Na $3H_2O$–$CO(NH_2)_2$. Bull Chem Soc Jpn 56 (1983):1223–1226

Wang JF, Xie HQ, Xin Z (2010) Enhancing thermal conductivity of palmitic acid based phase change materials with carbon nanotubes as fillers. Sol Energy 84:339–344

Wei J, Kawaguchi Y, Hirano S, Takeuchi H (2005) Study on a PCM heat storage for rapid heat supply. Appl Therm Eng, 25:2903–29320

Xia L, Zhang P, Wang RZ (2010) Preparation and thermal characterization of expanded graphite/ paraffin composite phase change material. Carbon 48(2010):2538–2548

Xu X, Zhang YP, Lin KP, Di HF, Yang R (2005) Modeling and simulation on the thermal performance of shape-stabilized phase change material floor used in passive solar buildings. Energy Build 37(10):1084–1091

Yin H, Gao X, Ding J, Zhang Z (2008) Experimental research on heat transfer mechanism of heat sink with composite phase change materials. Energy Convers Manage 49(2008):1740–1746

Yinping Z, Guobing Z, Ruib Y, Kunpinga L (2006) Our research on shape-stabilized pcm in energy-efficient buildings. In: Proceedings of ECOSTOCK, 10th international conference on thermal energy storage, Stockton, USA
Zalba B, Marin JM, Cabeza LF, Mehling H (2003) Review on thermal energy storage with phase change: materials, heat transfer analysis and applications. Appl Therm Eng 23(2003):251–283
Zhang Z, Fang X (2006) Study on paraffin/expanded graphite composite phase change thermal energy storage material. Energy Convers Manage 47:303–310
Zhang M, Medina M, King J (2005) Development of a thermally enhanced frame wall with phase-change materials for on-peak air conditioning demand reduction and energy savings in residential buildings. Int J Energy Res 29(9):795–809
Zhao CY, Zhang GH (2011) Review on microencapsulated phase change materials (MEPCMs): fabrication, characterization and applications. Renew Sustain Energy Rev 15(8):3813–3832
Zhisheng L, Dongmei L, Lijuan L, Guangqiang Z, Jianlong L (2006) Comparison of building energy efficiency and life span for different envelopes. In: Proceedings of the sixth international conference for enhanced building operations, Shenzhen, China, 6–9 Nov 2006
Zuo J, Li W, Weng L (2011) Thermal performance of caprylic acid/1-dedocanol eutectic mixture as phase change material (PCM). Energy Build 43:207–210

Chapter 4
Laboratory Thermal Testing of PCM-Enhanced Building Products and Envelope Systems

4.1 Introduction

In buildings, both sensible and latent energy accumulation can occur in the external envelope as well as in the internal building fabric. As discussed in earlier chapters, since late 1940s, latent storage using phase-change materials (PCMs) has been the focus of multiple research studies.

Researchers have found that in some novel building envelope applications their high latent energy capacity per volume can be more beneficial than sensible storage available in traditional construction materials. Building components with phase-transition capability allow large amounts of heat to be stored or released due to the latent heat of fusion or solidification. PCM-enhanced building envelopes are widely considered as prospective building technologies to achieve peak-hour load and HVAC energy consumption reductions. The majority of PCMs used in building envelope applications need to go through a complete phase transition during 24-h time periods in order to be fully effective. This is why it is crucial that the temperature in the locations where PCM is installed cycle above and below the temperature range in which phase transition occurs. For the correct design of a PCM-enhanced building envelope, as well as for the accurate whole building energy predictions, it is essential to characterize the thermo-physical and rheological properties of the PCMs. These properties include phase-transition enthalpy, melting and freezing temperatures, hysteresis and supercooling effects, thermal conductivity in the solid and liquid phases, viscosity of liquid PCM and its density as a function of temperature.

A great variety of organic and inorganic PCMs with stable thermo-physical properties are available and have been used worldwide in heat storage applications. These applications include heat sinks in electronics and hybrid vehicles, food packaging, solar thermal systems, and transportation of medical products. One of the major barriers reported, however, for expanding the use of PCMs is their relatively high cost. It is suggested that one effective way for cost reduction of PCM applications in buildings is optimization of the PCM location and its quantity.

J. Kośny, *PCM-Enhanced Building Components*, Engineering Materials and Processes, DOI 10.1007/978-3-319-14286-9_4

As expected, this requires an excellent understanding of thermal performance characteristics. That is why the development is needed for relatively inexpensive and easy-to-use testing methods for the determination of the key thermal characteristics of PCM-based products or systems. This chapter is mainly focused on laboratory-scale thermal-testing methods for PCMs, PCM-enhanced building products, and building envelope systems containing PCMs.

4.2 Scale of Thermal Performance Analysis and Testing Methods

After several decades of successful research, PCMs are now widely recognized as one of the most promising technical means of improving the energy efficiency and sustainability of residential and commercial buildings. In most of building applications, PCM-enhanced materials are utilized as lightweight thermal mass components controlling the transfer of exterior thermal loads to the interior space. The main benefit of using PCMs is the availability of significant amounts of latent heat during the phase-transition processes. In most PCM building applications, the latent heat storage is based on heat absorption or release during a change of state from solid to liquid and vice versa. The storage capacity of materials containing PCMs depends on the product of mass (m) and specific heat (c_p) followed by the enthalpy (h) curve and the functional temperature range (T_1, T_2), where the stored heat Q can be denoted as follows:

$$Q = \int_{T_1}^{T_2} mc_p \mathrm{d}T + mf_m \Delta h \tag{4.1}$$

where, f_m is the fraction of the material ongoing phase change.

Most of commercially used PCMs, when used in small temperature intervals around their phase-change temperature, show superior heat storage density (the amount of heat stored in a given material or system per unit volume or mass) compared to sensible heat storage capacity of conventional building materials, such as wood, concrete, or brick. Since, PCMs used in buildings represent a great variety of phase-change thermal characteristics, the exact shape of the enthalpy curve $h(T)$ has to be known to quantify their heat storage capacity and thus the prospective energy advantage of PCM-enhanced building systems. In Fig. 4.1, the enthalpy curves for melting and freezing of an organic PCM are depicted. An ideal PCM can be mathematically characterized by four scalar values, namely the specific heat, c_p, of solid and liquid phases, the phase-change enthalpy Δh, and the melting temperature T_m. A real PCM is better represented by the function $h(T)$. The precise determination of this function is the goal of PCM calorimetric measurements. As depicted in Fig. 4.1, enthalpy curves are not always identical for melting and

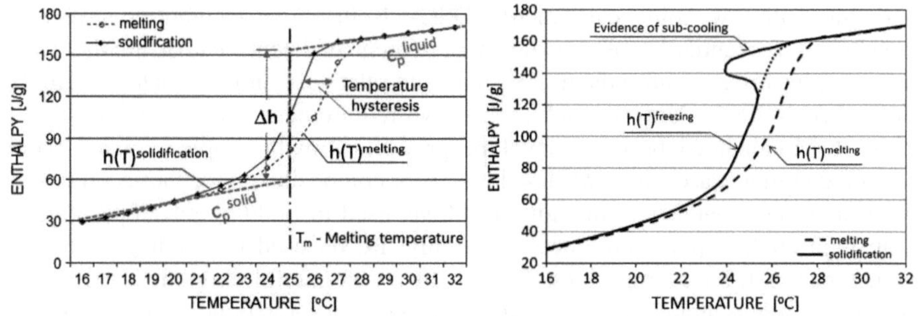

Fig. 4.1 Examples of enthalpy curves for melting and freezing processes showing temperature hysteresis (*left*) and supercooling effect (*right*)

freezing processes. It is often due to the temperature hysteresis phenomenon, which strongly depends on the used testing method and a type of equipment (see IEA 2011). In addition, the notable supercooling effect can be observed in some organic and majority of inorganic PCMs. It is associated with impaired PCM nucleation during the freezing process. In many PCM heat storage applications, supercooling may be a source of numerous thermal performance and durability problems. There are many publications available, which are focused on theoretical and experimental analysis of phase-change processes including hysteresis or supercooling phenomena (see Demirel and Paksoy 1993; Martin et al. 2003; Wunderlich 2005; Bony and Citherlet 2007; Mehling and Cabeza 2008; Günther et al. 2007, 2009).

As stated earlier, proper thermal characterization of building materials containing PCMs is essential for their performance analysis, which is necessary for system designing, code approvals, and energy cost analysis. Theoretical and experimental analyses of building technologies can be performed at three levels as follows:

- Material scale
- System scale
- Whole building scale

In laboratory work, PCM-enhanced building products are usually limited to analysis at the material and systems scale. However, a limited number of small building/room experiments have been performed as well.

A selection of most widespread methods for the testing of PCMs, PCM-enhanced building products, and building enclosure systems containing PCMs is briefly discussed below. More detailed descriptions of these testing procedures can be found in the following paragraphs. The author is aware of a large number of additional testing methodologies which can be utilized for measurements of dynamic thermal characteristics of PCM-enhanced building envelope products and systems (Wunderlich 2005). However, only mostly used and most widely accepted methods are discussed in this publication.

For a long time, the only readily existing thermal test methods, available for material-scale evaluations of PCMs, were the differential thermal analysis (DTA) and differential scanning calorimetry (DSC). Both above techniques have been considered as standard tools for the measurement of latent heat of PCMs. These methods, however, require the use of small and relatively uniform material samples (Wunderlich 2005). Unfortunately, the above requirement is rather unrealistic for most PCM-enhanced building envelope products. Small samples are not representative in most cases of PCM-based blends or composites, since these composites are usually heterogeneous. In addition, PCMs used in building envelopes are seldom high-quality, pure substances. Typically, PCMs used in buildings contain a wide range of different impurities. That is why a very careful sampling is required to obtain the representative test-generated characteristics for building products containing PCMs. It is also important to remember that in building envelopes, PCMs are commonly used in the form of packaged or encapsulated products. It is well known that a size of the PCM encasement, as well as thermal conductivity of a shell material, may affect PCM's phase-transition characteristics. Lastly, heat exchange between individual PCM packages, capsules, or arrays of pouches has three-dimensional character. Summarizing, because most of the PCM-enhanced building envelope products and PCM-insulation blends are non-homogenous and because they often use packaged or encapsulated PCM, the test data generated using DSC or DTA may be a source of significant inaccuracies. This limits the usefulness of DSC and DTA testing procedures to characterize PCM building applications.

In order to allow testing of larger PCM samples and heterogeneous PCM products, Zhang (1999) and Zhang et al. (2007) proposed the T-history method, as an alternative to DSC or DTA testing. It is an inexpensive and easy way to measure the phase-change enthalpy of PCM products using test samples significantly larger from what is required for DSC testing. According to Peck et al. (2006), the T-history method can be used for measuring the thermal properties of PCM composites and PCM blends. However, the accuracy of the T-history method is restricted by the size of the PCM sample due to potentially uneven temperature distribution and temperature stratification within the test sample. Paksoy (1996) introduced the twin water bath method, which determines the thermal storage capacity of PCMs or PCM-based composites. The water bath method is relatively inexpensive, simple, and fast, making it widely used today. The method enables evaluation of phase-change temperature and the thermal storage capacity of PCM for wide range of temperatures. In addition, in early 1990-s Zhang designed a simple bath calorimetric device which made possible test the specific heat at different temperatures and the solid–liquid phase-change enthalpy of ice (see Zhang and Li 2008). This method was useful for evaluating the thermal physical properties of PCM within a room temperature range.

Since, most often, PCMs used in building envelopes are difficult to analyze complex heterogeneous composites or structures with multidimensional heat transfer effects, Kośny et al. (2007b) introduced a new test method utilizing a conventional ASTM C518 heat flow meter apparatus (HFMA). This method determines the amount of phase-change energy available in heterogeneous PCM-enhanced materials to provide thermal storage (Kośny et al. 2009a; Shukla et al. 2013). Following this development, ASTM Committee C 16 approved a new test standard (ASTM C1784 2013)—*Standard Test Method for Using a Heat Flow Meter Apparatus for Measuring Thermal Storage Properties of Phase Change Materials and Products.* Considering the limitations of this new testing method when applied to building applications using arrays of PCM pouches or containers, a new combined experimental-analytical protocol was developed by Kośny et al. (2009b) for testing building envelopes containing complex three-dimensional geometries.

Experimental analysis of PCM-enhanced building envelopes can be also performed in full scale with the use of the dynamic hot-box testing. In 2006/2007, Kośny et al. (2007a) performed in Oak Ridge National Laboratory (ORNL), USA, a series of full-size wall tests, utilizing ASTM C1363 hot-box apparatus (see ASTM 2011b). The dynamic hot-box method enabled assessment of the phase-change energy storage capacity and determination of the system thermal response characteristics for full-scale wall and attic/roof assemblies containing PCM. During these dynamic tests utilizing thermal excitations on the climate side, temperature, heat flux, air velocity, and electrical power were recorded. Furthermore, with the use of an array of heat-flux transducers, the energy storage effect could be determined. With this methodology, direct side-by-side performance comparisons of two wall assemblies were possible. In 2010, similar full-scale dynamic wall testing was performed by the Norwegian University of Science and Technology (NTNU), Trondheim, Norway (see Cao et al. 2010).

A full-scale experimental investigation of a test room with walls containing PCM-enhanced wallboard was performed in laboratory conditions by Kuznik et al. (2007) at the Université de Lyon, France. The experimental apparatus, which is called MICROBAT, was composed of two identical test cells. Each test cell was a cubical enclosure with an internal dimension of 0.50 m. This research allowed energy performance analysis of a PCM wall system in a whole room scale. Results from this testing were later utilized to validate the TRNSYS Type 260 subroutine, containing a novel PCM algorithm (Kuznik et al. 2010). In this numerical procedure, a one-dimensional finite difference model is utilized to simulate the thermal performance of wall systems containing a layer of a PCM-enhanced material (this development will be discussed in more detail in the following chapter).

4.3 Differential Thermal Analysis (DTA) (Material-Scale Testing)

The main characteristic of PCM is the enthalpy–temperature relation that determines the heat storage capacity. This relation can be determined experimentally using different calorimetric methods. DTA is one of the earliest and most studied thermal measurement methodologies. It is a thermal measurement methodology well known in analytical chemistry for identifying and quantitatively analyzing the chemical composition of substances by observing the thermal behavior of a sample as it is heated. In DTA, the test sample and a reference material are subject to identical heating profiles. At the same time, temperature differences between the test sample and a reference are recorded. This differential temperature is then plotted against time, or against temperature (DTA curve or thermogram). Changes in the sample either exothermic or endothermic can be detected relative to the inert reference. Thus, a DTA curve can provide performance data on a number of physicochemical processes such as glass transitions, crystallization, melting, solidification, and sublimation. In order to obtain high sensitivity during DTA measurements, they required uniform heating rates, sample symmetry, and weight equivalence of the test sample and the reference.

The DTA method can also be used to measure the specific heat capacity and phase-change enthalpy. In this case, the DTA apparatus is utilized to measure the temperature difference between PCM and a reference material when they are both subjected to the same heat load. During this measurement, the PCM and the reference material absorb the same quantity of heat and their temperatures are usually different because of their different thermal characteristics within the phase-transition temperature range. That is why thermal performance of PCMs is evaluated by measuring the temperature difference between the analyzed PCM and the reference sample.

During the heating process, a PCM sample of the mass "m" and specific heat "c_p" absorbs energy and its physical state can change due to a phase transition, or just the internal energy change. For the case without phase transition, the received heat "δq" is proportional to the temperature increase "δT" as follows:

$$\delta q = mc_p \delta T \tag{4.2}$$

For the temperature range (T_1, T_2), the enthalpy "H" change can be noted as follows:

$$\Delta H = \int_{T_1}^{T_2} C\,\mathrm{d}T \tag{4.3}$$

where "C" is the heat capacity at constant pressure.

The melting of the substance causes a change of the enthalpy ΔH_M. Usually, if a PCM sample is heated, its temperature slowly increases until the phase-transition temperature T_M is reached. In ideal PCMs, the temperature stays constant during the phase-transition process. Before, during, and after the phase-change process (melting), the enthalpy change can be denoted using the following equation:

$$\Delta H = \int_{T_1}^{T_M} C\,\mathrm{d}T + \Delta H_M + \int_{T_M}^{T_2} C\,\mathrm{d}T \tag{4.4}$$

During the cooling process, the same amount of latent heat is released. The temperature is decreasing before and after the phase-transition point, while remaining constant during the transition.

As shown in Fig. 4.2, the DTA apparatus usually consists of a sample holder comprising thermocouples, sample containers, and a ceramic or metallic block; a furnace; a temperature programer; and a data acquisition system. During the 1950s, the DTA-measuring devices were relatively simple. However, later these devices were subsequently automated and miniaturized. The key feature is the existence of two thermocouples connected to a voltmeter. One thermocouple is placed in a reference material (T_r) such as, for example, aluminium oxide (Al_2O_3), while the other is placed in a test sample (T_s). Thermal performance of the test sample and the reference material is determined from the measured temperatures. The amount of heat involved and temperature at which the material changes take place are characteristic of individual chemical elements or compounds. That is why the DTA technique has been widely used for identification of chemical elements, compounds, or in thermal investigations of their blends.

Figure 4.2 depicts a typical setup of a DTA apparatus. The substance under investigation and a reference material are heated under the same temperature conditions (T_F) in a furnace. The increase of the temperature in the furnace should be linear with respect to time. Both the test sample and the reference material thermocouples are connected together in series, which enables direct temperature difference measurements (ΔT). As shown in Fig. 4.3, the reading on the control unit is zero as long as there is no heat consumed or released, because there is the same temperature $T_S = T_R$ in the sample and in the reference material and hence $\Delta T = 0$. If there is an endothermic reaction in the measured sample, its temperature increases slower as the reference sample temperature ($T_{S1} < T_{R1}$) and measured temperature difference $\Delta T_1 = T_{S1} - T_{R1}$ is negative. If there is an exothermic reaction, the measured sample temperature increases faster as the reference sample temperature

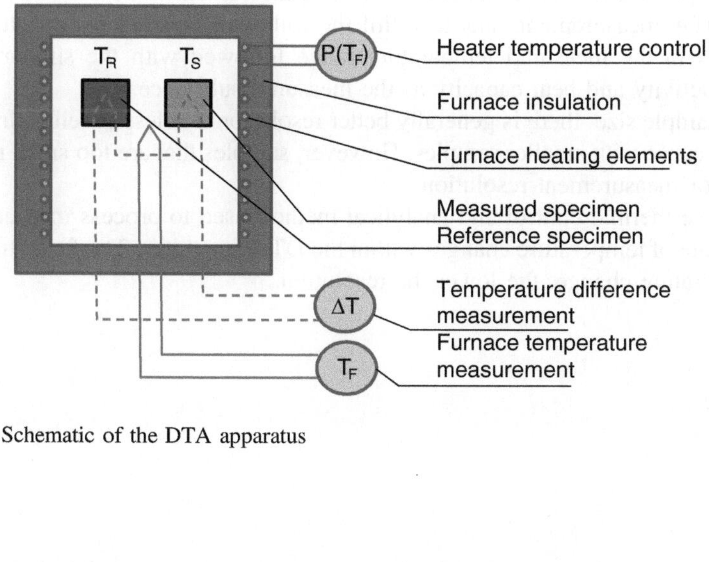

Fig. 4.2 Schematic of the DTA apparatus

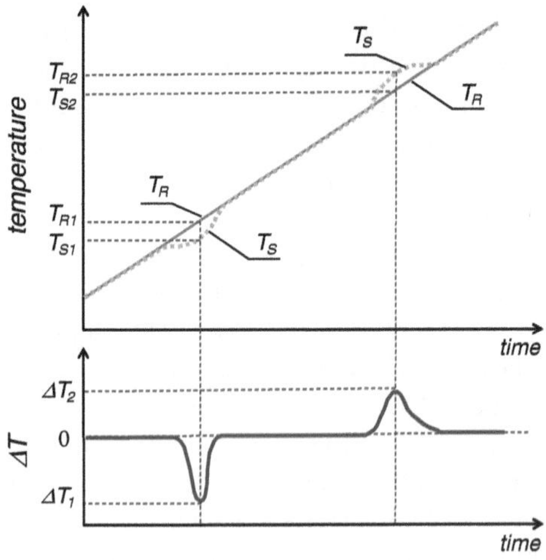

Fig. 4.3 DTA temperature–time curves for the test material and reference and temperature difference–time curve

($T_{S2} > T_{R2}$) and measured temperature difference $\Delta T_2 = T_{S2} - T_{R2}$ is positive. The DTA curve can be determined by subtraction of the curve T_S from the curve T_R.

During measurements of the PCM phase transition, the holding time in temperature–time curves is often used as a semi-quantitative evaluation for the phase-change enthalpy. This time is equal to the time from the beginning of the thermal process to the time of the peak maximum. During the phase transition, the temperature remains constant for this time. The peak area is a useful value for the determination of heat capacities. In a DTA diagram, the heat of reaction ΔH is proportional to the area between baseline and the DTA curve.

According to Gabbott (2008) and Xie et al. (2013), the main factors influencing the accuracy of the DTA enthalpy measurements are as follows:

1. Configuration of the DTA apparatus, including the heating method and the type and location of used thermocouples.
2. The type of a reference material used in measurements. A reference substance for DTA measurements has to fulfill the following conditions: no phase transitions in the measured temperature range followed with the similar thermal conductivity and heat capacity as the measured substance.
3. The sample size: there is generally better resolution and less baseline drift of the DTA curve with smaller samples. However, samples that are too small may lead to poor measurement resolution.
4. Furnace thermal control and analytical method used to process measured data.
5. The rate of temperature changes within the DTA apparatus. The faster the rate of temperature change, the lower the resolution.

4.4 Differential Scanning Calorimetry (DSC) Material-Scale Testing

DSC is one of the most widely used PCM measurement methods, because of the ease with which it can provide large amounts of thermodynamic data. Instead of measuring temperature changes during the heating process (as in DTA), DSC measures the amount of heat that is needed to increase the temperatures of the test sample and the reference material across a temperature interval. The DSC method was developed and patented in early 1960s by Watson and O'Neill (1966). DSC apparatus measures the change of heat flow rate difference which is normally released due to the change of the sample temperature. The test sample and reference material are kept at the same temperature as it changes across the specified temperature interval. The differing amounts of heat required to keep the temperatures identical are measured. As the test sample undergoes phase transitions, more or less heat is needed, which allows for energy diagrams to be created from the measured data. Additionally, specific heat, glass transition temperature, crystallization temperature, melting temperature, and oxidative/thermal stability, among other properties, can be measured using a DSC. That is why, DSC measurements are often used today in industrial manufacturing, ensuring sample purity and confirming compositional analysis. A DSC is also used in materials research to provide information about material properties and enables material compositions to be determined. In case of PCM measurements, a single DSC test yields qualitative and quantitative information on the phase transitions of a sample, such as transition temperature, enthalpy, heat capacity, specific heat, and heats of transition.

During DSC measurements, heat only flows if the temperature difference exists between tested material and a reference. The DSC apparatus measures energy directly and allows precise measurements of heat capacity and thermal conductivity (see Klančnik et al. 2010). DSC is today the widest used and one of the most powerful techniques to thermo-physically analyze a PCM, because it provides the melting/solidification enthalpies, the melting/solidification temperatures, and the specific heat of the materials under study. However, it presents several following limitations when it comes to testing of PCM-enhanced building products: a small size of the test sample, followed by the sample's mass influence on the thermal response characteristic (Günther et al. 2009), the response dependence on the heating rate (Castellón et al. 2008), and finally, results repeatability may not be reached when analyzing heterogeneous materials, blends, or composites (Barreneche et al. 2012). Figure 4.4 shows the schematic of a DSC apparatus which can be used for PCM.

Despite many similarities, there are several differences between DTA and DSC testing methodologies. First of all, during DTA testing, the temperature difference between the reference and the sample is recorded as a function of temperature or time, while in case of DSC measurements, the difference in heat absorbed by the reference material and the tested PCM sample is analyzed. In addition, during DTA test, the temperature difference between the reference and sample is recorded. The DSC apparatus, however, generates test charts where heat flow rate difference is

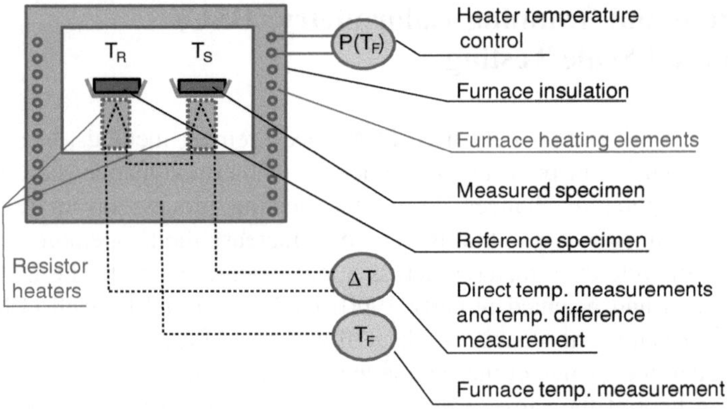

Fig. 4.4 Schematic of DSC apparatus

shown as the ordinate and temperature or time as the abscissa. Keep in mind that in DSC testing, the temperature difference between the two materials is zero, usually leading to more accurate results than in the case of DTA.

A DSC allows relatively fast measurements of phase-transition heats (heat flow rates) and their changes at characteristic temperatures for small sample masses (in milligrams and in grams) in wide temperature ranges with high accuracy (Höhne et al. 2003).

DSCs are mainly classified as

- Power compensation DSC (PC-DSC), or
- Heat-flux DSC (HF-DSC)

The primary measurement signal for both these types is temperature difference and the heat flow rate, which is proportional to the temperature difference (see Mehling and Cabeza 2008).

4.4.1 Power Compensation DSC (PC-DSC)

The PC-DSC belongs to the heat-compensating calorimeter family of measurements methods. The PC-DSC consists of two identical micro-furnaces, one for the test sample and the other for the reference with individual heaters. The sample furnace is heated following a programed temperature–time schedule, while the reference furnace follows this temperature schedule. It results in increase and decrease of the temperature in the reference furnace following a reaction. In this case, the compensating heating power is measured, which is actually the heat flow difference. The major advantage of the PC-DSC is a low weight of the individual furnace ($\sim$1 g), while in the HF-DSC apparatus it is usually significantly heavier (up to 200 g). That fact favors extremely short responding time in the PC-DSC, where the heating and cooling rates can be up to 500 °C/min. When an exothermic (heat releasing, i.e., solidification) or endothermic (heat absorbing, i.e., solidification) reaction appears, the energy is accumulated or released to compensate the energy change in both furnaces. The power required to maintain the system in equilibrium is proportional to the energy changes occurring in the sample (see Mehiling and Cabeza 2008; Ford and Mann 2012; Barreneche et al. 2012).

Power compensation DSC measures the power difference (ΔW) needed to maintain the reference material and the PCM test sample at the same temperature. This can be described by the following equation:

$$\Delta W = \frac{\mathrm{d}Q_S}{\mathrm{d}t} - \frac{\mathrm{d}Q_R}{\mathrm{d}t} = \frac{\mathrm{d}H}{\mathrm{d}t} \tag{4.5}$$

where Q_S and Q_R represent heat absorbed by the sample of the tested material and a reference, respectively.

Measured enthalpy change per unit time dH/dt can be described as a product of the total current I_T and voltage difference ΔV across the resistance heater as follows:

$$\frac{\mathrm{d}H}{\mathrm{d}t} = I_T \Delta V \tag{4.6}$$

Measured heat is compensated during the PC-DSC testing with electric energy, by increasing or decreasing an adjustable Joule's heat.

4.4.2 Heat-Flux DSC (HF-DSC)

A very common method used for calorimetric measurements of PCMs is the heat-flux DSC (HF-DSC), where heating or cooling can be performed at a constant rate (dynamic method). The HF-DSC is also known as the Boersma DTA (Klančnik et al. 2010). This instrument belongs to the class of heat-exchanging calorimeters in which a measurement of the heat flow rate between test sample and surroundings is performed and heat exchange with the environment takes place via a well-defined heat conduction path with known thermal resistance (Höhne et al. 2003).

The HF-DSCs instruments are further classified into three basic types such as:

- Disk-type measuring system,
- Turret-type measuring system, and
- Cylindrical-type measuring system

Since the DSC allows measuring the heat of fusion and its changes at characteristic temperatures with high accuracy, it could be useful to obtain thermo-physical properties of PCMs. These properties are specific heat (c_p) for solid and liquid phases, melting temperature (T_m), and phase-change enthalpy (Δh) (Höhne et al. 2003; Castellón et al. 2008).

The HF-DSC measures the temperature difference between the reference material and the PCM sample with equal heat input. The HF-DSC determines the amount of heat absorbed by a sample during a temperature change. The temperature within the sample placed in the furnace is compared with temperature of the reference. In the HF-DSC, a temperature sensor is placed on the surface of the furnace. Heating of the test sample leads to a deviation of the sample temperature from the reference temperature. This temperature difference is monitored and used to determine the heat flux between the sample and the furnace. From this indirectly measured heat flux, the specific heat $c_p(T)$ can be obtained as a function of temperature. Following this calculation, the enthalpy (Δh) (the storage capacity) can be derived from the specific heat values by integration over temperature (T). The following equation can be used for this purpose (Wunderlich 2005; Castellón et al. 2008).

$$h(T) = \int_{T_0}^{T} c_p(\tau)\mathrm{d}\tau \tag{4.7}$$

Temperature T_0 used in the above equation as the start point of the integration range can be chosen freely to normalize $h(T)$ curves. In HF-DSC, temperature difference ΔT between the reference and the test sample is proportional to the heat-flux difference. At the same time, the rate of heat flux $(\mathrm{d}H/\mathrm{d}t)$ is proportional to the sample specific heat capacity. That is why the PCM specific heat capacity c_p can be computed by the following equation:

$$c_p = \frac{1}{m}\left(\frac{\mathrm{d}H}{\mathrm{d}t}\right) : \left(\frac{\mathrm{d}T}{\mathrm{d}t}\right) \tag{4.8}$$

where c_p is the PCM specific heat capacity, $(\mathrm{d}T/\mathrm{d}t)$ is the temperature change per unit time, and m is the mass of the PCM sample.

4.4.3 Different Modes of DSC Operation

DSC can operated using (i) a dynamic ramp mode or (ii) an isothermal step-testing mode (Höhne et al. 2003; Castellón et al. 2008). These two test modes are discussed in detail in the following sections.

4.4.3.1 DSC Dynamic Ramp Mode

The DSC dynamic ramp mode is the most widely used for PCM testing technique. It consists of heating and cooling segments performed at steady temperature change

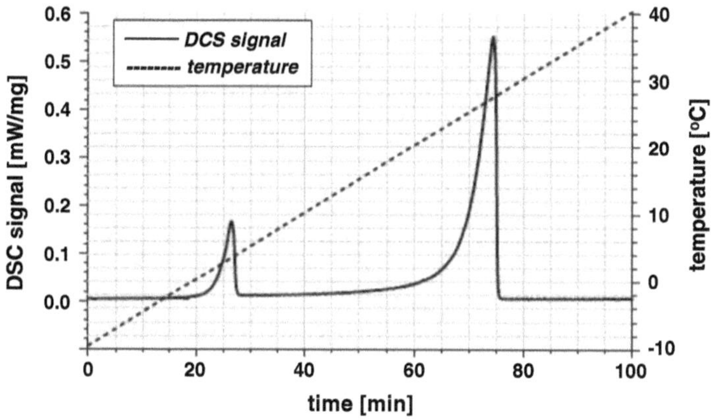

Fig. 4.5 Chart showing a typical heat flow and temperature evolution during a dynamic DSC measurement with constant heating rate (see Günther et al. 2009)

rates (Höhne et al. 2003; Castellón et al. 2008). Since the method involves continuous heating or cooling processes, the duration of this type analysis can be relatively short. An example of a typical temperature ramp during the heating process and measured DSC signal is depicted in Fig. 4.5.

With use of the DSC apparatus, the PCM enthalpy (h) can be experimentally determined by performing the following three dynamic tests:

1. The reference sample is tested against the empty crucible to determine the thermal response of the crucible.
2. The reference needs to be compared with the standard calibration material (usually sapphire).
3. Finally, the reference is tested against the PCM material of interest.

Then, the resulting specific heat of the sample as function of temperature $c_p(T)$ is given by the following formula (see Günther et al. 2009)

$$c_{p_{\text{sample}}}(T) = c_{p_{\text{preference}}}(T)\left[\frac{U_{\text{sample}} - U_{\text{empty}}}{U_{\text{reference}} - U_{\text{empty}}}\right](T)\frac{m_{\text{sample}}}{m_{\text{reference}}} \tag{4.9}$$

where,

m masses of the sample and reference material
U voltage signals of empty, sample, and reference run

Since the DSC dynamic ramp involves continuous heating or cooling processes, there are some related uncertainties to the dynamic mode which are discussed next.

The effect of different heating/cooling rates and the sample size can be clearly seen in Fig. 4.6 (Günther et al. 2009), which shows that the sample weight, as well the speed of the heating ramp, may significantly affect the shape of the enthalpy curve. Please also keep in mind that in DSC analysis it is assumed that the tested

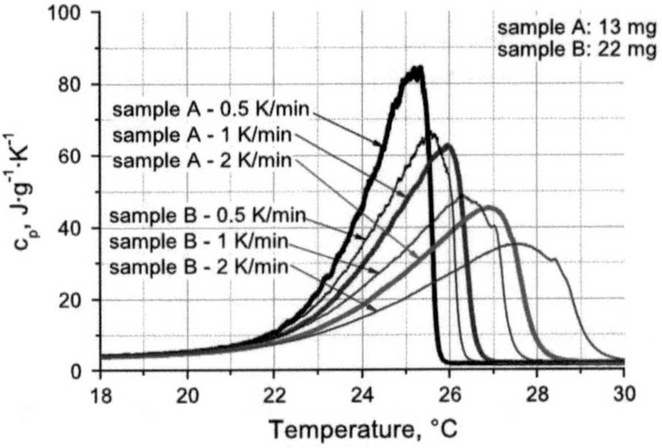

Fig. 4.6 Effect of different sample masses and different heating rates (see Günther et al. 2009)

sample has to be in isothermal stage at the beginning and at the end of the test. If the sample is not in the isothermal state during these two periods, these thermal stages cannot be properly characterized and the heat supplied cannot be accurately determined, which may compromise the overall measurement accuracy (see Höhne et al. 2003). When the PCM sample is exposed to the dynamic ramp, with continuous heating (or cooling) applied, the temperature gradient across the sample originates. This gradient varies consequently as the heating or cooling rates vary, or the sample mass changes. As a result, the measured values during the heating process can be overestimated or similarly underestimated during the continuous cooling. As a consequence, significant inaccuracies can be generated during these tests.

Another potential source of inaccuracies during the DSC testing of PCMs is nonlinear temperature distribution within the test sample. In a DSC apparatus, the temperature sensors are placed on the surface of the DSC furnace. However, as shown in Fig. 4.7, the measured temperature at the surface of the furnace is higher than the average sample temperature for heating and lower than the average sample temperature for cooling runs. Thus, the effect of continuous heating and cooling of the dynamic mode may lead to temperature variation across the sample thickness (see Wunderlich 2005; Mehling and Cabeza 2008; Castellón et al. 2008).

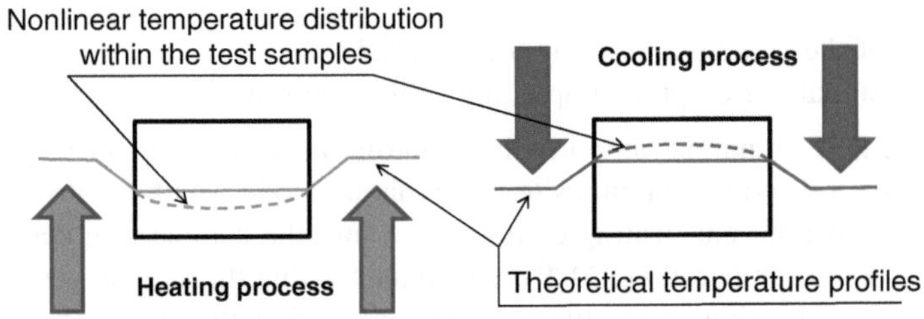

Fig. 4.7 Nonlinear temperature distribution within the test sample which may appear during the continuous heating or cooling testing

That is why the resulting enthalpy values generated by DSC during the heating and (or) cooling ramps can be systematically shifted to higher and/or lower temperature regions. This effect can be observed in Fig. 4.6. Theoretically, during the dynamic ramp DSC testing, the temperature profile within the tested material should be linear (solid lines—Fig. 4.7). In reality, due to the transient heat transfer character, the temperature profile across the sample is not linear (dashed lines). Therefore, in case of PCMs, the speed of heating or cooling needs to be slow enough to assure thermodynamic equilibrium within the test sample. If not, the heat supplied at each data recording ($dH = dQ$) cannot be assigned to the temperature measured at this time, because it may lead to notable errors in estimation of the heat stored by the test sample (see Höhne et al. 2003). Due to this reason and because the exact form of the gradient is not known, it cannot be assumed that the real value is indeed at the center of the test sample. Therefore, the distance between heating and cooling curves may be used to evaluate the uncertainty of this type of measurements (Mehling and Cabeza 2008; Castellón et al. 2008).

4.4.3.2 DSC Step-Testing Method

The DSC step-testing procedure was designed to ensure that measurements are made during the time when thermal equilibrium is present in the test sample (see Günther et al. 2009). In this method, continuous heating or cooling is not applied to the sample. Instead, a series of isothermal step temperature changes, followed by the time periods of constant temperature, are applied to the test sample. Since at the beginning and at the end of each heating or cooling ramp thermal equilibrium can be achieved, measurements are significantly more accurate than for the dynamic ramp method. The final resulting signal is the sum of these heating/cooling ramps. Using this procedure, it is possible to increase the temperature resolution of the stored heat measurement. Hence, the method is more precise than the dynamic ramp method because the uncertainty in the temperature is confined to the step size. In this method, the resolution of the temperature is proportional to the magnitude of the temperature steps. An example of a temperature history for the DSC step method and the resulting measured signal is presented in Fig. 4.8.

During the DSC step testing, the sample is heated (or cooled) in stepwise intervals in the predetermined temperature range. These steps are followed by the predetermined time delay without heating (or cooling) such that the heat exchange with the test sample and the reference is zero at the end of each step and the measured signals return to zero.

In order to evaluate the accuracy of this test method, pure PCM samples are used for calibration of the DSC apparatus. Calibration of the DSC apparatus is usually performed by dynamic performance comparisons against enthalpy data available from the literature for chemically pure PCM. To ensure accuracy of measurements, instrument calibration needs to be repeated periodically.

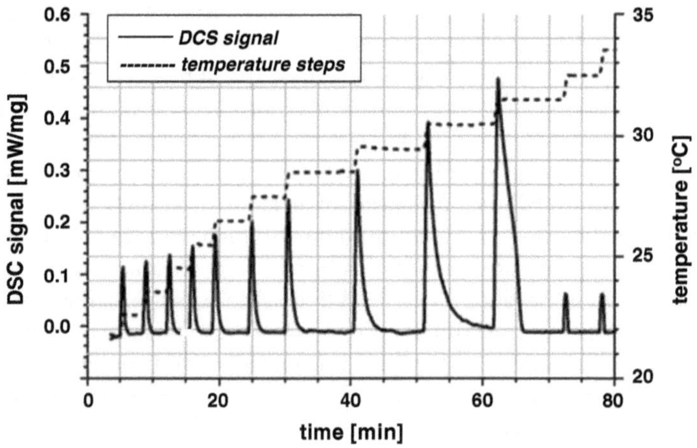

Fig. 4.8 Graph showing a typical heat flow and temperature recorded during an isothermal step DSC measurement (see Günther et al. 2009)

4.4.4 DSC Accuracy Studies for Dynamic PCM Testing

It is well known that good understanding of thermo-physical characteristics and overall system performance is essential for proper design of building envelope technologies using PCMs. In this case, a determination of PCM enthalpy as a function of temperature is an important enabling task, which is usually necessary in the early design stages. Please keep in mind that PCM thermal characteristics provide only preliminary information about dynamic thermal behavior of a building system containing PCM. During the subsequent building design tasks, associated with thermal analysis, energy performance optimization, and overall building design, thermal characteristics of the entire system containing PCM are much more useful.

As stated earlier, DSC testing is the most widely used experimental methodology for thermal performance analysis of PCMs and homogenous PCM-based products. The accuracy of DSC measurements is discussed in detail by Richardson (1997), Rudtsch (2002), Mehling and Cabeza (2008), and Günther et al. (2009). In general, DSC with standard heating rates is not optimal for high-resolution measurements of PCMs, since in enthalpy determination its accuracy is highly dependent on the heating/cooling rates. The fact is that during a dynamic DSC measurement, the PCM sample can be often far away from thermal equilibrium while changing its phase. Further, most PCMs used in buildings demonstrate a relatively low thermal conductivity. That is why, during DSC testing, in contrast to materials without a phase change or with high thermal conductivity, a significant temperature gradient is usually created inside the PCM sample. This can lead to significant inaccuracies in estimations of the heat storage capacity with respect to temperature. The magnitude of these errors depends primarily on the sample size and heating/cooling rates, followed by the sample's heat storage capacity and thermal conductivity.

The IEA Task 42/Annex 24—Energy Conservation through Energy Storage and Solar Heating and Cooling Programme "Compact Thermal Energy Storage" performed a series of DSC tests using different types of test equipment. The main objective of this work was to compare measurement results from different laboratories and to determine major sources of errors in estimations of PCM melting enthalpy as well as in analysis of melting and solidification behaviors. The other factors studied at that time were the influence of the DSC apparatus configuration, the impact of the sample preparation, and the size and shape of the test crucibles. Results show that good agreement for enthalpy measurements can be achieved in heating mode, but high discrepancies were observed in case of cooling mode measurements (see IEA 2011).

4.5 T-history Method

In 1980s, a number of research methods were developed, which allowed experimental analysis of larger test samples. The following test methods are considered today as predecessors of the T-history method.

- Vlasov et al. (1980) proposed an experimental method for determination of thermal diffusivity and thermal conductivity in both the liquid and solid phases. This method also allows determination of latent heat during the phase-transition processes. During the testing, a cylindrical sample of PCM is heated until it is completely melted. An array of thermocouples is used to measure temperature distribution within the sample during the test (Oró et al. 2010).
- The test method introduced by Carré and Delaunay (1985) allows the simultaneous measurement of thermal conductivity, specific heat, and the latent heat of PCM. This instrument utilizes the guarded hot-plate apparatus as a major component. Thermo-physical properties of PCMs are measured in three different regimes as follows: (1) steady-state intervals to measure thermal conductivity, (2) transient period between two stationary regimes to measure the variation of the specific heat as a function of temperature, and (3) test interval where the two phases coexist to measure latent heat of melting. The authors compared results generated by their instrument against the DTA testing, using commercial wax samples.
- An improved method for testing the thermal performance of PCMs was proposed by Demirel and Paksoy (1993). They tested the thermal performance of encapsulated heat storage materials using two identical temperature-controlled baths, which were heated by electrical resistance heaters. This method allowed measurements of heating and cooling enthalpies and heat transfer characteristics between the test sample and the different types of working liquids. During these experiments, temperatures of the working liquids and the test samples could be set between 0 and 200 °C. Assuming that the electrical energy input to the heating resistor is totally converted to heat, the power inputs of the baths were easily calculated. In order to evaluate accuracy of their measurements, the authors used DSC measurements for comparison.

Following the above developments, the T-history test method was introduced to measure the latent heat, specific heat, and thermal conductivity of a PCM (Zhang 1999). Simple, general-use laboratory equipment was utilized for this purpose. In addition to enthalpy measurements, the T-history method enabled analysis of the effect of super-cooling. The samples containing PCMs were required to melt to the liquid state and have a Biot number less than 0.1 (temperature distribution within the test sample was considered as uniform). The Biot[1] number is defined as $\mathrm{Bi} = hr/2k$ where r is the test tube radius, k is thermal conductivity of the PCM, and h is the free convection heat transfer coefficient. The lumped heat capacity method was used to analyze heat transfer, and the results obtained were accurate enough for engineering needs. The main advantages of the T-history method are as follows:

- Due to the larger sample sizes, it can be used for testing of encapsulated or composite PCMs and PCM-based products.
- The ranges of heating and cooling rates and temperature are large enough to fit PCM for many different applications.
- Instrumentation configuration is relatively simple and inexpensive.

PCM testing with the use of the T-history method involves recording of temperature change during the time of phase transition. It also requires for performance comparisons with a well-characterized reference, usually distilled water. This method is mainly used to determine the PCM freezing point, specific heat, latent heat, thermal conductivity, and heat diffusion coefficient. After its introduction in 1990, the T-history method became essential for thermal analysis of a large number of non-uniform PCM products. Because of larger size of the test samples, the T-history method fulfilled the industry's stipulation to characterize a large number of commercial products containing PCMs. In case of PCM-enhanced building technologies, this method helped in generation of more realistic results, which better reflected PCM's thermal behavior in real-life, whole building conditions. Keep in mind that sample size is an important factor when experimentally determining thermo-physical characteristics of PCM-enhanced building materials. A small sample usually results in a decrease in the response time of the test apparatus, and hence leading to the higher temperature resolution. However, since the building products containing PCMs are usually non-homogenous, the larger the test sample the more representative test results are. In that light, thermal performance data obtained for small quantities may be sometimes misleading.

As shown in Fig. 4.9, test equipment used for the T-history method testing is very simple. The experimental device usually consists of the heating equipment (flume) that provides a constant temperature water bath to absorb heat from, or release heat to the test samples. One of major advantages of the T-history testing

[1]The Biot number is a dimensionless quantity, which determines a magnitude of the temperature differences across an analyzed sample, while it is heated or cooled over the time. It was named after the French physicist Jean-Baptiste Biot (1774–1862).

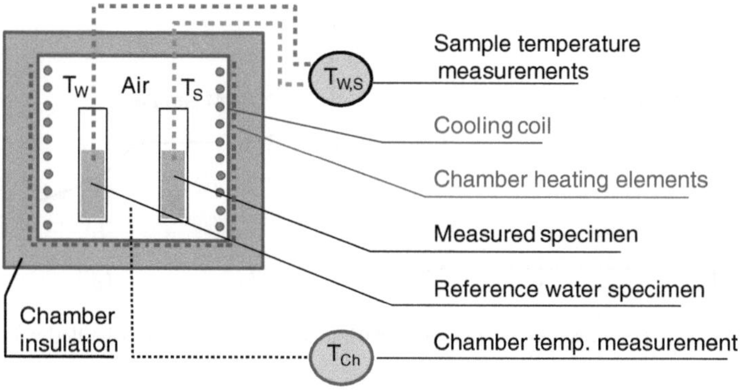

Fig. 4.9 Schematic of the T-history measurement device

methodology is that it allows measurements of several materials at the same time. In this case, a number of tested samples are only limited by the size of the water bath and a number of available data acquisition channels.

As shown in Fig. 4.9, during the T-history method measurements, tested materials are placed in laboratory glass tubes. At least one glass test tube is filled with tested PCM, while a second one contains a reference material (usually purified water). It is critical that both samples had the same mass.

Figure 4.10 shows an example of T-history curves which can be generated for distilled water and PCM samples during dynamic cooling ranging over the PCM solidification temperature. In case of the freezing process measurements, test samples need to be heated first to a temperature T_0, where $T_0 > T_m$ (T_m is the phase transition temperature). When the temperature of the two materials reaches T_0, the test tubes need to be rapidly chilled down to temperature $T_a (T_a < T_m)$ and allowed to stabilize their temperatures. It is important to remember that uneven distribution of the tested material inside the test tube may lead to uneven temperature distribution affecting measurement accuracy.

Dynamic measurements performed with the use of the described above equipment enable detailed analysis of material thermal characteristics during the phase-transition processes. In this case, thermo-physical properties can be obtained from the thermodynamic equations that describe basic phase-transition curves (see Fig. 4.10). If PCM is one of the tested materials, a typical T-history curve (a relation between temperature *T* and enthalpy *H*) can be generated for the phase-transition process. Since water is used as a reference, the T-history curve for the water sample is created as well. For PCMs without supercooling effect, the energy equations that describe this can be noted as follows:

$$m_s H_m + m_0 C_{p0}(T_{m1} - T_{m2}) = hBA_2 \tag{4.10}$$

$$m_0 C_{p0} + m_w C_{pw}(T_0 - T_{m1}) = hBA_1^w \tag{4.11}$$

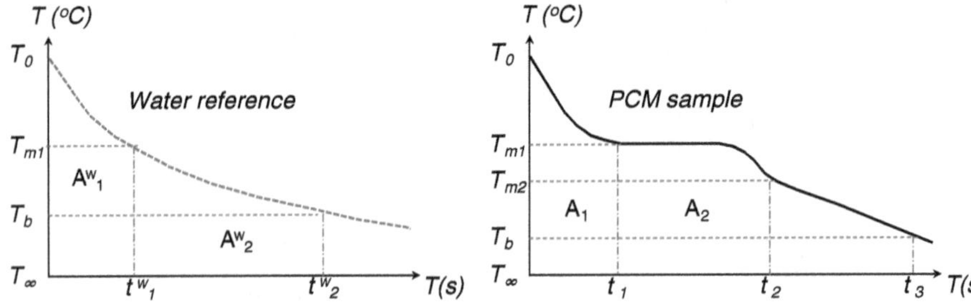

Fig. 4.10 Typical T-history curves generated during cooling process for distilled water and PCM samples

where,

$$A_1^w = \int_0^{t_1^w} (T - T_\infty)\mathrm{d}t \tag{4.12}$$

$$A_2 = \int_{t_1}^{t_2} (T - T_\infty)\mathrm{d}t \tag{4.13}$$

In equation (4.10), the latent heat H_m of the PCM sample can be described with use of the following formulas:

$$H_m = \frac{m_0 C_{p0} + m_w C_{pw}}{m_s} \tag{4.14}$$

where, m_0, m_s, and m_w were the mass of the glass tube, tested PCM, and water, respectively; c_{p0} was the specific heat of the test tube, c_{pw} was the specific heat of water, and B was the convection heat exchange area of the test tube.

Similar equations can be derived for PCMs or PCM composites demonstrating supercooling effect (Zhang 1999). Marin et al. (2003) proposed an improved mathematical formula for enthalpy analysis as a function of temperature.

Since 1990s, many researchers have been successfully using the T-history method to analyze variety of PCMs and PCM-based products. Zeng et al. (2006) tested the latent heat of n-octadecane ($C_{18}H_{38}$) using T-history and DSC testing. They reported less than a 5 % difference between DSC and T-history method measurements. However, testing performed on emulsions containing micro-encapsulated n-octadecane resulted in notable larger differences. Günther et al. (2009) used a T-history apparatus to test PCM samples approximately 1000 times larger than the typical DSC samples. Next, they used for comparison the DSC dynamic ramp mode and DSC isothermal step mode. It was found that the step method and the dynamic ramp method (with the heating rate 0.5 K/min) yielded the temperature measurement uncertainty of about 2 and 1 K, respectively. During this series of tests, the T-history method yielded similar measurement precision as DSC

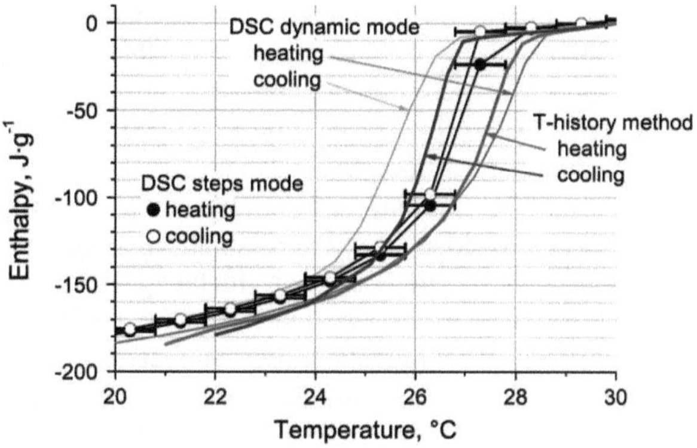

Fig. 4.11 Experimental comparisons between T-history method DSC thermal ramp testing and DSC isothermal step procedure (see Günther et al. 2009)

testing with significantly smaller sample sizes. The test results obtained by using the above three methods are depicted in Fig. 4.11 (see Günther et al. 2009). Test results comparisons of the enthalpy determinations of bulk PCM samples with the use of the DSC heat ramp method, DSC isothermal method, and T-history testing are shown there. Both heating and cooling curves are presented for each method.

In addition, Rady et al. (2010) evaluated the T-history method for characterization of granular PCM composites. Furthermore, Li et al. (2012) reported that the maximum discrepancies between T-history measurements of eutectic salt mixtures and the results of DSC testing were less than 8 %.

Although DTA and DSC methods are well developed, their shortcomings in case of testing of PCM-enhanced building materials are quite obvious. First of all, the samples required by these instruments are relatively small (1–10 mg) so that the test-generated thermo-physical properties are not usually representative of the bulk materials used in real building applications. A major reason for this difference is that PCM-enhanced building products are typically non-homogenous. In addition, the size of the test sample may impact phase-transition characteristics. For example, for liquid–solid phase-change processes of most composite PCMs, especially salt hydrates, if the phase-change material is put into a small container, the degree of supercooling of the PCM is increased while the degree of phase segregation can be decreased.

4.6 Dynamic Heat Flow Meter Apparatus (DHFMA) Testing Method

Different types of PCMs have been tested as dynamic components in buildings for at least four decades. PCMs in the building envelope and internal fabric are utilized to store energy, control temperature profiles, and mitigate temperature gradients throughout thermal envelopes. They are remaining at a nearly constant temperature during the melting and solidifying processes. A large number of laboratory and field experiments performed worldwide have already demonstrated that PCM-enhanced composite boards, membranes, arrays of PCM containers, and blends of insulation and PCM can be utilized as lightweight thermal mass components in low-energy buildings. It is expected that these types of dynamic thermal systems will contribute to the reduction of overall energy usage and peak load shifting in future buildings. Today, detailed thermo-physical data for these novel building materials are necessary for whole building simulations, energy analysis, and energy code work. The transient characteristics of PCM-enhanced products depend on the PCM content and quality of the PCM carrier. In the past, the only existing readily available methods of thermal evaluation utilized DSC testing, or sometimes T-history method measurements for limited number of materials. Unfortunately, the DSC method required small, relatively uniform test samples. This requirement is unrealistic in the case of many PCM-enhanced building envelope products. Small samples are not representative in case of PCM-based blends or composites, since these materials are not homogeneous. That is why a new laboratory-scale testing procedure enabling testing of significantly larger samples was introduced in late 2000 for the analysis of dynamic thermal characteristics of PCM-enhanced materials (Kośny et al. 2006, 2009a, b). It is using the conventional HFMA, which has been already widely used for steady-state thermal conductivity measurements according to the ASTM C 518-04 standard—*Standard Test Method for Steady-State Thermal Transmission Properties by Means of the Heat Flow Meter 4Apparatus* (ASTM 2011a). The DHFMA method is an upgrade to the previously developed rapid temperature ramp methodology that was developed to test PCM-enhanced fiber insulations and allowed for more accurate test data analysis (see Kośny et al. 2009a; Kośny et al. 2012; Shukla et al. 2013).

As mentioned previously, a conventional HFMA is typically used to measure steady-state thermal properties of thermal insulations and building materials. In general, a HFMA consists of two isothermal plates with one or more heat-flux transducers bonded to each plate. The plate temperatures are usually controlled using thermoelectric elements and water chillers. In principle, the dynamic heat flow meter apparatus (DHFMA) test method is similar to the step mode DSC testing, but it is applicable to larger-scale samples making possible the testing building products. At the beginning of DHFMA test, the top and bottom plates are set to the same temperature, unlike the conventional HFMA where top and bottom plates are set to different temperatures to impose a temperature gradient on the sample. DHFMA analysis utilizes temperature and heat-flux analysis to determine the dynamic thermal characteristics of PCM-enhanced building components. These upgrades improved the accuracy of the results with minimal modification to the existing equipment and requiring no costly hardware upgrades. Based on the described above research and development work, a new testing standard has been developed in 2013—ASTM C1784-13 *Standard Test Method for Using a Heat Flow Meter Apparatus for Measuring Thermal Storage Properties of Phase Change Materials and Products* (ASTM 2013).

In principle, a phase-transition event in a PCM-enhanced material system can be studied using the heat-balance equation for the material system with the consideration of temperature-dependent specific heat. The one-dimensional heat transport equation for such a case is as follows:

$$\frac{\partial}{\partial t}(\rho h) = \frac{\partial}{\partial x}\left[\lambda \frac{\partial T}{\partial x}\right] \tag{4.15}$$

where, ρ and λ are the material density and thermal conductivity, respectively, T and h are temperature and enthalpy per unit mass, respectively.

Considering a constant pressure during the thermal event, the effective heat capacity, c_{eff}, is defined as the derivative of the enthalpy (including latent and sensible heats) with respect to the temperature:

$$c_{\text{eff}} = \frac{\partial h}{\partial T} \tag{4.16}$$

For most PCMs, enthalpy profile with temperature depends on the direction of the phase-change process, and often, enthalpy profile during the melting process is different than that during solidification. Therefore, it is important to consider separate temperature-dependent specific heat functions for melting and solidification in the thermal design of the PCM-enhanced material. Effective heat capacity for a material which is a blend of a solid carrier and PCM may be expressed as follows:

$$c_{\text{eff}} = \frac{(100-\alpha)}{100} c_{\text{carr}} + \frac{\alpha}{100} c_{\text{eff PCM}} \tag{4.17}$$

where, α denotes the weight percent content of PCM, c_{carr} is the specific heat of the carrier material without PCM, and $c_{\text{eff PCM}}$ is the effective heat capacity of PCM.

During dynamic step testing with the use of the HFMA, temperature profiles and top and bottom plate heat flow rates, QT and QB, are recorded at time interval, τ, by thermocouples and heat-flux transducers, respectively, for each plate (see Kośny et al. 2006; Alderman and Yarbrough 2007). Each temperature step is allowed to continue until the thermal equilibrium condition is reached. Considering a constant pressure, enthalpy, H (in terms of heat per unit square of surface area), is determined by integrating heat flow rates over time as follows:

$$H = \sum \{H_i + [(QT_i - QT_{\text{final}})S_T + (QB_i - QB_{\text{final}})S_B]\tau\} \tag{4.18}$$

where, T_i and T_{final} are beginning and ending temperatures, respectively, for each temperature step, QT_{final} and QB_{final} are the residual heat flow signals from the upper and lower plates, respectively, at equilibrium that are subtracted from the signal of interest to eliminate drift caused by usually small flanking heat losses, and S_T and S_B are the calibration factor for top and bottom plates, respectively.

A series of dynamic thermal measurements were simultaneously performed with the use of DSC and DHFMA in order to estimate the accuracy of the DHFMA test method (see Fig. 4.12). A test sample of the bio-based shape-stabilized PCM (ss-PCM) was used for this purpose, because it can be synthesized both as small pellets and as larger-scale flat sheets, allowing a direct comparison between the above two methods (Shukla et al. 2013). As depicted in Fig. 4.13, a peak of the

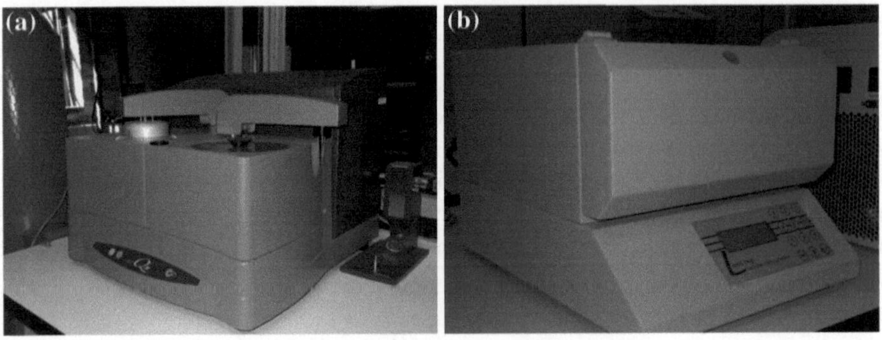

Fig. 4.12 Examples of experimental apparatus employed to determine dynamic thermal properties of PCM-enhanced materials: **a** differential scanning calorimeter (DSC), and **b** heat flow meter apparatus (HFMA)

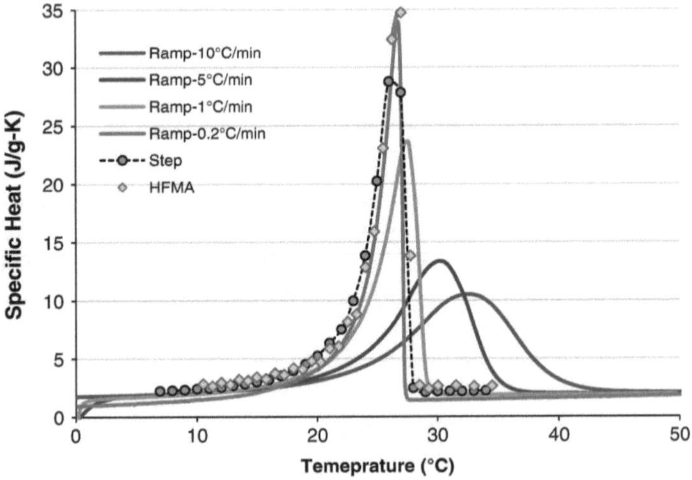

Fig. 4.13 Specific heat as a function of temperature measured for shape-stabilized (ss) PCM with the use of the DSC and DHFMA testing methods (Shukla et al. 2013)

specific heat data, generated during the DSC testing for PCM melting as a function of temperature, is shifting to lower temperatures with slower temperature changes. Correspondingly, for the solidification event, it is shifting to higher temperatures. This is due to the fact that DSC sample has a finite thermal mass, causing its thermal response lag. Slower ramps allow test sample more time to reach thermal equilibrium. This statement can be verified by comparing the DSC data recorded during the slowest test ramp (0.2 °C/min) against the DSC step test data. Next, DHFMA tests were performed on 0.4-cm-thick sheet of the ss-PCM. Two separate measurements for melting and solidification processes were performed using 1.5 °C temperature steps. As shown in Fig. 4.13, specific heat as a function of temperature measured by the DHFMA method was found to be in very good agreement with DSC's slowest ramp results (Shukla et al. 2013).

4.7 Dynamic Hot-Box Testing Method

The hot-box testing methodology has been used and perfected for decades. Historically, the hot-box test method was developed to measure the overall steady-state heat transfer through the full-size, system-scale building envelope assemblies. However, some hotboxes with cooling capability of the meter side can be used to obtain transient experimental data. There are two major types of hot-box testing devices—the guarded hotbox (GHB—an absolute test method) and the calibrated hotbox (CHB—a secondary method). The CHB apparatus is similar in form to the GHB; however, it requires calibration with the use of the calibration sample of well-characterized thermal conductance. The main difference between these two methods is that the GHB apparatus has a metering box inside a guard box (see Fig. 4.14)[2]. This reduces the heat loss through the metering box to a minimum as the metering box and the guard box temperatures are kept the same. The first GHB was built in the late 1930s (see Zarr 2001). Hot-box measurements are usually performed to measure the thermal performance of opaque and transparent building envelope components such as walls, floors, roofs, and windows. The hot-box test method can be used to measure either thermal transmittance (or U-value) which determines the

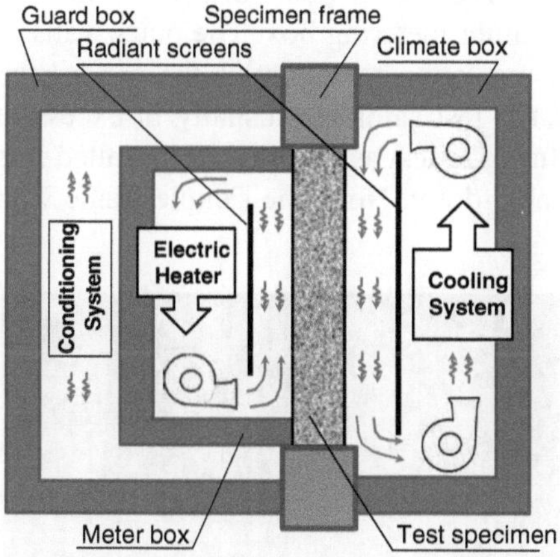

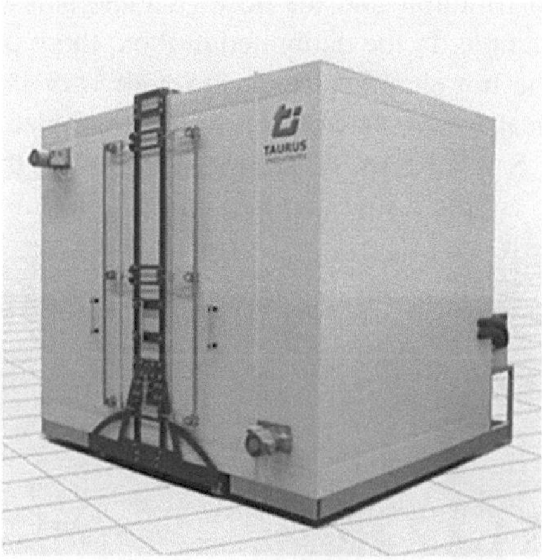

Fig. 4.14 A schematic of the guarded hotbox on the *left side* and a photograph of the commercially available hotbox on the *right side*

[2]http://www.taurus-instruments.de/en/products/hotbox-test-stations/taurus/tdw-4040-thermal-resistance-of-brickwork.html?cat=19&cHash=9272b7b5e8ee54ec1ae8903418149687

overall heat transfer through a structure; or thermal resistance (*R*-value), which is associated with heat flow within the structure.

Hotboxes, used in full-scale laboratory experiments worldwide, are usually designed in accordance with ASTM C1363-11 (2011b), ISO 8990 (1994), or ISO 12567-1 (2000). As shown in Fig. 4.14, the GHB test rig is mainly composed of two chambers: (i) climate chamber and (ii) guarded chamber with meter box placed inside. Climate chamber is sometimes called the "cold box" and is equipped with cooling units and fans to control the required low temperature inside. For PCM measurements, this part of the hotbox needs to be capable to rapidly change the internal temperature. During the conventional steady-state testing, the temperature in the cold box is kept at a constant value of sometimes −20 °C and on the opposite side at around +30 °C (this gives about 50 °C temperature difference across the test sample). The warm side of the GHB apparatus usually comprises the meter box which is surrounded by the thermal guard. The meter box functions as the main measurement cell, inside which electric heaters and fans are used to provide required temperature. The electric heating elements in the thermal guard chamber are used to control and maintain the air temperature on the same level as for the air inside the meter box. This ensures no heat transfer through the metering box walls. According to the energy balance during the steady-state measurements, the amount of heat which is generated inside the meter box needs to be equal to the heat transferred through the test sample. In the most advanced GHB test chambers, usually the meter side may have additionally cooling capability, which can be necessary for measurements of highly insulated building envelope assemblies (since electric fans may produce too much heat). On both sides of the hotbox, electric fans and radiant screens need to be designed in such way that uniform temperature distribution and air flow characteristics are achieved on both surfaces of the test sample. In the calibrated hotbox, there is no inner metering box. The outer walls of the hot chamber are made with very thick insulation to minimize the conduction heat losses. In hot-box testing, the size of the test sample is usually in excess of 1.5 m × 1.5 m. As shown in Fig. 4.15, the full-size test assemblies are installed into a sample frame that is mounted on a moveable dolly. Next, the sample frame with

Fig. 4.15 Full-size wall assemblies installed into the hot-box frames which are supported by the moveable dolly—Oak Ridge National Laboratory, USA, testing (authors archive)

test wall assembly is inserted between two chambers of identical cross-sections. Since the test wall assembly is located between the meter and climate chambers, their temperatures can be controlled independently.

Thermal characteristics of roof, attic, and floor assemblies can be tested in horizontal hotboxes. In this case, test sample is installed on the horizontal frame. Figure 4.16 shows the large scale climate simulator (LSCS) used by ORNL, USA. The LSCS apparatus allows relatively large horizontally mounted roof and floor systems to be tested at extreme environmental conditions. In addition, this hot-box apparatus contains solar simulator, which makes possible testing of roof assemblies in summer conditions.

Dynamic thermal characteristics for building envelope components (such as heat storage capacity or time of thermal response) can be generated during dynamic hot-box tests or by using analytical calculation methods. One of the first dynamic test methods for determining transfer function coefficients for complex wall samples using calibrated hotbox was developed by National Institute of Standard and Technology, USA (NIST) (see Burch et al. 1990). Following this work, the dynamic heat transmission characteristics of seven generic walls were measured by NRC Canada using a guarded hot-box procedure—ASHRAE Research Project 515-RP (NRC Canada 1991). First dynamic hot box tests of PCM-enhanced building envelopes were introduced by Oak Ridge National Laboratory (ORNL), USA. For this purpose, a combined experimental and theoretical method for analyses of dynamic thermal characteristic of building envelope assemblies was developed (see Kossecka and Kośny 1997; Kośny et al. 1998a). A series of dynamic hot-box tests have been performed on complex massive wall configurations (Kośny et al. 1998a, b).

A dynamic hot-box test typically consists of the following three basic stages:

- steady-state stage (steady temperatures on both sides of the wall),
- thermal ramp (rapid change of the temperature on the one side of the wall), and
- stabilizing stage (wall is kept under the second set of steady boundary temperatures until steady-state heat transfer occurs).

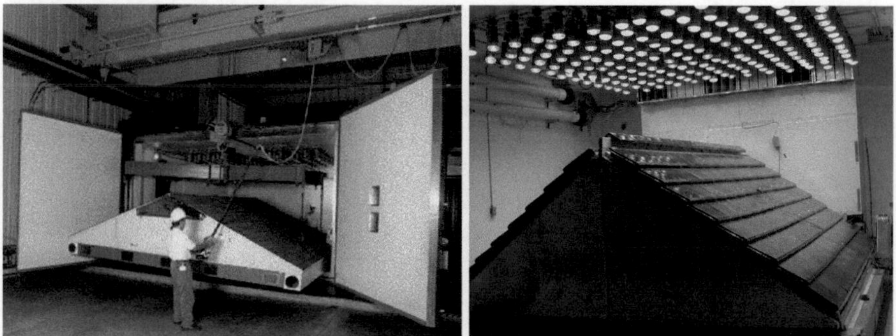

Fig. 4.16 Horizontal hotbox with solar simulator—large-scale climate simulator (LSCS) used by Oak Ridge National Laboratory, USA

This unique experimental/theoretical measurement technique can be utilized to generate dynamic thermal performance characteristics of building envelope assemblies, including response factors, structural factors, z-transfer function coefficients. A precise knowledge of these dynamic thermal characteristics is critical for accurate whole building energy simulations. Numerous wall assemblies containing conventional thermal mass have been studied using dynamic hot-box testing and transient heat conduction simulations (see Kossecka and Kośny 2001; Kośny and Kossecka 2002). In case of building envelope assemblies containing PCM, usually the following transient thermal characteristics are of major interest:

- total sensible heat storage capacity of assembly,
- phase-transition heat storage capacity of assembly,
- phase-change temperatures,
- thermal response time,
- peak load shifting characteristics, and
- damping rates for thermal excitations, etc.

One of the first tested by the author material configurations was gypsum-based stucco containing 20 % by weight of micro-encapsulated paraffinic PCM. The 2.4-m-by-2.4-m test wall, which contained about 15.9 kg of PCM (in a 1.9-cm-thick layer of the composite stucco), was constructed and tested with the use of the ORNL hotbox (Kośny et al. 2007a). Typical wood framing and fiberglass wall cavity insulation were used in this research.

Subsequently, a novel wood frame wall assembly used the multilayer insulation, containing several layers of the PCM-enhanced open-cell polyurethane foam, and aluminized plastic foil was evaluated by the same research team (see Fig. 4.17). In this research, the major role of the plastic foil sheet was to control moisture

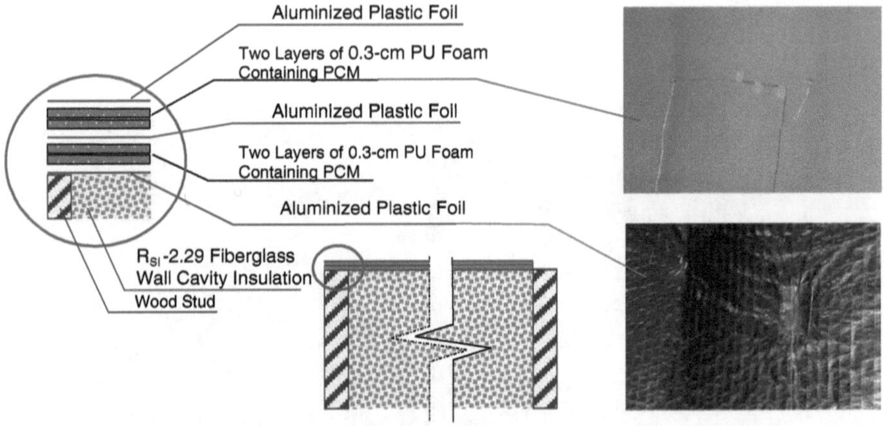

Fig. 4.17 Configuration details for the dynamic hot-box test of the wood-framed wall containing multilayer PCM-enhanced PU foam insulation and reflective foils (Kośny et al. 2007a)

migration inside the wall cavity and additionally a limited mitigation of radiation heat transfer, considering existence of small air spaces between insulation layers and a notable optical depth of the fiberglass insulation—as measured by Pelanne (1968, 1978). In total, the polyurethane foam contained about 0.49 kg/m^2 of PCM for the surface area (see Kośny et al. 2007a). The PCM melting point was about 25.5 °C, and its total phase-change enthalpy was about 140 J/g. From six identical wall cavities, three were insulated with unfaced R_{SI}-2.29 [m^2 K/W] fiberglass batts. In three other cavities, the R_{SI}-2.29 fiberglass batts were combined with the multilayer insulating facing. All cavities used conventional 1.3-cm oriented strand board (OSB) sheathing on one wall side. On the opposite side, 1.3-cm-thick gypsum board was installed. The part of the wall containing PCM (of surface area ~6 m^2) had a total heat storage capacity of about 202.6 kJ. Keep in mind that, because two different wall assemblies were tested next to each other during this experiment, major heat flow measurements had to be performed with the use of heat-flux transducers. This dynamic hot-box experiment was initiated with about 60 h of steady-state heat flow in the wall, and it was followed with series of temperature ramps performed on both wall surfaces. All major temperature changes performed on the cold and warm sides, during the described above dynamic hot-box experiment, are listed in Table 4.1.

Since the PCM wall warmed much more slowly than the conventional wall, the temperature difference between the hot-box meter side air and the surface of the PCM wall was higher compared to the conventional wall (Kośny et al. 2007a). Therefore, the heat flux on the warm side of the PCM wall was significantly higher as well. This difference in heat fluxes is presented in Fig. 4.18 as a "Cooling potential of the PCM wall." A side-by-side thermal performance comparison of the PCM part of the test wall containing multilayer insulation facing and the traditional wood frame wall assembly demonstrated a potential for steady-state and dynamic energy savings. After integration over the time, the heat fluxes measured on the cold side of the wall during the time just after the thermal excitation on the warm side demonstrated a difference of about 40 %. This value translates directly to a potential of 40 % reduction in the wall-generated peak-hour cooling load. Measurements of heat fluxes during periods of time with a steady-state heat flow enabled comparisons of the *R*-values of both parts of the test wall. The measured steady-state heat-flux difference was about 20 %, which was caused by the additional thermal resistance provided by the PCM–foam–foil facing.

Table 4.1 Temperature profiles of the dynamic hot-box test of the wood-framed test wall containing DRI and traditional fiberglass batt insulation

	Initial steady-state period (°C)	Ramp on the cold side (°C)	Rapid warmup ramp (°C)	Cooldown ramp (°C)
Warm side	23	26	35	20
Cold side	−6.6	19	20	19

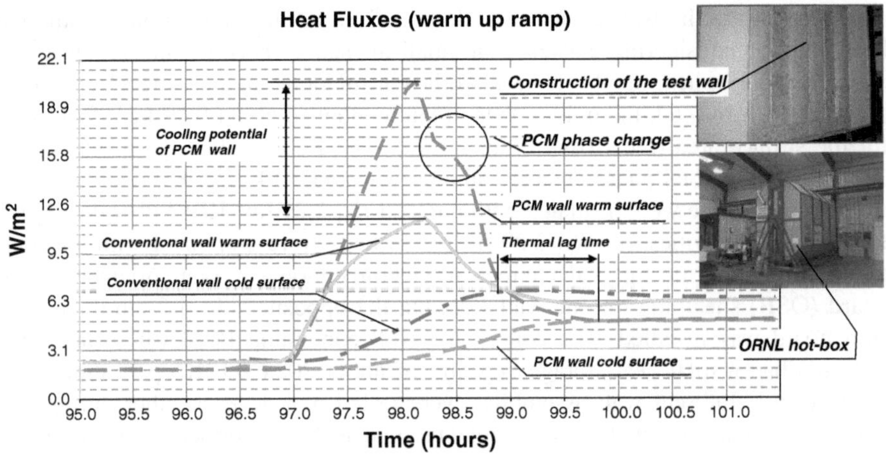

Fig. 4.18 Heat flows measured on the PCM and non-PCM sides of the wood-framed test wall, during the dynamic hot-box testing (Kośny et al. 2007a)

Following the dynamic measurements described above, the ORNL research team performed a series of similar dynamic hot-box experiments where thermal performance of PCM-enhanced cellulose fiber, used as a wall cavity insulation, was measured (Kośny et al. 2007b). A similar experimental procedure was used for testing of the wood frame wall assembly containing PCM-enhanced fiberglass (Kośny et al. 2010).

Analogous dynamic wall experiments were performed with the use of the guarded hotbox by NTNU, Trondheim, Norway (Cao et al. 2010). In this work, a series of measurements of a well-insulated wood frame wall with and without PCM layer was studied with respect to different interior conditions (generated heat and air velocity along the interior surface of the tested wall). Commercially available composite board product, containing shape-stabilized PCM, was used in these tests. Emphasis was put on the fluctuations in the interior air and wall surface temperatures. The PCM boards were installed just under the gypsum boards on the interior wall side. Temperature, heat flux, air velocity, and electrical power were recorded during thermal excitations. By applying well-distributed thermocouples, the influences of the PCM layer on the interior temperatures can be shown. As in ORNL testing, heat-flux transducers were used to measure the energy storage effects in test walls.

In addition to dynamic wall testing, dynamic hot-box experiments were performed on a residential attic module insulated with PCM-enhanced cellulose (Kośny et al. 2007b). The attic module was tested under periodic temperature changes in the ORNL LSCS (see Fig. 4.18). Two concentrations of micro-encapsulated PCM were used in these experiments (5 and 20 wt%). The main goal of dynamic attic tests was the evaluation of thermal response time of the PCM-enhanced attic floor insulation. During summer months, charging of the PCM material is not a problem, because of the intensive fluctuations of the attic air temperature (caused during the day by a rapid increase in temperature caused by the sun and during the night by intense irradiation from the roof surface to the clear sky). In a well-designed PCM application, 100 % of the PCM material should be able to fully discharge its energy before daytime operation the next day.

During these dynamic LSCS tests, the model of the residential attic was subjected to periodic changes of temperature (18 °C for about 16 h, rapid temperature ramp to 49 °C, and exposure to 49 °C for about 4 h, followed by natural cooling back to 18 °C). Please notice that during this test, hot-box climate side was warmer from the meter side, because of unique capabilities of the ORNL LSCS. The array of thermocouples installed at app. 2.5-cm intervals was used to monitor temperature distribution across the attic insulation. Figure 4.19 shows temperature profiles recoded with the use of the attic module during dynamic hot-box test in the LSCS. Please notice that recorded temperature fluctiations within the LSCS were slightly different from initial assumptions.

One of the interesting findings from the analysis of temperature fields within the PCM-enhanced attic floor insulation was that only layers located higher than 10 cm from the bottom of the attic were involved in the phase-changing process. It took about 6–8 h to fully discharge the energy stored in these layers. It is interesting that analysis of the temperature profiles demonstrated visual evidence of charging and discharging of PCM even in case of attic insulation containing only 5 wt% of PCM.

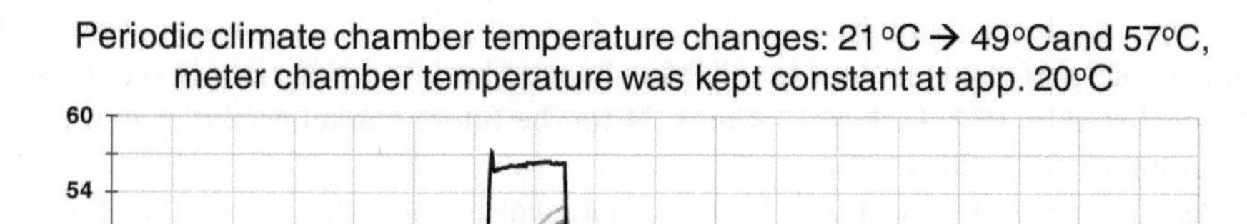

Fig. 4.19 Temperature profiles recorded inside the PCM-enhanced attic floor insulation during the dynamic full-size attic hotbox test performed with the use of the ORNL large-scale climate simulator (Kośny et al. 2007b)

4.8 A Need for Development of Testing Standards for Dynamic Thermal Analysis of PCM-Enhanced Building Materials and Systems

Thermal control and energy storage applications in buildings including PCM-based building envelopes are important technologies for energy efficiency. PCMs in buildings improve their performance by adding thermal inertia to the building structure and controlling transient heat flows through the envelope. Since an application of the PCM-enhanced building technologies means intense use of the melting and freezing enthalpies to store or to release thermal energy, it is crucial to well characterize all used building materials containing PCM. Proper design of building envelopes utilizing PCMs is essential for their effectiveness, cost, and long-term performance. That is why it is so important to be able to accurately characterize their thermo-physical characteristics (phase-change enthalpy, thermal conductivity in solid and liquid phases, and density, etc., as functions of temperature). Consider that access to the trustful thermal performance data is necessary for PCM-enhanced building envelope materials and systems to be used by designers and builders. As discussed in previous sections, for the purpose of design, code work, and energy analysis of the whole building PCM applications, dynamic thermal characteristics need to be primarily developed for PCM-enhanced building envelope products, both in material and system scales. Lack of respected and comparable performance data for PCM-enhanced building materials is one of key barriers listed by IEA Task 42—Annex 24 for the future market acceptance of these technologies (IEA 2011).

DSC is one of the most widely used analytical methods for thermal analysis of PCMs. However, a requirement of small and uniform samples makes difficult testing of PCM-enhanced building envelope materials. That is why a new laboratory-scale testing procedure for significantly larger samples was developed for the analysis of dynamic thermal characteristics of PCM-enhanced materials (see Kośny et al. 2006, 2009b; Shukla et al. 2013). It is using the conventional HFMA. In 2013, ASTM introduced ASTM C1784 test standard for dynamic HFMA measurements (ASTM 2013). Following this work, a similar standard method needs to be developed now for analysis of the long-term thermal performance of building materials, insulations, and board products containing PCM. In addition, dynamic thermal characteristics for full-size building envelope assemblies can be generated during the dynamic hot-box tests. Although hotboxes are used worldwide for full-scale steady-state experiments, an official standard method for hot-box measurements of dynamic thermal characteristics of full-size building envelope assemblies containing PCM has not been developed yet. This would be an indispensable step enabling, in the future, commercialization of many PCM-enhanced building envelope technologies.

References

Alderman RJ, Yarbrough DW (2007) Use of phase-change materials to enhance the thermal performance of building insulations. In: Proceedings of the international thermal conductivity conference, ITCC29 and ITES17 conference. Birmingham AL, 24–27 June

ASTM C518-04 (2011a) Standard test method for steady-state thermal transmission properties by means of the heat flow meter apparatus. ASTM Int Am Soc Test Mater 04.06:153–167

ASTM C1363-11 (2011b) Standard test method for thermal performance of building materials and envelope assemblies by means of a hot box apparatus. ASTM Int Am Soc Test Mater 04.06:717–768

ASTM C1784-13 (2013) Standard test method for using a heat flow meter apparatus for measuring thermal storage properties of phase change materials and products. ASTM Int Am Soc Test Mater. see: http://www.astm.org/Standards/C1784.htm

Barreneche C, Solé A, Miró L, Martorell I, Fernández AI, Cabeza LF (2012) New methodology developed for the differential scanning calorimetry of polymeric matrices incorporating phase change materials. Meas Sci Technol 23:5

Bony J, Citherlet S (2007) Numerical model and experimental validation of heat storage with phase change materials. Energy Build 39(10):1065–1072, 0378–7788 (Oct 2006)

Burch DM, Zarr RR, Licitra BA (1990) A dynamic test method for determining transfer function coefficients for a wall sample using calibrated hot box. ASTM—insulation materials, testing and application, ASTM 1030, pp 345–361

Cao S, Gustavsen A, Uvsløkk S, Jelle BP, Gilbert J, Maunuksela J (2010) The effect of wall-integrated phase change material panels on the indoor air and wall temperature—hot box experiments. In: Proceedings of renewable energy research conference 2010, Trondheim, Norway, 7–8 June

Castellón C, Günther E, Mehling H, Hiebler S, Cabeza LF (2008) Determination of the enthalpy of PCM as a function of temperature using a heat-flux DSC-A study of different measurement procedures and their accuracy. Int J Energy Res 32:1258–65

Carré P, Delaunay D (1985) Simultaneous measurement of the thermal properties of phase-change material and complex liquids using a nonlinear thermal-kinetic model. Meas Tech Heat Mass Transf Code 7413

Demirel Y, Paksoy HÖ (1993) Thermal analysis of heat storage materials. Thermo Chim Acta 213:211–21

Ford JL, Mann TE (2012) Fast-Scan DSC and its role in pharmaceutical physical form characterisation and selection. Adv Drug Deliv Rev 64(5):422–430

Gabbott PE (2008) Principles and applications of thermal analysis. Blackwell Publishing Ltd, Oxford ISBN-13: 978-1-4051-3171-1

Günther E, Mehling H, Hiebler S (2007) Modeling of subcooling and solidification of phase change materials. Model Simul Mater Sci Eng 15(8):879

Günther E, Hiebler S, Mehling H, Redlich R (2009) Enthalpy of phase change materials as a function of temperature: required accuracy and suitable measurement methods. Int J Thermophys 30:1257–1269

Höhne GWH, Hemminger WF, Flammersheim HJ (2003) Handbook of differential scanning calorimetry, 2nd edn. Springer, Berlin

IEA (2011) Development of a test-standard for PCM and TCM characterization part 1: characterization of phase change materials. Technical report. IEA—Solar Heating and Cooling/ Energy Conservation through Energy Storage programme—Task 42/Annex 24: Compact Thermal Energy Storage: Material Development for System Integration. http://www.iea-eces.org/files/a4.3.a2_appendix_wga2_1.pdf

ISO 8990 (1994) Thermal insulation—determination of steady-state thermal transmission properties—calibrated and guarded hot box. International Organization for Standardization, Geneva

ISO 12567-1 (2000) Thermal performance of windows and doors—determination of thermal transmittance by hot box method—part 1: complete windows and doors. International Organization for Standardization, Geneva

Klančnik G, Medved J, Mrvar P (2010) Differential thermal analysis (DTA) and differential scanning calorimetry (DSC) as a method of material investigation. RMZ—Mater Geoenvironment 57(1):127–142

Kossecka E, Kośny J (1997) Effect of insulation and mass distribution in exterior walls on the dynamic thermal performance of a whole buildings. In: Proceedings of the DOE, ASHRAE, ORNL conference-thermal envelopes VII, Clear Water, Dec 1998

Kossecka E, Kośny J (2001) Conduction Z-transfer function coefficients for common composite wall assemblies. DOE, ASHRAE, ORNL conference-thermal envelopes VIII, Clear Water, Dec 2001

Kośny J, Kossecka E, Desjarlais AO, Christian JE (1998a) Dynamic thermal performance of concrete and masonry walls. In: Proceedings of the DOE, ASHRAE, ORNL conference-thermal envelopes VII, Clear Water, Dec 1998

Kośny J, Christian JE, Desjarlais AO, Kossecka E, Berrenberg L (1998b) Performance check between whole building thermal performance criteria and exterior wall measured clear wall R-value, thermal bridging, and air tightness. ASHRAE Transactions 1998, vol 104, pt 2

Kośny J, Kossecka E (2002) Multidimensional heat transfer through complex building envelope assemblies in energy simulation programs. Energy Build 34:445–454

Kośny J, Yarbrough DW, Wilkes KE, Leuthold D, Mohiuddin SA (2006) PCM-enhanced cellulose insulation thermal mass in lightweight natural fibers. In: ECOSTOCK—IEA international conference on thermal energy storage—Stockton College of New Jersey. May 31–June 02 2006

Kośny J, Yarbrough DW, Miller W, Petrie T, Childs P, Syed AM, Leuthold D (2007a) Thermal performance of PCM-enhanced building envelope systems, thermal performance of the exterior envelopes of whole buildings X. In: Proceedings of the ASHRAE/DOE/BTECC conference, Clear Water Beach, 2–7 Dec 2007, pp 1–8

Kośny J, Yarbrough DW, Mohiuddin SA (2007b) Performance of thermal insulation containing microencapsulated phase change material. In: Proceedings of the international thermal conductivity conference, ITCC29 and ITES17 conference, Birmingham, 24–27 June

Kośny J, Kossecka E, Yarbrough DW (2009a) Use of a heat flow meter to determine active PCM content in an insulation. In: Proceedings of the 2009 international thermal conductivity conference (ITCC) and the international thermal expansion symposium (ITES), Pittsburgh, Aug 29–Sept 2, 2009

Kośny J, Stovall T, Yarbrough DW (2009b) Dynamic heat flow measurements to study the distribution of phase-change material in an insulation matrix. In: Proceedings of the 2009 international thermal conductivity conference (ITCC) and the international thermal expansion symposium (ITES), Pittsburgh, Aug 29–Sept 2, 2009

Kośny J, Yarbrough D, Miller W, Shrestha S, Kossecka E, Lee E (2010) Numerical and experimental analysis of building envelopes containing blown fiberglass insulation thermally enhanced with phase change material (PCM). In: Proceedings of the 1st central European symposium on building physics, Cracow, Sept 2010

Kośny J, Kossecka E, Brzezinski A, Tleoubaev A, Yarbrough D (2012) Dynamic thermal performance analysis of fiber insulations containing bio-based phase change materials (PCMs). Energy Build 52:122–131, Sept 2012

Kuznik F, Virgone J, Noel J (2007) Optimization of a phase change material wallboard for building use. Appl Therm Eng 28:1291–1298

Kuznik F, Virgone J, Johannes K (2010) Development and validation of a new TRNSYS type for the simulation of external building walls containing PCM. Energy Build 42(7):1004–1009

Li Y, Zhang Y, Li M, Zhang D (2012) Testing method of phase change temperature and heat of inorganic high temperature heat change materials. Experimental thermal and fluid science 2012 http://dx.doi.org/10.1016/j.expthermflusci. 2012.09.010

Marín JM, Zalba B, Cabeza LF, Mehling H (2003) Determination of enthalpy–temperature curves of phase change materials with the temperature-history method: improvement to temperature dependent properties. Meas Sci Technol 14:184–489

Mehling H, Cabeza LF (2008) Heat and cold storage with PCM. Springer, Berlin, Aug 15, 2008—ISBN 978 3 540 68556 2

NRC Canada (1991) Dynamic heat transmission characteristics of seven generic wall sample. ASHRAE Research Project 515-RP, IRC Contract CR5880

Oró E, deGracia A, Castell A, Farid MM, Cabeza LF (2010) Review on phase change materials (PCMs) for cold thermal energy storage applications. Appl Energy 99:513–33

Peck JH, Kim J-J, Kang C, Hong H (2006) A study of accurate latent heat measurement for a PCM with a low melting temperature using T-history method. Int J Refrig 29:1225–1232

Paksoy HO (1996) Determining thermal properties of heat storage materials using the twin bath method. Energy Convers Manage 37(3):261–268

Pelanne CM (1968) Experiments on the separation of heat transfer mechanisms in low density fibrous insulation. In: Proceedings of the eighth conference on thermal conductivity, Purdue University, USA, 7–10 Oct 1968

Pelanne CH (1978) Experiments to separate the effect of thickness from systematic errors in thermal transmission measurements. In: Presented at the ASTM C.16.00 thermal insulation conference, Tampa, 22–25 Oct 1978

Rady MA, Aquis E, LeBot C (2010) Characterization of granular phase changing composites for thermal energy storage using the T-history method. Int J Energy Res 34:333–44

Richardson MJ (1997) Quantitative aspects of differential scanning calorimetry. Thermochim Acta 300:15–28

Rudtsch S (2002) Uncertainty of heat capacity measurements with differential scanning calorimeters. Thermochim Acta 382:17–25

Shukla N, Cao P, Kośny J (2013) Lab-scale dynamic thermal testing of PCM-enhanced building materials. In: Proceedings of the ASTM symposium on next-generation thermal insulation challenges and opportunities, Jacksonville, 23–24 Oct 2013

Vlasov VV, Puchkov NP, Fedorov NP, Shatalov YS (1980) Automatic instrument for measuring thermophysical properties during a change of state. Heat Transfer-Soviet Res 14:48–54

Watson E, O'Neill M (1996) US Patent 3,263,484 (1966)

Wunderlich B (2005) Thermal analysis of polymeric materials. ISBN 3-540-23629-5 Springer, Berlin. Library of Congress Control Number: 2004114977

Xie J, Li Y, Wang W, Pan S, Cui N, Liu N (2013) Comments on thermal physical properties testing methods of phase change materials. Hindawi Publishing Corporation, Advances in Mechanical Engineering, vol 2013, Article ID 695762, 9p http://dx.doi.org/10.1155/2013/695762

Zarr RR (2001) A history of testing heat insulators at the national institute of standards and technology. ASHRAE Transactions 2001, vol 107, Pt 2, June 2001

Zeng RL, Chen BJ, Wang X, Zhang YP (2006) Two measurements of latent heat of microencapsulated phase change materials and suspensions. New Technol Dev Refrig, Nanjing

Zhang Y (1999) A simple method, the T-history method, of determining the heat of fusion, specific heat and thermal conductivity of phase-change materials. Meas Sci Technol 10(3):201–205

Zhang Y, Li QH (2008) Calibration of parameters and accuracy of an adiabatic water bath calorimeter. J Nanjing Norm Univ 8(1):47–50

Zhang Y, Zhou G, Lin K, Zhang Q, Di H (2007) Application of latent heat thermal energy storage in buildings: state-of-the-art and outlook. Build Environ 42(6):2197–2209

Chapter 5
Examples of Full-Scale Field Experiments—Test Huts and Whole Buildings Containing PCM-Enhanced Building Envelope Components

5.1 Introduction

Full-scale field testing of building envelope components containing PCMs is very often a critical last step before market implementation of new products utilizing latent heat. From the other side, a development of numerical tools, necessary for analysis of these technologies as well as validation of these tools is also used, is an important justification for full-scale field testing. Finally, energy code developments, energy performance demonstrations, and long-term durability questions for PCM-enhanced products may also require long-term full-scale field testing. This chapter will focus only on two basic forms of these experiments: test hut testing and whole-building experiments. It is good to mentions that for some internal building applications, many research centers have been using test rooms, which are not discussed herein. Only examples of the most current experiments are presented below. Due to a large number of commercial demonstration projects with use of PCMs, currently taking place around the world, only a limited selection of experiments and demonstration projects from different geographic regions are discussed in this publication.

5.2 Test Hut Field Testing

Test hut field testing is a convenient and relatively inexpensive way of getting system energy, thermal, and hygrothermal performance data for specific climatic conditions. Well-designed test hut experiments enable almost calorimetric type of dynamic measurements. In case of using a baseline hut and a test hut, usually it makes possible a separation of performance behavior effects generated only by a single system of interest, which is difficult in whole-building testing. Another advantage of hut experiments is a fact that due to limited sizes of these structures, construction and instrumentation costs are only a fraction of similar expenses in

J. Kośny, *PCM-Enhanced Building Components*, Engineering Materials and Processes, DOI 10.1007/978-3-319-14286-9_5

case of the whole-building testing. In case of an application of several test huts in a single experiment (to enable comparisons), it is critical that all construction and thermal parameters (i.e., geometry, dimensions, insulation levels, and air leakage), as well as utilized space conditioning systems are identical. Additionally, test huts allow testing of different building proportions, different types of roofs, attics, glazing, etc. Any necessary changes in orientation, tested system configuration, or instrumentation can be done very fast—within hours or days. Figure 5.1 presents two types of test huts used by Fraunhofer CSE for field testing of PCM-enhanced envelopes and other advanced building components. They are used to carry out long-term field evaluations in a semiarid climate of Albuquerque, NM, USA.

When passive solar systems, double facades, or solar fenestration systems are the main objective of research, very often southern opaque wall of the test hut is replaced by a large single window or glazing. At Concordia University, Canada an outdoor test facility has been developed with two configurations of the double façade. Athienitis et al. (1997, 2005), Charron and Athienitis (2003) performed an experimental analysis on a test room with a selection of passive and active building envelopes containing PCMs. Figure 5.2a shows the test hut with two configurations of the double façade with integrated photovoltaic panels and PCM heat storage. PV panels are forming the outside layer of the opaque wall section and airflow behind them. A similar rooftop test hut was developed by the Fraunhofer ISE in Freiburg, Germany,

Fig. 5.1 Fraunhofer CSE field testing with use of the test huts in Albuquerque, NM, USA

(a)

(b)

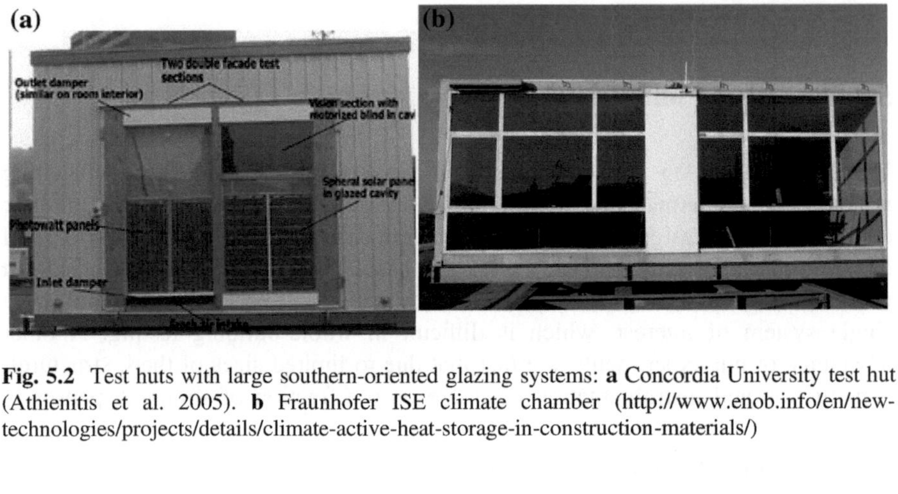

Fig. 5.2 Test huts with large southern-oriented glazing systems: a Concordia University test hut (Athienitis et al. 2005). b Fraunhofer ISE climate chamber (http://www.enob.info/en/new-technologies/projects/details/climate-active-heat-storage-in-construction-materials/)

for testing of climate-active heat storage in construction materials (see Fig. 5.2b). Field measurements for PCM-enhanced floors, walls, and ceilings were performed in Freiburg under controlled conditions. Integration included microencapsulated PCM that can be mixed with plaster, gypsum board, or other building products that can be applied to interior surfaces (Schossig et al. 2004).

5.2.1 Test Huts Built by the Alberta Research Council Inc., Canada

During 2000–2003, the Alberta Research Council Inc. (ARC), Canada, worked on the development of thermal solar systems that were designed for application in residential housing for the purpose of space heating. As a result of this research activity, a cost-effective thermal solar collector with direct heat storage (DHS) was developed and field tested (see Fig. 5.3). In the experimental system, three walls of the test hut (east, south, and west) were equipped with DHS panels. Plastic mats containing PCM of melting temperature around 29 °C and latent heat capacity of 190 kJ/kg were used for heat storage. During the 2004–2005 winter season, the prototype DHS solar system installed on the external walls of the test hut was extensively tested at different levels of the solar irradiation and at different ambient temperatures (Szymocha 2005). To perform an evaluation of the DHS system, two reference surfaces (black painted metal plate and standard vinyl siding) were also introduced into the testing system. The experimental huts were electrically heated and the internal temperature was controlled by thermostats set at 20 °C. It was observed that the temperature of the wall equipped with a PCM panel even on a short, semi-cloudy winter day can be kept at over 20 °C through most of the night (until 7 am of the next day). Recorded winter field test results have demonstrated that DHS solar thermal panels can reduce space heating energy consumption by about 50 % compared to traditional residential housing construction technology.

Fig. 5.3 Thermal components of the PCM test hut tested by the Alberta Research Council Inc. (ARC), Canada (Szymocha 2005)

5.2.2 University of Kansas Test Huts, Lawrence, KS, USA

A series of field experiments was performed using two small-scale test huts located on West Campus of the University of Kansas in Lawrence, Kansas, USA. The main focus of this research was different methods of inclusion of PCMs into the wood-framed walls and structural insulated panel walls (Zhang et al. 2005; Medina et al. 2008; Medina and Stewart 2008; Medina and Zhu 2008; Fang 2009). The PCM was integrated into test walls using direct mixing of the PCMs with cellulose insulation and macro-encapsulation. Two 1.83 m × 1.83 m × 1.52 m nominally identical test huts were built for these experiments using typical North American wood-frame wall construction practices (Fig. 5.4a, b). Both test huts were air-conditioned using fan coil units. A small refrigeration chiller produced cold water that was used for space conditioning (Zhang et al. 2005). Both test huts were instrumented to monitor and record space cooling and space heating energy consumption, indoor air and surface temperatures, indoor air relative humidities, outdoor surface temperatures, and wall heat transfer. One house was used as a control house and the other as a retrofit house.

In first Univ. of Kansas experiments, PCM was introduced to the wall using copper pipes (macro-encapsulation). The internal structure of the wall, showing the structural stud framing and PCM piping, is presented in Fig. 5.4c. A paraffin-based PCM with a phase transition range between 20 and 30 °C was selected for this research. During this experiment, 10 and 20 % percent PCM concentrations of the PCM were tested and evaluated during the summer season. Liquid PCM was encapsulated in copper pipes, which were later placed just behind the interior sheathing layer in all four walls (all orientations). Recorded date showed that the peak-hour heat fluxes through the PCM wall were substantially lower than for the conventional wood-framed wall. The overall cooling loads were reduced by 8.6 to 10.8 % for the 10 and 20 % PCM concentrations, respectively. The following research performed by the University of Kansas team, was focused on an application of the PCM-cellulose insulation mix (Fang 2009). From the experimental

Fig. 5.4 Field testing performed by the University of Kansas, USA with use of two test huts—*Source* Medina et al. (2008). **a** Test huts. **b** Wood-framing construction method used for walls. **c** Macro-encapsulation of PCM using copper pipes

results using a 30 % PCM concentration in the PCM-enhanced cellulose insulation (direct mixing), it was found that the PCM could lower and shift the peak-hour heat fluxes generated by the wall. On average, for the east wall, the absolute peak heat flux was reduced by 38.5 %; for the west wall, the reduction was 22.3 %; for the south wall, the reduction was 19.0 %; and for the north wall, the reduction was 27.4 %.

5.2.3 University of Auckland Test Huts—New Zealand

An array of identical wood-framed test huts was built on the Tamaki Campus, University of Auckland, New Zealand (Khudhair and Farid 2004, 2007; Tardieu et al. 2011). The single-story building dimensions were 2.60 × 2.60 × 2.60 m, giving floor area of 5.76 m^2 each. Their wooden frames were made of 9.8 × 6.3 cm solid pine wood profiles. Windows were installed on the north facing walls and doors on the east facing walls. The 1.25-cm-thick sheets of plywood were used as the exterior wall sheathing. The wall cavities were filled with fiberglass thermal insulations. As shown in Fig. 5.5, the test huts were elevated from the ground. In one of research projects, the PCM-enhanced gypsum boards were field tested and seasonal energy performance of this technology was analyzed using whole-building energy simulations (Tardieu et al. 2011). For the purpose of this project, gypsum boards were impregnated with 27 wt% of PCM with a melting range of 18–23 °C and a latent heat of fusion of 134 kJ/kg. The impregnation of the gypsum boards was realized by capillarity absorption of the PCM. Whole-building energy simulations using EnergyPlus software were conducted to predict the thermal their energy performance. The long-term measurements conducted for these test structures showed a good agreement with the simulation results. Both the simulation results and actual field test data have shown that the use of PCM-enhanced gypsum boards improves the thermal inertia of buildings. One of the observations was that the additional thermal mass of the PCM can reduce the daily indoor space temperature fluctuation by up to 4 °C on a typical summer day.

Fig. 5.5 Wood-framed test huts built for PCM testing on the Tamaki Campus, University of Auckland, New Zealand (Khudhair and Farid 2007)

5.2.4 Tianjin University Test Huts—Tianjin, China

Shilei et al. (2006) tested in field conditions of Tianjin, China, gypsum wall boards impregnated with organic PCM. In this experiment, a PCM-enhanced wall boards contained about 26 % PCM by weight. A mixture of 82 % of capric acid and 18 % of lauric acid of approximate melting point of 23 °C has total enthalpy around 150 kJ/kg. The field measurements of latent heat of PCM wallboards were performed in a 5 m × 3.3 m × 2.8 m test hut containing a 1.5 m × 1.5 m window in the south wall and a 1 m × 2 m wooden door in the north wall (Shilei et al. 2007). During the experiment, energy performance of conventional gypsum boards was compared to the performance of PCM-impregnated gypsum boards. 4.0-cm layer of glass wool was used to insulate the test hut. Compared with an ordinary room, it was found that the PCM wallboard can store about 5.7 % more energy compared to the conventional wallboards. It was also demonstrated that PCM room could greatly reduce the energy cost of HVAC systems and notable shift electric consumption peak.

5.2.5 University of Lleida Test Huts (Cubicles)—Lleida, Spain

Several interesting field studies investigating the inclusion of PCM into concrete, bricks, insulations, and other building envelope materials were performed by a research team of the University of Lleida, Spain. In the first study which was performed during 2005/2006, PCM-enhanced concretes were mostly analyzed (Cabeza et al. 2007). PCM concrete was used to build one of the two identically shaped test huts (see Fig. 5.6). The other test hut was built with conventional concrete for reference. Whole-building energy model TRNSYS using the PCM subroutine developed by the University of Lleida and validated based on the laboratory experimental results (Ibanez et al. 2005) was utilized for energy performance analysis. The paraffinic PCM with a melting point of 26 °C and a phase-change enthalpy of 110 kJ/kg was used in this work. The recorded results showed that PCM inclusion in the concrete helped to reduce the temperature on the test hut containing PCM. Temperature difference between a reference hut and PCM hut was about 4 °C. Also, the air in the reference hut reached the same day peak temperature (36 °C) about 2 h earlier comparing to the PCM hut.

In the following research in Lleida, PCM-enhanced bricks, alveolar bricks, and foam enhanced with macro-packaged PCM were analyzed (Castell et al. 2009, Castellón et al. 2010). These test huts were constructed using 4 concrete pillars, also with reinforcing bars, one in each edge of the test hut. The foundation was made of a concrete base (3 × 3 m) with crushed stone base and steel reinforcement. The walls were constructed of multicore bricks (29 × 14 × 7.5 cm) with a layer of insulating material (depending on the test hut) on the external side and plaster on the internal side. The external finish was made with hollow bricks and a cement

Fig. 5.6 Field testing of PCM-enhanced building materials performed by the University of Lleida, Spain, using an array of test huts (Cabeza et al. 2007). *Top* pictures show concrete test huts and concrete panels used for their construction. *Bottom* picture presents a panoramic view of the University of Lleida outdoor test facility

mortar finish. There was a 5 cm air cavity between the multi-core bricks and the hollow bricks there. The roof structure was made of the precast concrete beams and 5 cm of concrete slab. Plaster was used as the internal wall finish. Roof insulation was placed directly over the concrete. Roof slope was about 3 %. Roof surface was finished with cement mortar and double asphalt membrane from the top.

In another two test huts built in Lleida, 5-cm-thick spray polyurethane foam was used as a major insulation. One of these hut had an additional layer of paraffinic PCM located between the multi-core bricks and the polyurethane foam (on the southern and western walls and on the roof). An array of internal space temperature sensors was installed during this experiment. Recorded filed test results indicated a very strong relationship between the PCM melting point and internal space temperature set point. Best performance was recorded for the lowest temperature difference between PCM melting point and internal space temperature. This observation from Lleida confirms earlier findings from the field testing of the PCM-enhanced cellulose insulation (Kośny 2006, 2008).

5.2.6 *Oak Ridge National Laboratory Roofing Testing Facility—Oak Ridge, TN, USA*

An array of steep-slope attics with shed-type roofs was constructed by the Oak Ridge National Laboratory, USA (Fig. 5.7a) for a side-by-side field test performance comparison between different types of roof and attic assemblies (Miller et al. 2010a, b; Kośny et al. 2010). The test attics are adjacent to one another, and their

(a) (b)

Fig. 5.7 Roof testing facility at Oak Ridge National Laboratory, USA. **a** Array of test attics—south view. **b** Installation of foil pouches with inorganic PCM on *top* of the roof deck insulation—visible over-the-deck ventilation channel

shed roofs face directly south. The footprint for each attic is about 4.9 m long by 1.5 m wide. Roof slope was set at 1 cm of rise for every 3 cm of run (18.4° pitch). All test modules had ridge and soffit vents for ventilating the attic; the vent opening to the attic floor area was approximately 1–300 (Miller et al. 2007). A conventional attic covered with asphalt shingles of solar reflectance—0.093 and a thermal emittance of 0.89 were used as a control assembly for comparing the thermal performance of the prototype roof systems. All assemblies were equipped with heat flux transducers embedded in the roof deck and in the attic floor.

Since 2004, several generations of novel roof designs have been developed and tested by the ORNL research team (Miller et al. 2008). A variety of PCMs and PCM applications have been investigated in that time, starting from PCM-impregnated foams and fabrics, arrays of aluminum foil containers with inorganic PCM (Fig. 5.7b), and plastic foil mats containing organic PCM. In general, recorded test data demonstrated about 95 % reductions of peak-hour cooling loads generated by the attics and over 30 % overall cooling load reduction comparing with traditional roof designs of the same insulation R-value Kośny et al. (2010, 2012). In addition to PCMs and conventional insulations, some of these novel roof/attic systems utilized cool coatings, over-the-deck ventilation, and reflective insulation. Figure 5.7 shows a view of the ORNL attic testing facility and installation of foil pouches containing inorganic PCM on one of the test roofs.

5.2.7 Arizona State University Test Huts—Tempe, AZ, USA

Muruganantham et al. (2010) tested in field conditions several wall, floor, and attic systems containing bio-based PCM packaged in arrays of plastic foil containers. This experiment was performed by the Arizona State University during the entire 2008 calendar year in climatic conditions of Tempe, AZ, USA. The primary reason

Fig. 5.8 Arizona test huts field experiment (Muruganantham et al. 2010): **a** bio-based PCM packaged in arrays of plastic foil containers, **b** installation of PCM on the ceiling, and **c** east view of the test huts

for choosing Arizona location, as the testing location, is because of its abundant availability of solar insolation, dry almost desert weather, and very little precipitation during the year. The experimental setup consisted of two identical test huts (4.9 m × 3.7 m × 2.4 m) made of wood-framed construction as shown in Fig. 5.8. One of two test huts was traditionally insulated, while the second one had added plastic mats containing PCM. Fiberglass batt insulation was used for both attics and walls. The structures had enclosed attic spaces with R_{SI}-5.30 fiberglass batt between 0.6-m on center (o.c.) ceiling framing. Half-inch OSB roof sheathing was covered with roofing felt and standard three-tab fiberglass desert tan shingles. Walls were constructed with 2 × 4 (9-cm-thick) studs 0.4-m o.c. with R_{SI}-2.30 fiberglass insulation, composite wood siding, and 1.2-cm-thick finished gypsum board. A wooden door and a single-pane glass window were located on the east side of both huts. The test huts were equipped with identical heat pumps and connected to separate three-phase electricity meters in order to monitor the electric consumption by the HVAC system. In the test hut containing PCM, plastic foil mat with PCM containers with approximate heat storage density of 21 kJ/m^3 was installed in all walls between the fiberglass insulation and gypsum board. In addition, about 37 kJ/m^3 of PCM was installed in both the ceiling and floor of the PCM test hut. During the 2008 summer season, recorded cooling energy savings were oscillating between 25.9 % in September and 12.1 % in June. Similarly, the maximum and minimum heating energy savings were observed for November—29.3 % and March—9.2 %.

5.2.8 University of Twente Test Boxes—Enschede, Overijssel, The Netherlands

Entrop et al. (2011) and Prins et al. (2012) tested and analyzed PCM-enhanced floor systems in four plastic containers with dimensions approaching 1.0 m × 1.0 × 0.5 m see Fig. 5.9. These sizes made the test boxes approximate a scale of 1 to 5, compared

Fig. 5.9 University of Twente test boxes for PCM floor testing—University of Twente, Enschede, Overijssel, The Netherlands

to the reference Dutch residential city buildings. Inside of each experimental container, thermocouples were installed to monitor temperature profiles across building shell components. Two out of the four test containers had the floors containing microencapsulated paraffinic PCM, of melting temperature around 23 °C. Two levels of container thermal insulation were used (heavy and light insulation). The experiment has shown that indoor peak space temperatures can be reduced with up to 4.0 and 3.7 °C in the test containers using PCM-enhanced floors, for heavy and light insulated envelope configurations, respectively.

5.2.9 PCM Roof Testing at Kuwait University—Safat, Kuwait

Alqallaf et al. (2012) performed a field experiment with 1.2 × 1.2 × 1.8-m test huts dedicated to thermal analysis of a concrete roof containing PCM. The experiment was performed in Safat, Kuwait, climate, during May–September summer season. The experimental roof consisted of a 15-cm-thick concrete slab with vertical cylindrical holes filled with PCM to reduce the solar heat gain by absorbing the incoming energy by melting the PCM before it reaches the indoor space. Two different dimensions of the cylindrical holes were investigated with diameters of 10 and 14 cm and with holes' depths 15 and 7.5 cm, respectively. Paraffinic PCM with melting point approximately 39 °C was used in this experiment. The hole surfaces were sealed with metallic foil to prevent the leakage of the liquid PCM into the concrete slab. Thermocouples were placed and distributed along each cross section of the test roof. The thermal effectiveness of the analyzed roof PCM system was determined by comparing the heat flux at the indoor surface to a roof without the PCM. A parametric study was conducted to assess the effects of the PCM geometry. The results showed maximum 28 % reduction of the roof-generated heat fluxes for the roofs containing PCMs.

5.3 Whole-Building Field Testing

As mentioned above, field hut testing has a lot of advantages, mostly due to limited sizes of test structures allowing a lot of design flexibility and a relatively low cost. In case of projects where main interest lays in performance of building technologies containing PCMs, these types of measurements serve the best analysis performed in a system scale. However, one of the major benefits of using PCM-enhanced envelopes is their interaction with other building components (through thermal load mitigation and peak load shifting). Thermal response of PCM-enhanced building envelope is usually significantly slower comparing to, for example, loads generated by fenestration. So, it is very common that they may cancel each other, causing so-called free cooling effect (Kośny et al. 2012). Unfortunately, these dynamic interactions are very difficult to analyze without whole-building experiments. That is why, whole-building testing is critical for full performance analysis of many building envelope applications with use of PCMs. There is a large number of whole-building demonstrations in area of PCM-enhanced building envelops worldwide. The following section presents a selection of field test projects from different geographic regions. Most of these projects use either PCM-enhanced wall boards, internal plasters containing PCM, or ceiling panels with PCM.

5.3.1 BASF 3-L PCM House—Ludwigshafen, Germany

The 3-L house project was launched in 2001 in Ludwigshafen, Germany, as a housing unit for BASF employees.[1] The “3-Liter house” is a future-oriented, environmentally friendly house model, introduced by BASF. It maintains optimum temperatures of internal space with consumption of only 3 L of heating fuel per m^2 per year, utilizing energy-efficient construction technologies, highly efficient insulation materials, and a fuel cell system. The main project focus was whole-building energy retrofit of the building constructed during the 1950s and conversion into a low-energy multifamily house (Fig. 5.10). Initially, the main project goal was to lower the heating oil consumption to 7 L of heating oil per square meter per year. In practice, this energy-saving target was exceeded by an application of high R-value exterior opacified foam sheathing, installation of triple-glazed windows, passive solar heating from large newly glazed sunspaces, a controlled ventilation system with 85 % heat recovery, efficient heat and electricity generation by a new miniature power plant in the basement, and an application of PCM-enhanced gypsum boards or internal plasters. Three-year-long energy monitoring demonstrated that, an average consumption in the 3-L house is 2.6 L of heating oil per square meter per year. The 3-L house has since attracted considerable attention worldwide.

[1]http://construction.basf.us/files/resources/3LH_%20Brochure_e_EV_12_08_08.pdf.

Fig. 5.10 The 3-L house—Ludwigshafen, Germany—*Source* BASF

Fig. 5.11 BASF 3-L house—Kyonggi, Korea—*Source* BASF

5.3.2 BASF 3-L House—Kyonggi, Korea

A similar demonstration project was realized in Kyonggi, approximately 40 km from Seoul, Korea (Fig. 5.11). The two-story, approximately 125-m^2 house is the first "3-Liter house" in South Korea.[2] This low-energy building has been specially designed for the Korea's housing market. It introduces European energy saving know-how, novel materials, and innovative building technologies. The house was built as a result of collaboration between BASF Company Ltd., FuelCellPower Inc., and Daelim Industrial Co., Ltd (construction company). Low-energy consumption target was achieved by an application of high R-value wall thermal insulation, PCM-enhanced internal gypsum plasters, high-performance windows, thermal exchange ventilation system, and efficient fuel cell co-generation system.

[2]http://www.energyefficiency.basf.com/ecp1/Show-houses/show_houses_korea.

5.3.3 Haus der Gegenwart—Munich, Germany

In 2001, one hundred renowned architectural firms, mostly from Germany were invited to participate in an architectural competition organized by the German newspaper Süddeutsche Zeitung. The competition topic was to develop a modern-day family house (in German—Haus der Gegenwart)[3]—a home for four people for around EUR 250,000 (building costs only). House size was limited to around 200 m^2. It was suppose to be placed on a 500-m^2 flat lot. By the end of June 2001 deadline, 34 designs had been submitted. As a result of this competition, in January 2005, the Haus der Gegenwart was opened in Munich, Germany. It was built according to the plans from the Munich firm Allmann, Sattler, Wappner (ASW). This futuristic lightweight structure represents prize-winning floor plan with four autonomous living boxes joined by stairways. Three 32-m^2 wooden boxes are located on the ground floor. Each individual box can be subdivided and has its own entrance, its own bath and its own mini-garden extending to the boxwood hedge. Also, each box has its own passage upward to the 100-m^2 common kitchen/living room.

The individual boxes were constructed of wood using post and beam construction method. The walls and the ground of the largest box are insulated with 20 cm of mineral wool, while the roof required 40 cm. The walls of the three apartment boxes have the 15-cm-thick insulation and roofs—about 20-cm-thick layer of mineral wool insulation. The foundations of the apartment boxes are insulated with polystyrene foam. Windows' U-value was around 1.1 W/m^{2}K. Built in classical lightweight construction, the detached residential house incorporates a total of 600 m^2 of PCM-enhanced gypsum boards with melting temperature 23 °C to help with the internal space temperature management. This is equivalent to about 1800 kg of PCM, which yields a total PCM heat storage capacity of about 198 MJ after assuming phase-change enthalpy of 110 kJ/kg.

5.3.4 BASF House—Nottingham, England

The BASF House, located on the university campus in Nottingham, England, was built in 2008 with a UK's first-time home buyer's budget in mind. The BASF House is intended to demonstrate performance of new building materials and systems in full-scale field conditions and to showcase modern innovations and system integration. Lifetime cost and energy use analyses were performed in order to select building materials and to balance the first cost of the building with the requirement to make the house affordable to a first-time buyer. A use of alternative construction methods allowed reduction of cost, construction time, and eliminated a

[3]http://www.hausdergegenwart.de/.

need for expensive skilled labor. The house was designed using PassivHaus Standard to have near-zero carbon emissions, relying heavily on features such as passive solar energy design, a ground air heat exchanger, and a rainwater collection tank. The BASF house is part of the University of Nottingham's "Creative Energy Homes Project." It is a unique showcase of innovative state-of-the-art energy-efficient home of the future made with today's materials.

The BASF House is an 82-m^2 single-family home which can be extended to a row of terraces on demand. It has been inhabited by University of Nottingham's students and faculty since 2008. A low-carbon emission target was set for the house. The house complies with the PassivHaus standard requirement of energy consumption of 15 kWh/m^2. Actual measured energy consumption is 12.5 kWh/m^2/year, including hot water. The BASF House can be called a 1.5-L house—an average consumption of 1.5 L of heating oil equivalent per square meter per year.

A combination of insulation materials and products has been used on this project to demonstrate their potential for an application in affordable low-energy buildings. In order to minimize energy losses and infiltration through an opaque building envelope, structural insulated panels (SIPs) with closed-cell polyurethane foam core and insulated concrete forms (ICFs) were selected for the first floor and ground floor exterior walls, respectively. Similarly SIPs were used for construction of the roof. Thermal insulation on the ground floor was provided for by the use of ICFs made of opacified expanded polystyrene foam produced by BASF. ICF blocks are filled with lightweight concrete mixture, which generates tiny air cells during first hours of concrete setting, resulting in improved concrete thermal resistance. All of this helped to achieve overall R_{SI}-value of about 6.7 m^2 K/W for the walls and roof. In addition, roof was covered with metal panels containing cool coating to reduce solar-generated cooling loads during the summer time.

The BASF House can naturally maintain comfortable temperature range for indoor space by combining solar gains, natural ventilation, and thermal mass provided by PCM-enhanced gypsum boards. PCM wallboards are produced with use of microencapsulated PCM. Melting temperature is 23 °C and phase-change enthalpy—110 kJ/kg. As shown in Fig. 5.12, there is a south-oriented, fully glazed, adjustable two-layer sunspace. The sun warms the air in the sunspace and acts as the primary heating source for the house. Windows between the solar area and the main part of the house can then be opened to enable the warm air to flow around the rest of the house. An affordable ground air heat and cooling exchange system and a biomass boiler are used to provide an effective, affordable heat, and cooling for the house as well.

5.3.5 Kingspan Lighthouse—Watford, Hertfordshire, UK

The Kingspan Lighthouse was built in 2007, at Building Research Establishment Ltd. (BRE) Innovation Park in Watford, Hertfordshire, UK. BRE Innovation Park is an urban field test facility with selection of experimental buildings. Designed by

Fig. 5.12 The BASF House, located on the University of Nottingham campus, England—*Source* BASF

Sheppard Robson, the Kingspan Lighthouse was a 100-m^2 single-family home built by Kingspan Off-Site and Arup. It was the first home to meet level 6 guidelines of the UK Code for Sustainable Homes (net zero carbon emission home).[4]

During five years long field demonstration the Kingspan Lighthouse served as a demonstration and prototype house for the standard which will become mandatory for all UK residential buildings in 2016. High-performance building materials were used for construction of the Lighthouse with low U-values for all building shell components—0.07 W/m^{2}K (R_{SI}-14.3 m^{2}K/W) for the windows, 0.11 W/m^{2}K for the opaque envelopes, with airtightness of 1.0 m^3/hr/m^2 at 50 Pa. The walls and roof were made of 2 layers of structural insulated panels. The windows were wood-frame gas filled with triple glazing. The home was built using recycled and reclaimed materials when possible. In addition, the ratio of glazing to the wall area in the Lighthouse was 18 % as opposed to 25–30 % in traditional UK houses.

The Lighthouse utilized variety of low and high technologies coupled with a strong integrated design approach.[5] In order to increase the thermal inertia of house and to achieve a low-energy and zero carbon emission status, PCM-enhanced wallboards using microencapsulated PCM have been included in suspended ceilings and south-facing walls. Melting temperature of microencapsulated PCM used in the Lighthouse was 23 °C and phase transition enthalpy was 110 kJ/kg. PCM-enhanced ceiling and southern wall surfaces helped absorb daytime heat and then give it up to cooler nighttime purge ventilation.

---

[4]http://www.planningportal.gov.uk/uploads/code_for_sust_homes.pdf.

[5]http://www.kingspanlighthouse.com/.

Fig. 5.13 Roof and arch ceiling construction details for the Crossway House—designer Richard Hawkes, Staplehurst in Kent, UK

5.3.6 Crossway Eco-House—Staplehurst in Kent, UK

Located near Staplehurst in Kent, England, the Crossway Eco-House was the first in the UK to use PV–thermal systems and has officially been named as Kent's first zero-carbon house. Since 2008, it has been a family home of its visionary architect, Richard Hawkes. One of several innovative energy-saving features in the project was the installation of shape-stabilized PCM boards manufactured by DuPont. For over 2 years, about 127 m^2 of PCM boards with melting point 21.7 °C have been tested in the Crossway House. The monitoring analysis performed by Cambridge University, UK, shows that an application of PCM boards helped in reduction of summer peak temperatures by on average 4 °C.[6]

Building probably the world's largest vaulted wood–brick arch incorporates 26,000 handmade red brick tiles, fabricated from local clay within 4 miles of the site. It spans 20 m and is 9 m high. Arch's thermal mass provides thermal inertia during summer months and the clay's hygroscopic properties help maintain a healthy internal environment. As shown in Figs. 5.13 and 5.14, the 12-mm-thick tiles are bonded together with a mortar without a formwork.

The Crossway House hosts an experimental 580 -L PCM heat storage unit linked to the UK's first PV–thermal system. Heat storage helps buffer the solar-generated thermal energy during periods of cloudy days. Crossway's annual PV electricity generation is about 5500 kWh. At the same time, 1800 kWh of solar thermal energy is produced each year by integrated solar thermal part of the system. The PV–thermal panels are supported by an 11-kW wood pellet boiler.

[6]http://energain.co.uk/Energain/en_GB/assets/downloads/documentation/references/ref_Daily_usage_of_Energain.pdf.

Fig. 5.14 Front and side elevations of the Crossway House—designer Richard Hawkes, Staplehurst in Kent, UK

Four-year-long monitoring by University of Cambridge has shown Crossway's primary energy consumption to be 54.59 kWh/m^2 per year with annual heating load of 14.82 kWh/m^2.

5.3.7 Academic Office Building, Charles Sturt University—New South Wales, Victoria, Australia

Charles Sturt University (CSU) located in New South Wales, Victoria, Australia, is a developing, dynamic model of how communities can address environmental concerns and sustainable living in inland Australia. The $4.2 mln. Academic Office Building at CSU Thurgoona Campus is the first building which was constructed in 2008 as a part of the Albury-Wodonga Campus relocation program (AWCRP).[7]

It was designed by Wayne McPhee Architects and constructed by Zauner Constructions. Architects and CSU worked with BASF to develop together a new approach for achieving super-low-energy office buildings, with use of microencapsulated PCM in floors and through an application of the PCM-enhanced gypsum boards. PCM ceiling boards with melting point of 23 °C have been used on the first- and second-floor ceilings. The PCM microcapsules have been embedded in the concrete floor throughout. Due to PCM applications which effectively doubled the buildings thermal storage capacity, operating temperatures inside the building were between 21 and 23 °C in winter and 23–26 °C in summer. It is good to mention that the use of microencapsulated PCM in the concrete floor was in 2008 one of the first world's applications of that type! This building had also a first in Australia office buildings' application of PCM in ceilings.

[7]http://news.csu.edu.au/uploads/documents/Briefing%20on%20Building%20AA3.pdf.

Both above PCM applications were integrated with the radiant space conditioning system. Hydronic tubing that has been imbedded into the concrete slabs is freezing PCM during summer nights using off-peak power for release during the warm days. During the warm days, PCM's role is to stabilize the interior space temperature. PCM-enhanced concrete floors and ceilings act as oversized radiators to cool or heat the internal occupied office spaces. During a summer's day, as the ambient air temperature increases, the temperature of the building components remains relatively constant at 23 °C due to PCM's heat storage capacity. No refrigerative cooling is used. Integrated evaporative cooling system chilling the water for hydronic system is working mostly during the nights to maximize efficiency of the system.

This 880-m^2 building was constructed using lightweight steel framing technology. That is why the low-mass building structure, thermal mass of massive concrete slabs, and precast walls were integrated together with PCM's heat storage capacity into a well-insulated building fabric, minimizing fluctuations of temperature and humidity (see Fig. 5.15). Building design team paid a lot of attention for reduction of potential energy losses through the building envelope. As a result, roof had R_{SI} of 5.6 m^2K/W and walls R_{SI}-3.6 m^2K/W. Natural ventilation was achieved through the use of windows to control heat and air flow. Window sizes and position have been optimized for daylighting; thermal performance; and the creation of a natural, healthy, and yet productive office environment.

A network of temperature sensors has been installed in the strategic positions in order to monitor actual building performance and long-term behavior of the various

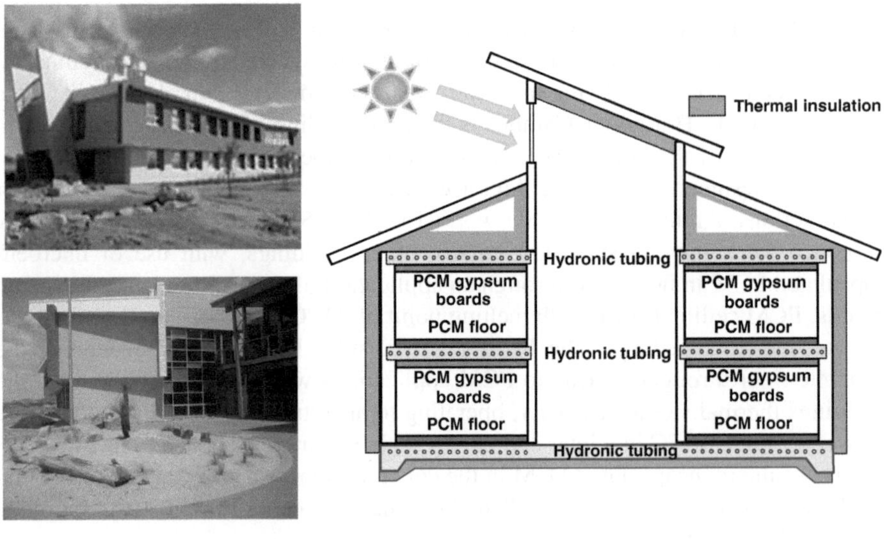

Fig. 5.15 Academic Office Building at CSU Thurgoona Campus, New South Wales, Victoria, Australia—*Source*—Charles Sturt University

Fig. 5.16 SARL Busipolis company building constructed in 2010, in Metz, France

materials and systems throughout the structure. In 2009, the Academic Office Building at the CSU Thurgoona site has been awarded six green stars and "world leader" status for its environmentally sustainable features by the Green Building Council Australia (GBCA). According to GBCA, the principles used in buildings at Thurgoona demonstrate a unique environmentally sensitive process that spans from site planning to selection of non-conventional materials and applications. One of the innovative aspects of the building that received special attention from GBCA was reduction of carbon dioxide production and energy consumption by about 65 %, compared to similar conventional office buildings in Australia.

5.3.8 SARL Busipolis Company Building—Metz, France

Since construction in 2010, the SARL Busipolis company building, in Metz, France, has received several awards for its excellent energy performance. This 1950-m^2 office building has a total energy consumption of 38 kwh/sqm per year. One of the awards was from the PREBAT competition for Lorraine region and the other from the "Lorraine Qualité Environnementale." This contemporary architecture building (Fig. 5.16) with steel and glass structure[8] incorporates more that 500 m^2 of shape-stabilized PCM boards with melting point 21.7 °C.[9] The PCM used in the Busipolis company building has been installed behind the gypsum wallboards and within the suspended ceilings in order to bring the required thermal inertia and increase the comfort in the building. It is manufactured by DuPont and has latent heat storage capacity of 515 kJ/m^2, for phase-change temperature range between 18 and 24 °C. Special ventilation system is used to help with discharging PCM panels.

[8] http://lqe.fr.

[9] http://energain.co.uk/Energain/en_GB/assets/downloads/documentation/references/ref_Busipolis_Metz.pdf.

Fig. 5.17 PCM test house—PCM-enhanced double-wall system and attic designed by Jan Kośny for the Oak Ridge field experiment (Miller et al. 2010a). **a** Framing work. **b** PCM house during construction. **c** Completed PCM house

5.3.9 Oak Ridge PCM House—Oak Ridge, TN, USA

Several high-performance building envelopes have been studied and monitored in actual homes as part of a collaborative research project sponsored by the US Department of Energy and Tennessee Valley Authority. This project was a partnership between Oak Ridge National Laboratory and building materials and construction companies to demonstrate and develop new energy-efficient technologies for US homes. Four homes were constructed in 2009/2010 using four different envelope systems in Oak Ridge, Tennessee, USA (Miller et al. 2010a). One of the test houses was designed to demonstrate thermal performance of high R-value walls and attic containing PCM-enhanced thermal insulation developed jointly by ORNL, Advanced Fiber Technologies (thermal insulation company), and Microtek (manufacture of PCM). The PCM house is a two story construction with total floor area of 253 m^2 (Fig. 5.17). The building is equipped with over 250 sensors and an on-site weather station. Airtightness of the building was 3.5 air changes per hour at 50 Pa. Windows in the PCM house are argon gas filled triple pane vinyl units that have a U-value of 1.25 W/m-K and a solar heat gain coefficient (SHGC) of 0.17. Inside surfaces of exterior and middle panes are low-e surfaces.

The house has a conventionally framed attic with dynamic solar ventilation and PCM-enhanced insulation installed on the attic floor and attic gable walls. The attic floor is insulated with 10.2-cm-thick layer of PCM-enhanced cellulose insulation on top of 25.4 cm of regular cellulose insulation. Approximately 9 cm of PCM-enhanced cellulose insulation was installed on the attic gable walls and all internal partitions. An infrared reflective painted metal shake was used as the roofing material with solar reflectance and thermal emittance of 0.34 and 0.85, respectively. A tapered EPS insulation was inserted under the metal shakes to minimize roof-generated peak-hour thermal loads. The reflective OSB boards were used on the roof deck. Low-e aluminum foil surface is facing attic space. Solar-powered gable ventilators are installed on the attic gables to reduce attic generated cooling loads.

Figure 5.17a, b shows the PCM test house during construction, while Fig. 5.17c shows completed PCM house. It has double walls filled with PCM-enhanced cellulose insulation. The wall studs are made of 2 × 4 laminated strand lumber (LSL)

and are spaced 0.60 m o.c. As shown in Fig. 5.18, the studs from one layer of framing are offset by 0.30 m from the other wall's studs. The interior framing is supported on top of the bottom plate that is fastened through floor sheathing and floor truss, while the exterior framing is supported on the sill plate and is fastened to the floor truss. A wide top plate was used to tie the two walls together for lateral strength. It is anticipated that this double-wall system functions as composite system. The interior frame will act as the structural wall component responsible for transmitting gravity and lateral force to the foundation. However, since the two double-top plates are mechanically connected and wall sheathing is provided in the exterior wall, a portion of lateral load is carried by the exterior walls as well. In double walls, thermal bridging at the corners is minimized by applying double-stud corner practice with insulation surrounding it (Fig. 5.18).

As depicted in Fig. 5.18, fiber fabric is stapled between the two sets of 2 × 4 studs to separate and hold two different types of blown cellulose insulation. The wall interior cavity is insulated with conventional dense-packed cellulose, while 20 % by weight of microencapsulated bio-based PCM was added to blown cellulose fibers in the exterior framed cavity (Miller et al. 2010a). Since 2002, the ORNL and Fraunhofer CSE research teams have analyzed several configurations of building envelope board products and insulations blended with microencapsulated and shape-stabilized PCMs (Kośny et al. 2006, 2007, 2012; Fallahi et al. 2012; Childs and Stovall 2012; Shukla et al. 2012).

5.3.10 TrekHaus—Portland, OR, USA

TrekHaus is a small building successfully completed in 2012 by a design-and-build team headed by architect Robert Hawthorne, of PDX Living, LLC, and builder Bart Bergquist of Willamette Valley Remodeling in Portland, Oregon, USA. The two adjacent homes in this project divide the space under the roof. The unit floor plans

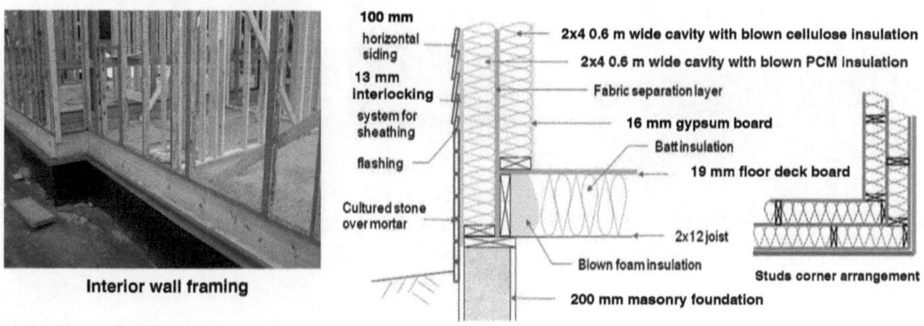

Fig. 5.18 Internal framing layer during construction and cross sections of double-wall assembly designed by Jan Kośny for the PCM house

are mirror images, each with about 155 m^2 of conditioned space, which includes three bedrooms and two baths, plus a 13-m^2 semi-conditioned workshop (see Fig. 5.19).

High-performance building envelope components are used throughout the TrekHaus to minimize winter heat losses and cooling loads during the summer. Thermal resistance is R_{SI}-6.7 for the floor slab insulation, R_{SI}-8.6 for the exterior walls, and R_{SI}-14.6 for the roof. Airtightness came in at 0.34 air changes per hour at 50 Pascals pressure difference. Triple-glazed Thermotech fiberglass-framed windows are used throughout. Mechanical systems in the TrekHaus units include a heat-pump water heater, a minisplit heat-pump for space conditioning, and a heat-recovery ventilation system. The TrekHaus is expected to operate at net zero energy with three people in each unit. Building is designed with two independent 4.14-kW roof-mounted photovoltaic systems, which are monitored for performance. Another energy-saving component of the building is BioPCM—a soy-based phase-change material. This PCM product is manufactured by Phase Change Energy Solutions from Asheboro, North Caroline, USA. Arrays of BioPCM pouches packed in plastic foil have been installed behind the drywall of one unit's interior walls and second-floor ceiling. Construction costs, excluding the land and PV system but including everything else, came to \$1.500 per m^2.

In order to monitor its energy performance, the TrekHaus building is equipped with a system of sensors and data acquisitions system installed by the Green Building Research Laboratory, of Portland State University. The overall energy performance of the building, the performance of the building envelope, and the building's mechanical systems will be monitored during coming years. In addition, the effects of occupant behavior on the building's performance will be analyzed.

Fig. 5.19 Construction details of the TrekHaus duplex building constructed in 2012 in Portland Oregon, USA—designer Robert Hawthorne, of PDX Living, LLC (http://www.trekhauspdx.com/home/reaching-for-net-zero-energy-2)

References

Alqallaf H, Abu-Arab T, Alawadhi E (2012) Experimental study of a roof with cylindrical holes containing PCM to reduce the heat gain. In: Proceedings of the international conference on solar energy for MENA region (INCOSOL), Amman, Jordan, 22–23 Oct 2012

Athienitis AK, Liu C, Hawes D, Banu D, Feldman D (1997) Investigation of the thermal performance of a passive solar test-room with wall latent heat storage. Build Environ 32 (5):405–410

Athienitis AK, Zhang J, Feldman D (2005) A study of double facades with phase-change storage and photovoltaics. In: Proceedings of international conference passive and low energy cooling for the built environment, Santorini, Greece, May 2005

Cabeza L, Castellon C, Nogues M, Medrano M, Leppers R, Zubillaga O (2007) Use of microencapsulated PCM in concrete walls for energy savings. Energy Build 39:113–119

Castell A, Medrano M, Castellón C, Cabeza LF (2009) Analysis of the simulation models for the use of PCM in buildings. In: Proceedings of Effstock 2009—the 11th international conference on thermal energy storage, Stockholm, Sweden, 14–17 June 2009

Castellón C, Medrano M, Roca J, Cabeza L, Navarro M, Fernández A, Lázaro A, Zalba B (2010) Effect of microencapsulated phase change material in sandwich panels. Renew Energy 35 (2010):2370–2374

Charron R, Athienitis AK (2003) Optimization of the performance of PV-integrated double façades. In: Proceedings of ISES solar world congress, Goteborg, Sweden

Childs K, Stovall T (2012) Potential energy savings due to phase change material in a building wall assembly: an examination of two climates—oak ridge national laboratory report—ORNL/TM-2012/6, Oak Ridge, Mar 2012

Entrop AG, Brouwers HJH, Reinders AHME (2011) Experimental research on the use of micro-encapsulated phase change materials to store solar energy in concrete floors and to save energy in Dutch houses. Solar energy. ISSN 0038-092X, 85, pp 1007–1020

Fallahi A, Shukla N, Kośny J (2012) Numerical thermal performance analysis of PCMs integrated with residential attics. In: Proceedings of SimBuild2012 conference, Madison, WI

Fang Y (2009) A comprehensive study of phase change materials (PCMs) for building walls applications. Ph.D. Thesis, Department of Civil Engineering, The University of Kansas

Ibanez M, Lazaro A, Zalba B, Cabeza LF (2005) An approach to the simulation of PCMs in building applications using TRNSYS. Appl Therm Eng 25:1796–1807

Khudhair AM, Farid MM (2004) A review on energy conservation in building applications with thermal storage by latent heat using phase change materials. Energy Convers Manag 45:263–275

Khudhair AM, Farid MM (2007) Use of phase change materials for thermal comfort and electrical energy peak load shifting: experimental investigations. In: Goswami DY, Zhao Y (eds) Solar world congress 2007. Solar Energy and Human Settlement, Beijing, China, 18–21 Sept 2007

Kośny J (2008) 2006/07 Field testing of cellulose fiber insulation enhanced with phase change material. Oak Ridge National Laboratory report—ORNL/TM-2007/186, Sept 2008

Kośny J, Biswas K, Miller WA, Kriner S (2012) Field thermal performance of naturally ventilated solar roof with PCM heat sink. Reference: SE3130—solar energy, 20 Jun 2012

Kośny J, Miller W, Zaltash A (2010) Dynamic thermally disconnected building envelopes—a new paradigm for walls and roofs in low energy buildings. In: Proceedings of DOE, ASHRAE, ORNL conference—thermal envelopes XI—thermal performance of the exterior envelopes of buildings, Clearwater, Florida, Dec 2010

Kośny J, Yarbrough DW, Wilkes KE, Leuthold D, Mohiuddin SA (2006) PCM-enhanced cellulose insulation thermal mass in lightweight natural fibers. In: Proceedings of ECOSTOCK—IEA international conference on thermal energy storage. Stockton College of New Jersey, New Jersey, 31 May–02 June 2006

Kośny J, Yarbrough DW, Mohiuddin SA (2007) Performance of thermal insulation containing microencapsulated phase change material. In: Proceedings of thermal conductivity conference ITCC29 and ITES17 conference, Birmingham, AL, 24–27 June 2007

Medina MA, Stewart R (2008) Phase-change frame walls (PCFWs) for peak demand reduction, load shifting, energy conservation and comfort. In: Proceedings of sixteenth symposium on improving building systems in hot and humid climates, Plano, TX, 16–17 Dec 2008

Medina MA, Zhu D (2008) A comparative heat transfer examination of structural insulated panels (SIPs) with and without phase change materials (PCMs) using a dynamic wall simulator. In: Proceedings of sixteenth symposium on improving building systems in hot and humid climates, Plano, TX, 16–17 Dec 2008

Medina MA, King JB, Zhang M (2008) On the heat transfer rate reduction of structural insulated panels outfitted with phase-change materials, energy. Int J 33(4):667–678

Miller WA, Childs PW, Desjarlais AO, Haun L, Jacobs J, Karagiozis AN, Kośny J, Petrie TW, Syed AM, Wilson J, Yarbrough DW, Youngquist A (2008) Next generation roofs and attics for homes. ACEEE summer conference, Pacific Grove, CA, Aug 2008

Miller WA, Desjarlais A, Shoemaker L, Kriner S, Dunlap R, Youngquist A (2007) Designing low-emittance low-slope roofs that comply with California title 24 requirements. In: Proceedings of ASHRAE thermal performance of the exterior envelopes of buildings X conference, Clearwater, FL, Dec 2007

Miller WA, Karagiozis A, Kośny J, Shrestha S, Christian J, Kohler C (2010a) Demonstration of four different residential envelopes. In: Proceedings of 2010 ACEEE summer study on energy efficiency in building, Pacific Grove, California, 15–20 Aug 2010

Miller WA, Kośny J, Shrestha S, Christian J, Karagiozis A, Kohler C, Dinse D (2010b) Advanced residential envelopes for two pair of energy-saver homes. In: Proceedings of 2010 ACEEE summer study on energy efficiency in building, Pacific Grove, California, 15–20 Aug 2010

Murugananthama K, Phelan P, Horwath P, Ludlam D, McDonald T (2010) Experimental investigation of a bio-based phase-change material to improve building energy performance. In: Proceedings of ASME 2010 4th international conference on energy sustainability, ES2010, Phoenix, Arizona, USA, 17–22 May 2010. http://proceedings.asmedigitalcollection.asme.org/volume.aspx?volumeid=15077

Prins J, Entrop A, Dorée A (2012) Phase change materials in Betonvloeren—Deel II modellering en Aanpassingen. TVVL Magazine, ISSN 0165-5523, 10, pp 20–23

Schossig P, Henning H, Haussmann T (2004) Microencapsulated phase change materials integrated into construction materials. In: Proceedings of Eurosun, European solar energy conference, Freiburg, Germany

Shilei LV, Guohui F, Neng Z, Li D (2007) Experimental study and evaluation of latent heat storage in phase change materials wallboards. Energy Build 39(2007):1088–1091

Shilei LV, Zhu N, Feng G (2006) Impact of phase change wall room on indoor thermal environment in winter. Energy Build 38(1):18–24

Shukla N, Fallahi A, Kośny J (2012) Performance characterization of PCM impregnated gypsum board for building applications. In: 1st international conference on solar heating and cooling for buildings and industry (SHC 2012), vol 30, pp 1–1434

Szymocha K (2005) Advanced thermal solar system with heat storage for residential house space heating. In: Proceedings of SESCI 2005 conference British Columbia Institute of Technology, Burnaby, British Columbia, Canada, 20–24 Aug 2005

Tardieu A, Behzadi S, Chen J, Farid M (2011) Computer simulation and experimental measurements for and experimental PCM-impregnated office building. In: Proceedings of building simulation 2011: 12th conference of international building performance simulation association, Sydney, 14–16 Nov 2011

Zhang M, Medina M, King J (2005) Development of a thermally enhanced frame wall with phase-change materials for on-peak air conditioning demand reduction and energy savings in residential buildings. Int J Energy Res 29(9):795–809

Chapter 6
Thermal and Energy Modeling of PCM-Enhanced Building Envelopes

6.1 Introduction

Energy storage in building envelopes can be accomplished by means of sensible and latent heat accumulation. Latent heat storage using PCMs has been the focus of multiple building studies, due to its relatively high thermal energy storage capacity per volume, compared to sensible storage available in conventional construction materials. PCM-enhanced building envelopes are widely considered as a prospective building enclosure technology which can help in the near future in peak-hour load reductions and cutting of HVAC energy consumption. Currently, there is a great variety of commercially available PCM products that vary in PCM types (salts, paraffins, fatty acids, etc.) and in packaging and encapsulation options (i.e., PCM pouches, containers, macro- and microencapsulation). In addition, PCMs can be differentiated based on their melting temperatures, which together with PCM location and its quantity, are essential for designing of PCM building envelope applications.

Most typically, PCMs used in building envelope applications, need to go through a complete phase transition during 24-h time periods in order to be fully effective. This is why it is critical that the temperature in the location where PCM is installed fluctuates (possibly daily) above and below the PCM functional temperature range. As depicted in Fig. 6.1, the PCM functional temperature range starts usually at the lowest temperature limit of the solidification process. Similarly, it ends at the highest temperature of the melting process.

Ideally, the entire heat transported should be available at approximately the same temperature for both, melting and solidification processes. However, in real live, this situation only occurs for the paraffinic PCMs. That is why, for well-designed PCM systems, it is crucial that this temperature range be as narrow as possible. One of the complications is an existence of the temperature hysteresis between PCM melting and solidification. Figure 6.1 depicts a situation when the melting point of a PCM-enhanced material is well above its solidification point. In addition, in most of inorganic and some organic PCMs the supercooling effect can be observed.

J. Kośny, *PCM-Enhanced Building Components*, Engineering Materials and Processes, DOI 10.1007/978-3-319-14286-9_6

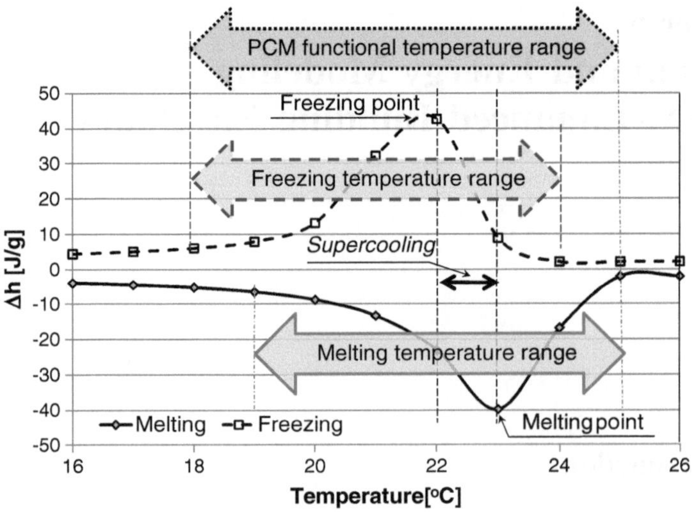

Fig. 6.1 Differential scanning calorimeter (DSC) data for microencapsulated organic PCM showing the temperature range necessary for PCM to be fully functional

The scale of supercooling can be significantly different for different types of PCMs. It can be also affected by the size of the PCM microcapsules, or through the usage of additives (like nucleation aid agents or fire retardants). In order to make the PCM presented in Fig. 6.1 fully efficient, a temperature in place of its location needs to cycle above 25 °C and below 18 °C. This simplified strategy can be used for initial selection of PCMs for specific building envelope applications.

During the last decades, numerous research studies focused on low-energy buildings utilizing latent heat storage took place. A large number of numerical heat transfer applications and whole-building energy programs enabling thermal and energy performance analyses of building fabric components containing PCM have been used. In a recent IEA Annex 23 work based on the survey of over 250 research publications, it was concluded that the general confidence in currently used numerical models is still too low to use them for designing and code purposes. In addition, it was found that in many cases, thermal characteristics of PCMs are poorly known, which creates inherent sources of inaccuracies in computer simulations (IEA Annex 23 2011). Please keep in mind that in building envelope applications, pure PCMs are rarely used. Instead, PCM-enhanced materials, microencapsulated PCMs, or arrays of small containers with PCMs are utilized, which yields a need for even more theoretical data with complex thermal characteristics, before numerical analysis can be implemented. Moreover, due to common complex geometries of arrays of PCM containers (which are sometimes used in building envelopes), heat flow is not one-dimensional in their vicinities. If combined with thermal bridging caused by structural and attachment details, most often, heat transfer in building fabric components containing PCM has a strong three-dimensional character (Kośny et al. 2009a; Childs and Stovall 2012). These facts make thermal and energy analyses of PCM-enhanced

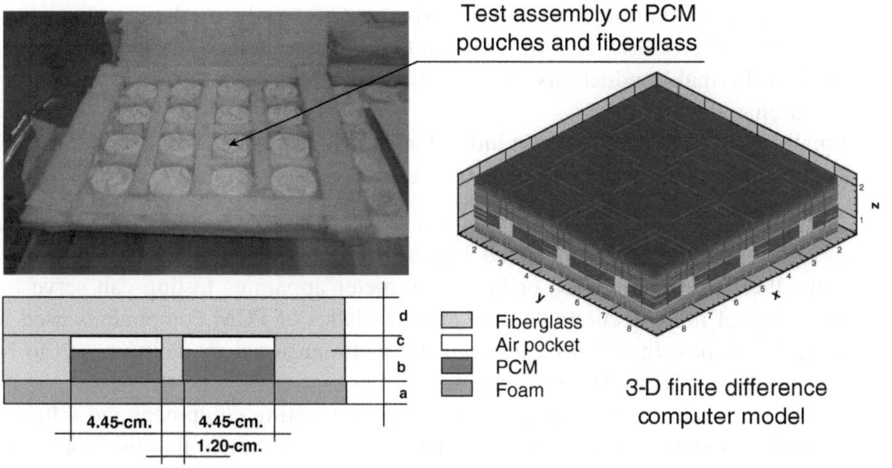

Fig. 6.2 Test specimen used in dynamic heat flow meter apparatus measurements and schematic of three-dimensional finite-difference computer model used by Kośny et al. (2009a)

building envelopes even more challenging. Figure 6.2 shows a complex array of PCM pouches packed in thermal insulation. This PCM assembly was dynamically tested with the use of the heat flow meter apparatus and simulated using three-dimensional finite-difference computer model (Kośny et al. 2010a). Good agreement was reported for comparisons between experimental and simulation results.

Another important factor effecting accuracy of thermal modeling in the case of PCM-enhanced building systems is quality of the enthalpy data used in the numerical analysis. Most often, enthalpy as a function of temperature is generated from DSC measurements for pure PCMs. If this data is used in thermal analysis of building envelopes or in whole-building energy simulations, it could be a source of significant inaccuracies, because of the following six reasons:

1. PCM-enhanced building products are not homogenous. This limits an application of DSC data for development of thermal characteristics, necessary for thermal simulations.
2. Due to cost limitations, organic PCMs used in building envelopes are seldom high-quality, pure substances. Typically, they contain a wide range of different impurities and may sometimes include additional PCMs of different melting temperatures.
3. Usually, PCM-enhanced building envelopes utilize PCM products which can be complex mixtures with fire retardants, nucleation additives, thickening agents, and thermal conductivity enhancers.
4. It is a common practice that PCMs used in building envelopes are mixed with porous structural materials (such as concrete, gypsum), thermal insulations, or other structural PCM carriers. This gives final products of significantly different thermal characteristics than the original PCM.

5. In addition, in building envelops PCMs are commonly used as packaged, or encapsulated products. It is well known that the size of the PCM encasement as well as thermal conductivity of a shell material can affect PCM's phase-transition characteristics.
6. Finally, heat exchange between individual PCM packages, capsules, or arrays of pouches has usually three-dimensional character.

All above factors show how important is to use only phase-change enthalpy data generated for actual PCM-enhanced building products. As mentioned in earlier sections, the T-history method or heat flow meter apparatus testing can serve for development of more accurate thermal characteristics of PCM components used in building envelopes. In addition, transient three-dimensional modeling needs to be applied in cases of complex envelope assemblies.

Several research centers, using different types of testing equipment and different sizes of PCM samples, have performed measurements to verify that the isothermal step method provides the enthalpy as a function of temperature and results of these measurements are independent of a rate of temperature changes and a size of a test specimen (Castellón et al. 2008; Günther et al. 2009; Lazaro et al. 2013; Shukla et al. 2012). It is also good to remember that, during the phase-change process, the latent heat is not evenly absorbed and released along the phase-change temperature range estimated during traditional way of DSC testing, utilizing simple heating or cooling ramps (see Fang and Medina 2009). Furthermore, PCMs in building envelopes will typically not be operating under theoretical assumption that the entire phase-change transition occurs during each diurnal cycle. In fact, typically the phase-change process will begin from a partially melted state, resulting from incomplete nighttime solidification, partial daytime melting, or temperature fluctuations throughout the day. These facts are not fully understood for PCM-enhanced building envelopes yet. More experimental and theoretical work is needed in this area.

One often reported problem is lack of validation for computer programs used in performance analyses of PCM-enhanced technologies. According to the IEA Annex 23, many numerical models have not been experimentally verified, which, in conjunction with often used too simplified modeling parameters, makes it difficult to realistically describe the complex phase-transition physics needed to predict the actual performance of PCMs added to a building (IEA Annex 23 2011).

Another factor, which is seldom taken into account during performance analysis of building envelope products containing PCMs, is their multi-functional nature. As mentioned in previous chapters, PCM-enhanced building envelopes are typically considered as thermal storage components and sometimes, as an engineering means of achieving local temperature control or thermal load shaving and shifting. However, more realistically, PCM-enhanced building envelopes act in a whole-building scale, helping in thermal load management for the entire internal space and involving dynamic interactions between many building fabric components (Kośny et al. 2013) (SE). This tells that limited system-scale thermal models should not be used in long-term energy performance analysis of these technologies. They can be, however, very helpful during analysis of test results, or they can support designing stage of the PCM-enhanced building components. The same conclusion refers to economical predictions, which should be based on whole-building energy consumption simulations.

In that light, a considerable portion of published literature, forecasting energy and economical performance of the PCM-enhanced building envelopes and based only on material- or system-scale analyses, can be incomplete and misleading. Consequently, conclusions regarding energy performance of PCM-enhanced building envelopes cannot be based only on numerical analysis of local thermal effects and later extended to the whole-building scale. According to IEA Annex 23 (2011), coordinated international activities need to be undertaken as soon as possible to create international standard model benchmarks for PCM-enhanced building technologies that will improve the confidence of the users of the numerical methods.

6.2 Heat Transfer Through a Building Envelope Assembly Containing PCM

The thermal energy storage systems containing PCMs have been widely recognized as one of the most advanced technical means of improving the energy efficiency and environmental impact of buildings. A very useful feature of PCMs is that they can store both latent and sensible energy. The systems incorporating PCMs benefit also from the isothermal nature of the phase-change process. The optimization of material properties within PCM-enhanced building envelopes and the analysis of their thermal performance are very important during their designing stage. At the same time, reliable whole-building energy models can numerically facilitate an optimized design of the PCM-enhanced building component without a need for time consuming and expensive whole-building field experiments. Heat transfer in PCMs during the phase-change processes is quite complex due to its nonlinear nature and, in addition, because of the fact that the simulated material has different thermo-physical characteristics for solid and liquid phases (see Lamberg et al. 2004). When the PCM undergoes a phase transition, both the solid and the liquid phases are present and they are separated by a moving interface. The difficulty in solving a phase-change problem is the presence of a moving boundary or region in which heat and mass balance conditions have to be met.

The solid–liquid phase-change problem can be described analytically with the use of partial differential equations for a particular kind of solid–liquid boundary definition, where the phase boundary can move with time. This physical process has been initially studied by Clapeyron and Lamé (1831) during analysis of the Earth's crust formation. An analytical solution for phase transition was first found by Franz Neumann, who introduced it in his 1835–1840 lecture notes (see Brillouin 1930). Later in 1889, this physical phenomenon was studied by the Slovenian physicist Jožef Stefan for the case of the water freezing (Stefan 1889). According to Stefan and followers, predicting of the thermal behavior in case of materials with phase transition is difficult due to its inherent nonlinear nature at moving interfaces, for which displacement rate is controlled by the latent heat lost or absorbed at the boundary. The Stefan problem can be described by the following set of equations for temperature distribution $u(x, t)$ (Stefan 1889; Tikhonov and Samarskii 1963):

$$c_s\rho_s\frac{\partial u}{\partial t} = k_s\frac{\partial^2 u}{\partial x^2} \quad \text{when:} 0 < x < \xi(t),\ t > 0; \tag{6.1}$$

$$c_l\rho_l\frac{\partial u}{\partial t} = k_l\frac{\partial^2 u}{\partial x^2} \quad \text{when:} 0 < x < \xi(t),\ t > 0 \tag{6.2}$$

with the following boundary condition: $u(o, t) = u_s = \text{const} > T$ and $t > 0$; and with the initial condition: $u(x, o) = u_l = \text{const} > T$ and $t \geq 0$ (where T is the temperature).

As shown in Eqs. (6.1) and (6.2), for the case of one-dimensional heat transfer, the interface between the solid and liquid phases of matter (during melting or solidifying processes) can be described by a function $\xi = \xi(t)$, combined with the temperature distribution, which is expressed as $u = u(x, t)$. An expression describing the law of motion of the solid–liquid interface and expressed as a law of energy conservation under phase transition is called the Stefan condition. The following theoretical formulas can be used for the phase-change process:

$$u[\xi(t) - 0, t] = u[\xi(t) + 0, t] \tag{6.3}$$

$$\lambda\rho_s\frac{\mathrm{d}\xi(t)}{\mathrm{d}t} = k_s\frac{\partial u[\xi(t) - 0, t]}{\partial x} - k_l\frac{\partial u[\xi(t) + 0, t]}{\partial x} \quad \text{for:} \xi(0) = 0 \text{ and } t > 0; \tag{6.4}$$

In Eqs. (6.3) and (6.4), λ represents the latent heat of fusion divided by unit of mass, ρ is the density, k is the thermal conductivity, and t is time. Finally, s and l indexes refer to solid and liquid phases.

Analytical solutions for the phase-change processes in pure PCMs are discussed in Evans (1951), Douglas (1957), Carslaw and Jaeger (1959), Friedman (1964), Rubibstein (1971), Crank (1984), and Hill (1987). Today, the solution of the Stefan problem is widely considered as one of the mostly used analytical solutions for one-dimensional solid–liquid phase transition (Ogoh and Groulx 2010).

6.2.1 Heat Transfer Model for PCM-Enhanced Porous Materials and Thermal Insulations

As mentioned above, PCMs used in building envelopes are seldom homogenous layers of pure substances. Very often porous building materials or thermal insulations are mixed with PCMs. The heat storage capacity for such PCM-enhanced product is a key factor that controls dynamic thermal performance. A theoretical model of the material with temperature-dependent specific heat can be used to calculate phase-change processes in most common materials (Heim and Clarke 2004; Kissock and Limas 2006).

The one-dimensional heat transport equation for such a case is as follows:

$$\frac{\partial}{\partial t}(\rho h) = \frac{\partial}{\partial x}\left[\lambda \frac{\partial T}{\partial x}\right] \tag{6.5}$$

where ρ and λ are the material density and thermal conductivity, T and h are temperature and enthalpy per unit mass. Heat flux q is given by

$$q(x,t) = -\lambda \frac{\partial T(x,t)}{\partial x} \tag{6.6}$$

The enthalpy derivative with respect to temperature (at constant pressure) represents the effective heat capacity, with phase-change energy being one of the components:

$$c_{\mathrm{eff}}(T) = \frac{\partial h}{\partial T}. \tag{6.7}$$

Only few types of PCMs demonstrate isothermal phase-change processes. For a large number of PCMs, variations of enthalpy with temperature depend to some extent on the direction of the phase-transition process and are different for melting and solidification.

Figure 6.3 depicts an example of temperature-dependent enthalpy and enthalpy differences curves for microencapsulated PCM, generated using a differential scanning calorimeter. In this material, the melting process takes place around 27 °C and solidification around 26 °C. The total phase-change enthalpy, H_m, within temperature interval [20°, 29°] is 122 J/kg.

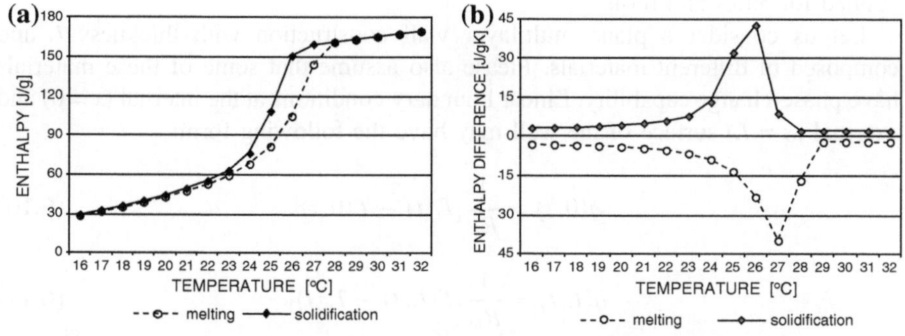

Fig. 6.3 Calorimetric data for the microencapsulated PCM: **a** enthalpy as a function of temperature, **b** enthalpy difference curves for melting and solidification processes

Effective heat capacity, c_{eff}, for a material which is a blend of insulation and PCM may be expressed as follows:

$$c_{\text{eff}} = (1 - \alpha)c_{\text{ins}} + \alpha c_{\text{effPCM}}, \tag{6.8}$$

where α denotes the weight fraction of used PCM, c_{ins} the specific heat of insulation without PCM, and c_{effPCM} is effective heat capacity of PCM.

For PCM-enhanced thermal insulations, thermal conductivity is the most important thermal parameter, which determines their thermal performance in different thermal conditions. Earlier research performed on PCM-enhanced cellulose insulations showed that an addition of microencapsulated PCM does not always result in a significant increase of thermal conductivity of the PCM–insulation blend (Kośny et al. 2009b). For cellulose fiber insulation, linear dependence of conductivity on temperature has been observed (Kośny et al. 2009a, b):

$$\lambda(T) = \lambda_0 + \beta \cdot T \tag{6.9}$$

where $\lambda_0 = 0.03575$; $\beta = 0.00013$.

Equation (6.9) can be used in numerical analysis of PCM-enhanced building envelopes, if numerical heat transfer algorithm allows variable temperature-dependent thermal conductivity.

6.2.2 *Thermal Balance for a Wall Containing PCM-Enhanced Material*

In 2009, a theoretical heat transfer model for PCM-enhanced building envelopes using insulation-PCM blends was introduced by Kossecka and Kośny (2009). The model describes heat transfer within building materials impregnated or mixed with either pure or microencapsulated PCMs. The one-dimensional model presented here is for a wall construction. However, a similar set of theoretical formulas can be applied for attics and roofs.

Let us consider a plane multilayer wall construction with thickness L and composed of different materials. Please also assume that some of these materials have phase-change capability. Linear boundary conditions at the internal ($x = 0$) and external ($x = L$) surface of the wall may have the following form:

$$q(0,t) = \frac{1}{R_{si}}[T_i(t) - T(0,t)] \tag{6.10}$$

$$q(L,t) = \frac{1}{R_{se}}[T(L,t) - T_e(t)] \tag{6.11}$$

where T_i and T_e are ambient temperatures, and R_{si} and R_{se} and surface film resistances.

In the following equations, R_{i-x} and R_{x-e} are used to denote the resistances for heat transmission from the point x in the wall to the internal and external environment, respectively. Similarly, R_u expresses the total resistance for heat transmission through the wall. For thermal conductivity along the wall's thickness x, which is represented by the function of $\lambda(x)$, the wall resistances can be described by the Eqs. (6.12)–(6.14):

$$R_{i-x} = R_{si} + \int_0^x \frac{\mathrm{d}x}{\lambda(x)} \tag{6.12}$$

$$R_{x-e} = \int_x^L \frac{\mathrm{d}x}{\lambda(x)} + R_{se} \tag{6.13}$$

$$R_u = R_{si} + \int_0^L \frac{\mathrm{d}x}{\lambda(x)} + R_{se} \tag{6.14}$$

By integrating a product of the Eq. (6.5) and the function R_{x-e} with respect to x over the thickness L of the wall, we can obtain the following formula:

$$\begin{aligned}\int_0^L \rho \cdot \frac{\partial h}{\partial t} \cdot R_{x-e}\mathrm{d}x &= \int_0^L \frac{\partial}{\partial x}\left(\lambda \frac{\partial T}{\partial x}\right) \cdot R_{x-e}\mathrm{d}x \\ &= \left(\lambda \frac{\partial T}{\partial x}\right) \cdot R_{x-e}\Big|_{x=0}^{x=L} + \int_0^L \frac{\partial T}{\partial x}\mathrm{d}x \\ &= -R_{se} \cdot q(L) + R_{0-e} \cdot q(0) + [T(L) - T(0)]\end{aligned} \tag{6.15}$$

In Eq. (6.15), temperatures $T(0)$ and $T(L)$ can be eliminated with the use of Eqs. (6.10) and (6.11). This leads to the following formula, representing the heat flux at the internal wall surface:

$$q(0) = \frac{1}{R_u}(T_i - T_e) + \int_0^L \rho \frac{\partial h}{\partial t} \cdot \frac{R_{x-e}}{R_u}\mathrm{d}x \tag{6.16}$$

Analogously, a formula for the heat flux at the exterior wall surface $q(L)$ can be obtained by integrating by parts a product of Eq. (6.5) and R_{i-x}.

As shown in Eq. (6.17), the impact of time-dependent variations of enthalpy h along the thickness of the wall, on the heat flux across the surface $x = 0$, depends on the factor R_{x-e}/R_u. This factor has values close to 1, in the vicinity of the wall internal surface $x = 0$, and low values, close to 0, in the vicinity of the opposite surface $x = L$. This means that in local heat-flux calculations, only time-dependent enthalpy variations near the considered surface play an important role. At the same time, thermal effects of enthalpy variations near the opposite wall surface are relatively small.

Let us assume now that densities and conductivities are temperature independent and thus constant in time. This has to be only understood as an approximation; especially for materials as described in Eq. (6.9). Next, Eq. (6.16) can be integrated with respect to time, over the time interval $[t_1, t_2]$. After interchanging of the integration order, the following formula can be obtained:

$$Q(0, t_1, t_2) = \int_{t_1}^{t_2} q(0, t)\mathrm{d}t \tag{6.17}$$

$$Q(0, t_1, t_2) = \frac{1}{R}\int_{t_1}^{t_2} [T_i - T_e]\mathrm{d}t + \int_0^L \int_{t_1}^{t_2} \rho \frac{\partial h}{\partial t} \cdot \frac{R_{x-e}}{R_u} \mathrm{d}t\mathrm{d}x \tag{6.18}$$

In Eq. (6.18), the time integral in the second term represents the enthalpy difference Δh, which can be expressed as follows:

$$\Delta h(x, t_1, t_2) = \int_{t_1}^{t_2} \frac{\partial h}{\partial t}\mathrm{d}t = \int_{t_1}^{t_2} \frac{\partial h}{\partial T}\frac{\partial T}{\partial t}\mathrm{d}t = \int_{T(x,t_1)}^{T(x,t_2)} c_{\mathit{eff}}(x, T)\mathrm{d}T \tag{6.19}$$

Keep in mind that integration of $c_{\mathrm{eff}}(x, T)$ over temperature can be conducted when the final temperature distribution in the plane wall is known. For particular heat transfer processes, it is possible to determine the enthalpy difference Δh without performing detailed simulations. Taking into account Eq. (6.18), the total heat flow in time interval $[t_1, t_2]$ across the internal wall surface $x = 0$ is given by

$$Q(0, t_1, t_2) = \frac{1}{R_u}\int_{t_1}^{t_2} \Delta T_{i-e}(t)\mathrm{d}t + \Delta H_{s0} \tag{6.20}$$

$$\Delta T_{i-e}(t) = T_i(t) - T_e(t) \tag{6.21}$$

$$\Delta H_{s0}(t_1, t_2) = \int_0^L \rho\, \Delta h(x, t_1, t_2) \cdot \frac{R_{x-e}}{R_u}\mathrm{d}x \tag{6.22}$$

In Eq. (6.22), ΔH_{s0} represents that part of the total enthalpy difference in a wall which is due to the heat flow across the surface $x = 0$. It is good to remember that in case of long-term cyclic processes, combined with an assumption of a perfectly performing

PCM (no loss of heat storage capacity with time), $\Delta H_{s0} = 0$, if time integration is over the integral number of cycles. In this case, the total heat flow depends mainly on the difference between internal set point temperature and mean ambient temperature.

6.2.3 *Time Constant of a Building Assembly Containing PCM*

One of the most important characteristics of PCM-enhanced building envelopes is their ability to shift peak thermal loads. It is expressed by a time constant τ_s, which can be easily computed for a large number of simple building envelope assemblies with one-dimensional heat transfer. The time constant represents a time of thermal response on the surface of the wall, roof, or slab after a step thermal excitation on the opposite side. The following section will bring a simplified methodology for estimating the time constants for simple building envelope assemblies containing PCM. A slab example is used for this purpose.

Let us assume that temperature on the exterior side of the slab was rapidly increased. The heat flux $q(t)$, and its changes across the unit surface area on of the opposite side of the slab, can be expressed as a function of time t. This process may be approximately described as a thermal relaxation of the slab (cooling), with the time constant τ_s. While considering a step function temperature excitation on the exterior side, the time constant τ_s can be approximated with the use of the following expression (Kossecka and Kośny 2008, 2009):

$$q(t) \approx \frac{\Delta T}{R}\left[1 - \exp\left(-\frac{t}{\tau_s}\right)\right] \tag{6.23}$$

The total heat flow in time interval [0, t] is given by

$$Q(t) \approx \frac{\Delta T}{R}\left[t - \tau_s\left(1 - \exp\left(-\frac{t}{\tau_s}\right)\right)\right] \tag{6.24}$$

For sufficiently long time t, as compared with τ_s, the steady state of heat flow is attained, and $Q(t)$ may be approximated by

$$Q(t) \approx \frac{\Delta T}{R}(t - \tau_s) \tag{6.25}$$

Comparison of Eqs. (6.20) and (6.23) yields

$$\tau_s = R \cdot \frac{\Delta H_{s0}}{\Delta T} \tag{6.26}$$

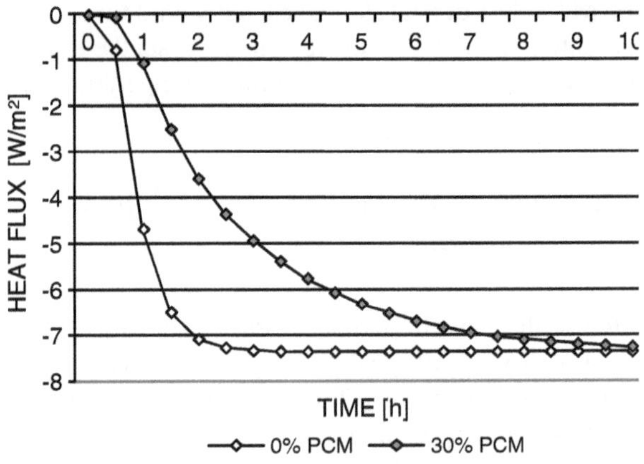

Fig. 6.4 Heat fluxes across the surface of an insulation layer after temperature jump at the opposite surface

In earlier work of Kossecka and Kośny (2008), a similar analysis was used for estimating the time constants for conventional non-PCM assemblies, for purposes of the hot-box test measurements. However, in the case of a PCM specimen, the time constant is strongly dependent on its enthalpy difference within the temperature interval taken into account.

In addition, Kossecka and Kośny (2009) performed analysis a conventional building envelope assembly containing two following types of thermal insulation: (I) conventional blown cellulose insulations and (II) PCM-enhanced cellulose insulation. Figure 6.4 depicts comparison of heat fluxes across surfaces of insulation layer of thickness 0.14 m. Thermal performance of insulation with no PCM is compared against thermal insulation with 30 % content of the microencapsulated paraffinic PCM. Thermal response after a rapid temperature jump, from initial temperature of 14 °C to final temperature of 44 °C, was analyzed in this case. The time constant, τ_s, estimated using Eqs. (6.19), (6.22), and (6.26), was close to 56 min for pure insulation and about 153 min for the same insulation with 30 weight % of PCM.

6.3 Numerical Performance Analysis of PCM-Enhanced Opaque Building Components

6.3.1 An Overview of Numerical Models' Developments

Energy performance improvements due to inclusion of PCMs into the building fabric depend on a number of factors, which include the amount of the PCM and its thermal characteristics, the location of the PCM, climatic conditions, overall design of the building, and details of the space-conditioning system. Therefore, a series of detailed energy simulations combined with analysis of associated local thermal effects within the building are necessary to evaluate the actual benefits of the PCM application. Dynamic thermal behavior of PCM-enhanced building envelopes can be numerically investigated using system-scale thermal models and whole-building energy simulations. Considering that PCMs placed in building components dynamically interact with the entire building fabric and the mechanical systems, it is essential that energy performance analysis of these technologies is performed with the use of a whole-building energy model.

A number of analytical solutions are available today for heat transfer with phase transition. They can be found in advanced heat transfer publications such as those by Evans (1951), Douglas (1957), Carslaw and Jaeger (1959), Friedman (1964), Rubibstein (1971), Crank (1984), Dalhuijsen and Segal (1986), Hill (1987), Alexiades and Solomon (1993), and Ozısık (1993). Please keep in mind that the analytical solution for a perfect case of isothermal phase transition may not provide a complete representation of PCM technologies used in buildings. PCMs in these systems are usually modified by a number of additives improving their durability, increasing thermal conductivity, or reducing flammability. In addition, these technologies may sometimes contain a mixture of different PCMs. These factors mean that phase-change processes in PCM-enhanced building envelopes are not isothermal and they usually occur across a temperature range. In that light, some of the above-referenced analytical solutions may provide useful guidance but may be impractical for actual PCM applications in buildings.

Approximate numerical solutions are usually used in practice to handle this class of heat transfer problems. The numerical methods for addressing phase-transition problems have been reviewed in the following publications: Basu and Date (1988), Idelsohn et al. (1994), Hu and Argyropoulos (1996), Voller et al. (2006), Dutil et al. (2011), Tabares-Velasco (2012), AL-Saadi and Zhai (2013). One of the earliest theoretical techniques for solving phase-transition problems is so-called the heat source method, introduced by Eyres et al. (1946) in mid-1940. In this method, the total enthalpy is split into the specific heat and latent heat and the latent heat is considered as a heat source. Most recently, Costa et al. (1998) implemented a similar numerical scheme to simulate the thermal performance of phase-changing thermal storage.

During the 1970s and 1980s, theoretical foundations for numerical analysis of a heat transfer problem with incorporated phase transition has been laid by Fox (1975), Furzeland (1980), Crank (1981), Voller and Cross (1983), Pham (1985), and others. In early theoretical analyses, solid–liquid phase-transition problem had been usually numerically solved using finite-difference or finite-element methods. Later, finite-volume and control-volume finite-element methods were also employed. The most common numerical methods in solving nonlinear phase-change behavior include the enthalpy method, the heat capacity method, and their modifications.

The enthalpy method deals with a total amount of energy required during the phase-change process, which includes both sensible and latent heat. The enthalpy method was first time introduced by Eyres et al. (1946) to deal with variations of thermal properties with respect to temperature. It is one of the most commonly used today fixed-domain methods for solving so-called the moving boundary problem, also known as the Stefan problem (Stefan 1889; Sharma et al. 2006, 2009;

Ogoh and Groulx 2010; Jayalath et al. 2012). This method presents heat capacity in terms of its integral form, H(T), with respect to the temperature. Alexiades and Solomon (1993) have presented several numerical methods for solving phase-transition problems that use the enthalpy method and using both linear and non-linear approaches. Similar enthalpy formulations have been extensively investigated by many authors for numerical analysis of building-related phase-change problems (Swaminathan and Voller 1993; Costa et al. 1998; Brousseau and Lacroix 1998; Chen and Sharma 2006; Chen et al. 2008a, b). In addition, Knoll and Keyes (2004) examined various novel approaches utilizing nonlinear solvers to solve the Stefan problem. In particular, Knoll introduced an algorithm to solve the Stefan problem using the Jacobian-free Newton–Krylov method for both ideal phase transition with isothermal melting and non-isothermal cases where melting process occurs over a range of melting temperatures. Furthermore, several alternative methods using the enthalpy approach and combined with an application of linear solvers, such as Gauss elimination or tri-diagonal algorithm or iterative methods and the Gauss–Seidel method, were proposed by Patankar (1980), Shamsunder and Sparrow (1975). Finally, an iterative Newton linearization scheme, using the solution process similar as that of Shamsunder, was introduced by Furzeland (1980). However, several authors reported that iterative methods such as Gauss–Seidel are too slow to be used in practical applications (AL-Saadi et al. 2013). Therefore, a number of fast numerical schemes have been introduced to improve the computational efficiency (see Pham 1985; Voller 1990; Swaminathan and Voller 1993). In addition, the ability of the enthalpy method to correctly account for the subcooling effect was discussed by Günther et al. (2007).

The heat capacity method describes the temperature change $T(x, t)$ using the heat capacity $Cp(T)$. Within the temperature range of a phase transition, this method deals with heat capacity as a function of temperature (between melting and solidification). It numerically imitates the effect of enthalpy (sensible and latent heat) by controlling the heat capacity value during the phase-changing stage. Two following theoretical approaches can be identified with the heat capacity method: (i) the apparent heat capacity (Idelsohn et al. 1994; Hu and Argyropoulos 1996; Voller 1997) and (ii) the effective heat capacity (Minkowycz et al. 2006; Muhieddine et al. 2009; Heim 2010; Poirier 1988). In the heat capacity method, the calculation is controlled for PCMs by temperature and total latent heat. The key in this approach lies in the heat capacity approximation that combines sensible and latent heat contributions.

The apparent heat capacity method was introduced by Hashemi and Sliepcevich (1967) to solve a one-dimensional heat transfer with phase change in a mushy region. This method is relatively popular today because during simulations the temperature is the only independent variable that needs to be solved in discretized form. In apparent heat capacity method, experimental results can be used to form an empirical expression to approximate the heat capacity. In order to generate high-quality enthalpy and heat capacity data necessary for numerical analysis, dynamic measurements can be performed for pure PCMs and homogenous PCM products, with a use of the differential scanning calorimeter (DSC) utilizing the step-testing method (Castellón et al. 2008; Günther et al. 2009; Lazaro et al. 2013). For example, Fang and Medina (2009) used DSC test results to develop a mathematical formula for the heat capacity of the paraffinic PCM. In cases of non-uniform PCM-enhanced materials and PCM composite blends, T-history method (Zhang 1999), or dynamic heat flow meter apparatus measurements (Kośny et al. 2009a, b; Shukla et al. 2012) need to be applied. Apparent heat capacity can be also estimated using numerical methods (Lemmon 1981; Voller 1997). A numerical technique was applied by Comini et al. (1974) to determine the heat capacity as a derivative of enthalpy with respect to temperature. Later and Morgan (1978) has adjusted this numerical technique to avoid the convergence problems. However, Voller (1997) found that the apparent heat capacity approximations, based on the direct test-generated relationships, are more accurate than the Morgan's approximation method.

The effective heat capacity method assumes that the PCM can be out of phase-change temperature range only in two cases: when it is fully charged or discharged (Heim 2010). During simulations with the use of the effective heat capacity method, the heat capacity value is usually computed for the specific node with a help of the discretization equations based on the corresponding nodal temperature. However, since the relationship between heat capacity and temperature is highly nonlinear during phase transition, an explicit scheme is often used. This may cause some inaccuracies in situation of sharp heat capacity changes with respect to temperature (Sadasivam et al. 2011).

In addition to the enthalpy and heat capacity methods, there is also a large group of modified methods using either components from one or both above-described numerical algorithms. For example, Pham (1985, 2006) introduced a new numerical scheme which combines the features of the enthalpy and heat capacity methods. This hybrid method is popularly called the quasi-enthalpy method. It consists of two following numerical steps: (i) a temperature prediction step and (ii) a temperature verification and correction step. In the prediction step, the nodal temperature is predicted based on guessed initial values. Next, the predicted temperatures are checked for consistency against the enthalpy–temperature curve and corrected if necessary. This method, after implementation of accuracy improvements proposed by Swaminathan and Voller (1993), has been recently introduced as one of the alternative PCM simulation algorithms in ESP-r, a whole-building energy simulation program (Sadasivam et al. 2011).

In order to overcome time and spatial limitations of the heat capacity method, a novel temperature transforming scheme was introduced by Cao and Faghri (1990). This simulation procedure was later called the improved temperature-based equivalent heat capacity method (Zhanhua and Yuwen 2006). Initially, this numerical algorithm was tested against several benchmark examples. Preliminary comparisons showed inconsistent results, especially when mass transfer through the PCM was considered. After a series of enhancements, this numerical model has been used in a number of projects (Zeng and Faghri 1994a, b; Zhanhua and Yuwen 2006). Today, this method is not in common use; however, it offers an alternative solution to the apparent heat capacity method. A list of numerical methods used in the most popular whole-building energy simulation tools, with capability of modeling phase transition, is given in Table 6.1.

Table 6.1 Numerical methods used in best-known numerical tools which can be used for phase-change simulations in buildings

Whole-building computer tool	Numerical method	Discrete form	Heat capacity computations	References
ESP-r	Effective heat capacity	Finite volume	Explicit	Heim and Clark (2004)
TRNSYS Type 204, combined with Type 56	Effective heat capacity	Finite element	Crank–Nicholson	Jokisalo et al. (2000)
TRNSYS Type 222, combined with Type 56	Indirect calculations			Ibanez et al. (2005)
TRNSYS Type 241, combined with Type 56	Enthalpy method	Finite difference		Schranzhofer et al. (2006)
TRNSYS Type 260, combined with Type 56	Effective heat capacity	Finite difference	Explicit	Kuznik et al. (2010)
TRNSYS Type 399, combined with Type 56	Two different methods available based on the heat capacity approach	Finite element	Crank–Nicholson	Dentel and Stephan (2013)
MATLAB	Enthalpy method	Finite volume	Crank–Nicholson	Sadasivam et al. (2011)
EnergyPlus	Enthalpy method	Finite difference	Implicit	Pederson (2007)

In building applications, PCMs are usually either encapsulated, or packed in very tight containers, which makes very common a situation where all locally enclosed PCM is going through the phase transition at the same time. In larger PCM containers, internal convection, during liquid phase, may impact overall system effectiveness. Internal axial convection within a PCM encasement was analyzed by Beasley and Clark (1984). A study by Ahmad et al. (2006) demonstrated that complexity of thermal model does not always yield improvements in simulation accuracy. In this work, numerical predictions generated by a complex thermal model considering a solid–liquid interface and results of a simpler model using the effective heat capacity algorithm were compared. Recorded results showed that a simpler heat capacity approach was more accurate. In another similar study, finite-element computer simulations were performed to compare accuracies of the enthalpy and heat capacity methods (Lamberg et al. 2004). When compared to experimental data, more accurate results were obtained with a use of the heat capacity method.

Today, only few commercially available computer models use separate enthalpy–temperature curves for melting and freezing. A large number of commercially available whole-building energy tools use simplified algorithms considering identical thermal characteristics for melting and freezing processes. Unfortunately, among commonly used PCMs, this simplification only works for paraffins, which nucleate without noticeable subcooling. Consequently, for a large number of non-paraffinic organic and inorganic PCMs, an application of simplified algorithms which do not include hysteresis may lead to substantial inaccuracies. A numerical study performed by Kuznik and Virgone (2009) concluded that separate enthalpy curves for PCM melting and freezing should be included in numerical simulations in order to improve their overall accuracy. In a separate research focused on a sandwich wall system, containing a layer of the PCM-enhanced lightweight concrete, PCM behavior was numerically analyzed with the use of COMSOL 4.0[1] and WUFI 5.0[2] simulations (Baghban et al. 2010). The authors have also found that it is essential to use separate melting and freezing characteristics, to achieve good agreements between experimental results and numerical predictions during PCM modeling.

6.3.2 Numerical Research Studies of PCM-Enhanced Building Products: Historical Outline

A large number of numerical studies, which have been recently performed in different countries, helped in better understanding of the physics behind the PCM-enhanced building products and their potential energy performance. For decades, different types of PCM-enhanced building boards and plasters have been the most popular objects of computer simulations. Earliest numerical studies started during late 1970s and had been continued till 1990s. They were mainly focused on gypsum wallboards impregnated with paraffin (Solomon 1979; Drake et al. 1987; Tomlinson and Heberle 1990; Kedl 1990; Stovall and Tomlinson 1995; Stetiu and Feustel 1996; Kissock et al. 1998). A combined experimental–numerical work was performed by Athienitis et al. (1997), who conducted an extensive field testing followed with one-dimensional numerical analysis of a full-scale outdoor test hut with PCM-enhanced gypsum board installed as an inside wall sheathing.

In more recent projects, PCM wallboards and plasters containing microencapsulated PCMs (Hawlander et al. 2002; Darkwa and O'Callaghan (2005), Schossig et al. 2005; Kendrick and Walliman 2007; Poulad et al. 2011) have been studied. In addition, thermal performance of shape-stabilized PCM board products has been analyzed using numerical methods (Kuznik et al. 2007; Virgone et al. 2009; Constantinescu et al. 2013). Due to flammability concerns about paraffinic PCMs, a number of numerical models have been utilized recently to analyze thermal performance of boards and insulation products thermally enhanced with bio-based alternatives to paraffins (Rozanna et al. 2005; Riza 2007; Kośny et al. 2009c; Dhanusiya and Rajakumar 2013).

In addition to the material research cited above, several numerical studies have focused on numerical optimization of PCM wall systems and their thermal characteristics. Numerical work with the use of one of the first PCM algorithms for TRNSYS[3]—a whole-building energy code, was performed in early 1990 by Stovall and Tomlinson (1995). It concluded that surface film coefficients on the wall

[1]http://www.comsol.com/products.

[2]http://www.wufi.de/index_e.html.

[3]http://www.trnsys.com/.

internal surface may notably impact the overall performance of walls containing PCM-enhanced gypsum boards. In a similar project performed about 15 years later, Liu and Awbi (2008) have compared natural convection on the interior side of a conventional wall and a wall containing PCM. The authors developed a numerical model to evaluate several convective heat transfer correlations available from the literature, for natural, mixed, and forced convection flows. In another project, Kuznik et al. (2007) have performed a thermal optimization of a wallboard containing PCM. They have concluded that a thickness of the PCM-enhanced board should be determined based on the time interval of thermal excitation and desired system time response. Similarly, Zhou et al. (2007, 2008) have analyzed the optimum location for the PCM-enhanced wall panels, concluding that the internal surface location is most thermally effective.

Since the majority of passive PCM ceiling applications use either PCM enhanced boards, membranes, or arrays of PCM containers (see Fig. 6.2), it was assumed that from the thermal perspective, they should work in a similar way to the internal PCM wallboard technologies. That is why analogous numerical models can be utilized for their performance analysis. Examples of transient thermal analysis of the passive ceiling systems containing PCM can be found in Stritih and Butala (2010), Oyeleke (2011), Jaworski and Poplawski (2013). Additional example of transient simulations, performed on the active ceiling technologies using night precooling, can be found in Corgnati (2007). Thermal performance of the PCM ceiling systems integrated with the hydronic heat exchangers was also investigated numerically by Schossig et al. (2003, 2005) and Kalz et al. (2007). Comparing to passive ceiling applications, active ceilings with PCMs require more complex numerical models, enabling analysis of mechanical systems and heat exchangers. In different reports, Antonopoulos and Tzivanidis (1997), and Antonopoulos et al. (1998) describe an application of the T-history experimental method and a numerical procedure to solve the three-dimensional transient problem of night precooling of the ceiling thermal system using PCM and hydronic tubing.

All of the cases of numerical analysis discussed above focused on building applications using relatively thin layers of PCM products. However, there is also a second group of thicker PCM applications utilizing either PCMs dispersed in porous materials, or PCMs blended with thermal insulations or composites. These applications started in 1970s from an introduction of the concrete and gypsum impregnated with salt hydrates or paraffinic hydrocarbons (Balcomb et al. 1983; Hawes and Feldman 1992). Since that time, PCM-enhanced wall masonry products have been extensively investigated around the world as dynamic building envelopes for residential and commercial buildings (Hawes et al. 1990; Salyer et al. 1995; Romanowska et al. 1998; Lee et al. 2000; Zhang et al. 2004; Bentz and Turpin 2007; Castell et al. 2009). A numerical study which utilized a new numerical subroutine enabling PCM modeling, combined with an existing software package, was performed by Ibanez et al. (2005). In this case, the TRNSYS program was used to simulate the energy effects of incorporating microencapsulated PCM into the concrete wall structures. Detailed configurations of test huts used in supporting field experiments and recorded test results for similar PCM technologies were presented in Cabeza et al. (2007) and Castell et al. (2010). A review of available literature on PCM applications in concrete building technologies was presented by Jayalath et al. (2012). They utilized finite-element modeling for analyzing the thermal performance of the PCM–concrete composite in building applications.

In addition, PCM-enhanced natural stones and aggregates have been used in buildings for external façades and internal space applications. Numerical thermal performance analysis of agglomerate floor tiles containing 20 wt% of encapsulated octadecane has been performed by Lee and Hittle (2005). In this work, dynamic whole-building energy simulations were performed using BLAST, a whole-building computer program.[4] Several research studies focused on the natural stone impregnated with PCM have been performed in southern Europe. Examples of experimental projects supported by the numerical performance analysis can be found in Katsourinis et al. (2010) and Romero-Sánchez et al. (2012). They used PCMs for thermal enhancement of a natural stone in order to improve its thermal performance as a building material.

Apart from direct building energy applications, PCMs in concrete can be used for durability reasons as well. The National Institute of Standards (NIST), Gaithersburg, MD, USA, investigated experimentally and numerically an application of the microencapsulated PCM as a concrete additive, reducing the temperature rise (and subsequent rate of temperature decrease) in large concrete sections during the concrete curing processes (Bentz and Turpin 2007). The main goal of this application was to minimize thermal cracking in concrete slabs. In addition, a potential reduction of a number of freeze/thaw cycles experienced by concrete decks exposed to a winter environment was investigated by the same authors using one-dimensional finite-difference computer model (Bentz and Turpin 2007).

As mentioned earlier, PCMs can be also blended with thermal insulations and composite materials. In building envelope applications, thermal insulations containing PCM help in reduction of space-conditioning loads and enable significant shifting of peak-hour thermal excitations, transmitted through the opaque building shell components (Halford and Boehm 2007; Kossecka and Kośny 2010; Tabares-Velasco 2012; Kośny et al. 2013a). One of the first experimental and analytical projects focused on PCM-enhanced building insulations was performed by Petrie et al. (1997). The PCM used was hydrated calcium chloride dispersed in the lightweight perlite aggregate working as dynamic ceiling insulation. Extruded polystyrene (XPS) foam was the insulation below and above a layer of the PCM–perlite blend. Finite-difference computer modeling was used in this study to enable comparisons to the behavior of the conventional insulation. Petrie reported that with PCM-enhanced perlite, the total daily energy flow into the conditioned space below the test cells was reduced and the peak heat flow rate was delayed in time and decreased in magnitude.

In building envelopes, plastic foam insulation can be used as a carrier for PCMs. Plastic foams can be applied in form of a rigid sheathing, thin flexible lining, or as a

[4]http://apps1.eere.energy.gov/buildings/energyplus/.

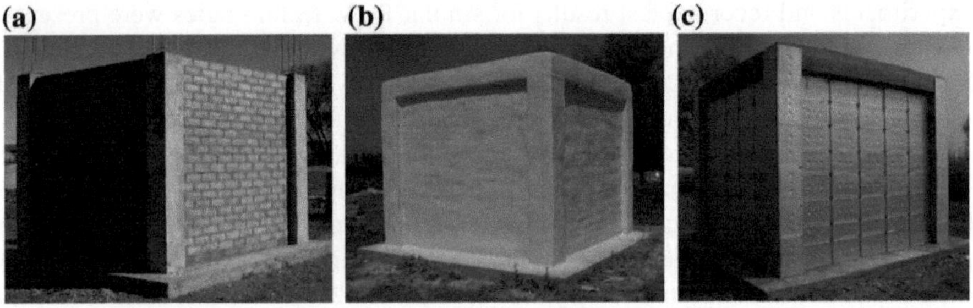

Fig. 6.5 Field testing of brick structures (**a**), insulated with polyurethane foam (**b**), and containing PCM panels attached to the exterior surface (**c**)—*Source* Castell et al. (2009)

layer of the sprayed foam product. As mentioned in earlier sections, the University of Lleida, Spain, constructed a series of test huts for field experiments with different PCM-enhanced building envelope technologies (Cabeza et al. 2007). Some of these technologies were also analyzed using numerical methods. A unique application of a thin PCM layer installed together with polyurethane foam was tested in field conditions of Lleida, Spain (Castell et al. 2009). In this case, PCM containers were directly attached to the sprayed polyurethane foam (see Fig. 6.5). Ceramic bricks were used as a base wall structure in this experiment.

Several configurations PCM-enhanced foams were successfully analyzed on wood frame walls by using dynamic hot-box experiments at the Oak Ridge National Laboratory, USA (Kośny et al. (2007a). In a separate project, small-scale field testing was performed on a novel retrofit strategy for masonry walls using PCM-enhanced foam insulation, reflective aluminum facing, and vented air cavity (Kośny et al. 2010a). A series of finite-difference simulations were used to select PCM and optimize the insulation configuration in this case. To enable direct performance comparisons, PCM wall insulation was installed next to the conventional foam insulation of equivalent *R*-value—as shown in Fig. 6.6. During the summer 2007, this design of the PCM-enhanced wall insulation yielded about 60 %

Fig. 6.6 Field wall testing with PCM-enhanced foam insulation containing reflective foil facing—*source* authors archive. **a** Installation of PCM–foam insulation with reflective surface, **b** conventional foam insulation (*left*) and PCM-enhanced foam insulation (*right*), and **c** *front* view of completed conventional foam insulation section (*left*) and PCM-enhanced foam insulation section (*right*)

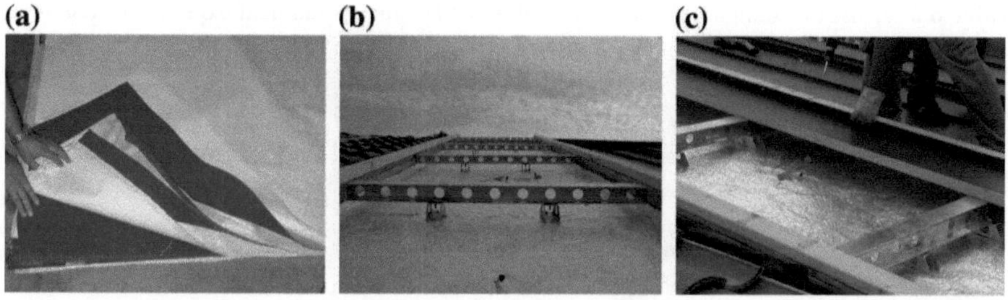

Fig. 6.7 Field testing of the vented metal roof assembly containing two different PCMs—*Source* authors archive. **a** Multilayer sandwich of PCM-enhanced foam and fabric, **b** vented air cavity with reflective surface providing thermal break and enabling removal of excess heat during the summer, and **c** installation of the metal panels on top of the roofing assembly

reduction of the peak-hour cooling load, comparing to the conventionally insulated south-oriented concrete masonry wall of the same *R*-value (Kośny et al. 2008a, b). Following the wall testing, thermal performance of one of the first PCM roofing applications, utilizing materials of different melting temperatures, was analyzed in field conditions by Kośny et al. (2008b) and Miller et al. (2010). A multilayer sandwich made of the PCM-impregnated foam and fabric had two different melting temperatures. As shown in Fig. 6.7, it was a ventilated roofing system with over-the-deck heat sink. Layers of foam and fabric containing PCMs were installed in conjunction with the air-vented cavity and were surfaced with reflective insulation. A series of finite-difference computer simulations were utilized for numerical performance comparisons of the above PCM application with other conventional sloped roof configurations without PCM (Kośny et al. 2010b). In addition, whole-building energy analysis of building envelopes with thin layers of building products containing PCM was performed with the use of the EnergyPlus computer program[5] by Kośny et al. (2010a).

Between 2002 and 2013, a number of novel PCM blends with thermal insulations were introduced, tested, and numerically analyzed in the USA. Different arrangements of PCM–cellulose (Kośny 2008) and PCM–glass fiber insulation (Kośny et al. 2010c) have been used by the Oak Ridge National Laboratory, USA, as a cavity insulation in framed walls and for attic floor insulation. In addition, several methods for mixing the microencapsulated and shape-stabilized PCMs with aerogel insulation have been analyzed by Aspen Aerogels and Fraunhofer SCE during a US Department of Defense research project.[6]

Examples of finite-difference numerical analysis of the wall systems containing fiber insulations blended with PCM can be found in Kośny et al. (2010c, 2012a, b), Kośny and Kossecka (2013) and Childs and Stovall (2012). In addition,

[5]U.S. Army Corps of Engineers Engineer Research and Development Center, Phase I, Contract # W9132T-12-C-0012.

[6]http://www.esru.strath.ac.uk/EandE/Web_sites/10-11/Schools/Model.html.

Table 6.2 A list of example references available for laboratory and field experiments combined with thermal modeling of PCM-enhanced thermal insulations

	Test method	References: Testing	References: Simulations
1	Small material-scale testing using dynamic heat flow meter apparatus	Kośny et al. (2007a, b, 2012b), Alderman and Yarbrough (2007), Shukla et al. (2012)	Kośny et al. (2009a, c)
2	System-scale laboratory testing with the use of the dynamic hot-box experiments for walls	Kośny et al. (2007b, 2008b)	Kośny et al. (2008b, 2010b, c)
3	System-scale laboratory testing with the use of the dynamic hot-box experiments for attics and roofs	Petrie et al. (1997), Kośny et al. (2010b)	Petrie et al. (1997), Kośny et al. (2009b)
4	System-scale field testing of walls	Kośny (2008), Kośny et al. (2010c), Medina and Stewart (2008)	Kośny et al. (2010c, 2012b), Fallahi et al. (2012), Childs and Stovall (2012), Tabares-Velasco (2012)
5	System-scale field testing of roofs and attics	Miller et al. (2010), Kośny et al. (2010a, 2012a)	Kośny et al. (2010b, c), Fallahi et al. (2012)

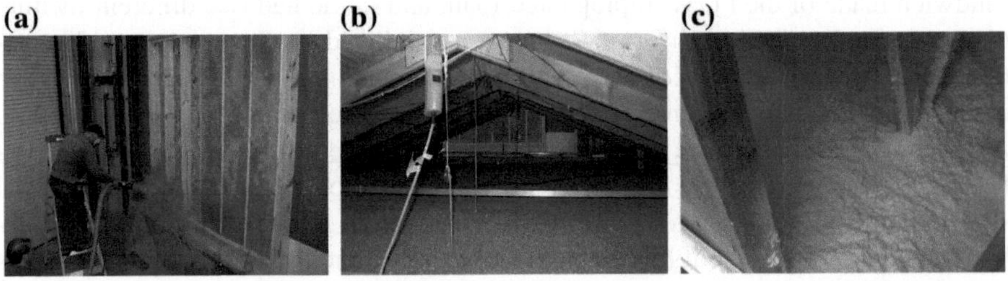

Fig. 6.8 Examples of system-scale test assemblies, constructed by the Oak Ridge National Laboratory, USA, containing PCM-enhanced insulations—*Source* Author's archive. **a** Installation of the PCM-enhanced cellulose inside the test frame, before dynamic hot-box wall testing, **b** test attic with PCM-enhanced cellulose insulation installed on the attic floor, prepared for climatic chamber testing, and **c** PCM-enhanced fiberglass insulation installed on the attic floor of the test attic, prepared for field testing

whole-building computer models like EnergyPlus and ESP-r[7] have been utilized for detailed system optimization and for energy consumption analysis in buildings containing PCM-enhanced wall systems (Kośny et al. 2010a, b; Tabares-Velasco (2012). Table 6.2 summarizes major US references available for testing and modeling of PCM-enhanced thermal insulations.

Figure 6.8 presents examples of laboratory and field experiments with the use of PCM-enhanced fiber insulations, performed by Oak Ridge National Laboratory, USA. In late 2000, a new dynamic testing method was introduced for thermal

[7]http://www.esru.strath.ac.uk/EandE/Web_sites/10-11/Schools/Model.html.

performance analysis of non-uniform composites and blends containing PCM (Kośny et al. 2007a, 2009a; Alderman and Yarbrough 2007; Kossecka and Kośny 2009; Shukla et al. 2012). In these tests, finite-difference simulations have been often used to support transient analysis of PCM-enhanced insulations (Kośny et al. 2010a). Analytical formulas for analysis of PCM content in PCM-enhanced thermal insulations were given by Kossecka and Kośny (2009).

6.4 System-Scale Thermal Modeling

In buildings, a scale of thermal behavior improvements attributable to an introduction of PCM-enhanced building envelopes depends on a number of factors. This includes, for example, the amount of PCM and its physical properties, climate conditions, and design of the building. Therefore, a comprehensive numerical analysis of thermal and energy effects in buildings using PCMs is essential for evaluation their energy benefits. Detailed numerical models, developed for system-scale thermal simulations of PCM-enhanced envelopes, may capture more physics of heat transfer process and with higher accuracy than less accurate algorithms used in most of whole-building energy simulation tools. However, simplified models can be sometimes preferred because of their ease to use and due to the fact that they can yield quick estimates of the PCM's thermal performance.

With respect to PCM-enhanced building envelopes, thermal design and optimization processes require detailed performance simulations. Numerical system-scale modeling is usually a first step in the process of designing of the material configuration for the PCM-enhanced building system. This process frequently includes decisions about the following: (i) sequence of materials, (ii) location of PCM, and (iii) amount of PCM. Thermal design decisions also include estimation of the optimal phase-change temperature range, which leads directly to selection of a specific PCM. One-dimensional simulations work well for a large number of PCM-enhanced building products, including boards, foils, bulk insulations, or composites containing PCMs. Multilayered and sandwich systems are good examples of building envelopes for which one-dimensional analysis can be used. Please keep in mind that one-dimensional models are only useful for analysis of simple building envelopes without thermal bridging generated by structural members. In case of buildings having framing members, or other highly conducting structural components imbedded into the building envelopes (i.e., wood and steel framing walls, post-and-beam concrete and steel structures), three-dimensional simulations would be preferable. However, there is also a second group of the PCM-enhanced systems utilizing arrays of small-size PCM containers or PCM pouches packed in plastic or aluminum foils. These assemblies require detailed multi-dimensional thermal simulations, because of the complex character of their internal heat transfer. In such situations, structural thermal bridging may only add further complication to the entire thermal analysis. It is good to remember that building envelopes need to be designed from the hygrothermal perspective as well.

Areas with internal thermal shorts or spaces with lower local temperatures, like in case of some PCM applications, can be susceptible to water condensation which may lead to durability problems. That is why, a process of thermal optimization of PCM-enhanced building envelopes needs to be most often accompanied with hygrothermal simulations.

A second important area for system-scale numerical applications is associated with laboratory thermal testing of PCM-enhanced building products and system. In this case, numerical analysis can be useful for numerous purposes including (i) PCM optimization, (ii) accuracy analysis, (iii) estimation of actual amounts of PCM in tested assemblies, or (iv) analysis of recorded test data.

Reliable whole-building-scale models can help in thermal and energy performance analyses without a need for time consuming, full-scale experiments. In whole-building energy simulations, numerical analysis of PCM components is quite challenging due to discussed earlier difficulties associated with multi-dimensional heat transfer effects. Please bear in mind that in whole-building energy simulations, complex geometries of conductive components, which are common in building envelopes, need to be accurately represented by one-dimensional whole-building algorithms (see Kossecka and Kośny 1997, 2001; Lamberg et al. 2004; Regin et al. 2009). Numerous analytical solutions, which have been developed for commercially available whole-building energy models, deal only with one-dimensional geometries and simple boundary conditions. In 2001, ASHRAE evaluated a theoretical procedure for conversions of complex building components with three-dimensional heat transfer, into thermally equivalent one-dimensional representations, which can be used in whole-building energy simulations (see ASHRAE RP-1145 2001).

As discussed earlier, the Stefan problem is one of the best-known formulations for one-dimensional solid–liquid phase transition (Stefan 1889; Sharma et al. 2009; Ogoh and Groulx 2010; Jayalath et al. 2012). Typically, phase-change heat transfer problems are numerically solved with the use of finite-difference, finite-element, finite-volume, or control-volume algorithms. The enthalpy and heat capacity methods are the most common computational methods used today for nonlinear phase-transition behavior. It is also good to remember that, from all PCMs used in buildings, only paraffins exhibit either no or insignificant subcooling. This allows for systems using paraffins, an application of only a single set of phase-change characteristics in transient whole-building simulations. However, great majority of non-paraffinic organic and almost all inorganic PCMs show notable hysteresis, often combined with supercooling. That is why more advanced computer models usually enable incorporation of subcooling effect and enthalpy hysteresis (Bony and Citherlet 2007; Bedecarrats et al. 2009a, b).

A large number of numerical models have been used worldwide for analysis of PCM-enhanced building envelopes. In addition, quite a few computer models developed for non-building applications have been adopted for analysis of PCM-enhanced building components as well. In research project sponsored by the US Department of Energy, a comprehensive comparison of functionality, computational speed, accuracy, and various other features of about 20 building energy simulation tools was presented (Crawley et al. 2005). In addition, several other authors discussed a number of research and commercial models, coming from various engineering fields and having a potential for applications in buildings area (see Zalba et al. 2003; Caldwell and Kwan 2004; Verma et al. 2008; Regin et al. 2008; Kośny et al. 2009a, b, c, d; Agyenim et al. 2010; Dutil et al. 2011; AL-Saadi and Zhai 2013). Among a large number of currently available numerical tools for thermal and energy simulations in building components; HEATING, RADCOOL, COMSOL, METLAB, BSim, DeST, IDA ICE, ANSYS Fluent, PowerDomus, and SUNREL are considered as capable to deal with phase-transition problems (see Crawley et al. 2005; Pedersen 2007; Rose et al. 2009). The following sections will discuss examples, of most referenced commercially available computer models, used for this purpose.

6.4.1 PCM Simulations with the Use of Finite-Difference and Finite-Element Computer Codes

In past decades, several novel designs of PCM-enhanced building envelopes have been introduced worldwide. Usually, before full-scale application, these technologies require system-scale analysis to ensure that design thermal performance parameters are met. This task can be performed either through experimental work, or with the use of system-scale numerical analysis. Frequently, both methods are simultaneously used. Experimental results are often utilized for validation of computer models, which can later help with system optimization. An application of numerical analysis is usually a cost-saving approach comparing to expenses associated with experimental analysis. While majority of numerical tools developed for PCM-enhanced building envelope allow only one-dimensional analysis, more detailed, two- and three-dimensional heat transfer approaches have been utilized as well. It is critical that thermal models used in PCM analysis provided quality approximations of the system geometry and physics in phase-transition processes. However, very often a trade-off between speed of calculation and the accuracy has to be made. Several commercially available computer tools, that have been created using well-established and validated numerical algorithms, are available today for thermal analysis of PCM technologies. This group offers good accuracy and modeling flexibility combined with acceptable speed of computations. Below, some of the most referenced commercial analysis tools are discussed and examples of system-scale numerical analyses are presented.

Starting from 1980s, several research projects focused on phase-change latent heat systems have been performed by the Oak Ridge National Laboratory, USA. Some of these projects involved numerical analysis. Since, starting from 1940, Oak Ridge, TN, USA, has been a national center for nuclear research, very often computer models used for analysis of PCM building technologies, utilized numerical algorithms developed earlier for thermal analysis of nuclear systems. For example, in the early 1990, A. Solomon et al. used WALL88, a computer code for two-dimensional transient simulations of the gypsum board impregnated with PCM (Tomlinson and Heberle 1990). In this computer code, the PCM-enhanced building products could be either represented as a blend with the matrix material, or in the form of discrete PCM pellets distributed randomly throughout the solid support matrix with the use of a random number generator. This computer model utilized earlier developed numerical algorithms for transient thermal transport and storage of both sensible and latent heat (Solomon et al. 1984, 1986, 1990). The WALL88 numerical model was validated against analytical solution and experimental laboratory results. The model was found to give an excellent agreement with experimental results only after allowing the PCM to melt over a temperature range rather than at a specific temperature.

HEATING[8] is the second computer program, which has been initially developed for designing of the equipment for nuclear facilities, by the Oak Ridge National Laboratory. It has been later adopted for analysis of building envelope technologies including PCM-enhanced systems. During the last two decades, different research groups from the U.S. have used this program for numerical analysis of PCM-enhanced building technologies. HEATING (Heat Engineering and Transfer in Nine Geometries) has been extensively validated by the author against dozens hot-box tests of lightweight and massive wall systems (Christian and Kośny 1995). It served for development of the ASHRAE modified zone method for *R*-value calculations of steel-framed wall systems (Kośny and Christian 1995). In addition, ASHRAE project RP-1145 performed by Enermodal, Canada, Oak Ridge National Laboratory, and Polish Academy of Sciences, demonstrated with use of HEATING an excellent agreement between hot-box test results and numerical heat transfer predictions of the wall assemblies with complex geometries (ASHRAE RP-1145 2001; Kośny and Kossecka 2000). In this project, the equivalent wall method—developed earlier, numerical procedure enabling conversion of complex three-dimensional wall assemblies into the thermally equivalent one-dimensional multilayer configurations, was utilized in terms of application into whole-building energy simulations (Kossecka and Kośny 1996, 1997; Kośny et al. 1998).

HEATING is a multi-dimensional, general-purpose, finite-difference, heat transfer code, which solves steady-state and/or transient problems in one-, two-, or three-dimensional Cartesian, cylindrical, or spherical coordinates. Transient heat-conduction problems can be solved in HEATING using several numerical techniques including the following:

1. The implicit technique, which can range from Crank–Nicolson to the classical implicit procedure. In implicit scheme, system of equations is solved by point-successive-over-relaxation iteration.
2. The classical explicit procedure using the Levy explicit method.

Heating simulations may include multiple materials, and the thermal conductivity, density, and specific heat of each material may be both time- and temperature-

[8]http://www.oecd-nea.org/tools/abstract/detail/psr-0199/.

dependent. Thermal conductivity may be anisotropic. The mesh spacing may be variable along each axis. Heating allows up to five changes of phase for transient calculations. Boundary temperatures may be either time-, or position-dependent. Boundary conditions include specified temperatures or any combination of prescribed heat flux, forced convection, natural convection, and radiation. The boundary condition parameters may be time- and/or temperature-dependent. In HEATING, the heat-generation rates may be dependent on time, temperature, or position.

One of the first research projects, where HEATING was utilized for analysis of the PCM-enhanced attic floor insulation, was performed by Petrie et al. (1997). The PCM used was hydrated calcium chloride dispersed in the lightweight perlite aggregate working as a dynamic insulation. In order to specify the configuration of PCM–perlite blend and conventional insulation that achieves optimum thermal performance enhancement, a transient numerical model had to be used to perform parametric variations. The purpose of this modeling was to reproduce the experimentally measured temperature profiles in several configurations of the test cells and their heat fluxes, thereby verifying understanding of the heat transfer and phase-change phenomena occurring during this experiment. In addition, three-dimensional HEATING simulations were used to enable thermal behavior comparisons with conventional insulation. Petrie reported an excellent agreement between numerical predictions and experimental data.

HEATING was also used for development of the dynamic heat flow meter apparatus testing (DHFMA) procedure for laboratory thermal measurements of PCM-enhanced building products and PCM blends with thermal insulations (Kośny et al. 2009a). A series of the DHFMA tests were performed on PCM-enhanced fiber insulations. Experimental results were compared against the HEATING predictions with a very good agreement (Kośny et al. 2009a, b). Developing a numerical model in two- or three-dimensional domain is complicated and difficult to be generalized for different geometries, PCM applications, and physics. A series of three-dimensional HEATING simulations were employed to analyze the results of the DHFMA tests of the complex arrays of the PCM pouches packed in the plastic foil (Kośny et al. 2009d). Results of this analysis were later used for development of thermally equivalent one-dimensional numerical model of the wall assembly containing arrays of the PCM pouches. This simplified model was utilized in the whole-building energy simulations with the use of the EnergyPlus whole-building energy code (Kośny et al. 2010a). In 2012, as a part of the research sponsored by the U.S. Department of Energy, Kośny et al. (2013) and Childs and Stovall (2012) used HEATING transient simulations for the thermal performance analysis of the PCM-enhanced thermal insulations. In addition, HEATING has been used as a standard benchmark numerical package to validate the finite-difference algorithm used for PCM modeling in EnergyPlus (Tabares-Velasco 2012).

A number of existing simulation packages such as COMSOL[9] or FEMLAB (COMSOL's predecessor) has been lately used for analytical and design work.

[9]http://www.comsol.com/.

COMSOL is a finite-element simulation package that allows multi-physics modeling for many engineering applications. Although COMSOL is widely known as a highly flexible numerical model, it has not been fully explored yet as a tool for heat transfer processes with latent heat storage.

A research team from the Helsinki University of Technology—Espoo, Finland, has used FEMLAB commercial package (in later version—COMSOL multiphysics) to develop a numerical model of the wall assembly containing PCM (Lamberg et al. 2004). In this project, phase-transition simulations were performed using both the enthalpy method and the heat capacity method. Simulation results generated with the use of both numerical procedures were compared against experimental results. It was found that both used methods yielded good approximations of the phase-transition processes. However, it was also found that for narrow phase-change temperature ranges, the heat capacity method was found more precise.

In another numerical study performed by the Norwegian University of Science and Technology, Trondheim, Norway, COMSOL has been used to analyze several configurations of multilayer building envelope systems containing PCM (Baghban et al. 2010). A three-layer sandwich structure contained two layers of lightweight concrete with and without PCM and a thin layer of silica aerogel were used in numerical analysis. In this work, the numerical results generated by COMSOL were compared with WUFI-5.0[10] simulations performed for identical structures. WUFI family of hygrothermal programs was developed by the Fraunhofer IBP, Germany. WUFI-5.0 allows realistic calculation of one- and two-dimensional heat and moisture transport in multilayer building components exposed to natural weather conditions. It also permits a use of materials with temperature-dependent heat capacity. Good agreement was reported for results generated by both simulation programs, for identical configuration of the PCM-enhanced building envelope.

6.4.2 PCM Simulations with the Use of MATLAB/Simulink

MATLAB/Simulink has recently become the universal numerical tool in large number of universities and research centers worldwide. In terms of phase-change material analysis, METLAB simulation environment allows a development of either self-standing numerical models, or relatively simple simulation components collaborating with more complex whole-building energy models like ESP-r or TRNSYS. Several advantages of the MATLAB/Simulink modeling platform, which can be useful in parametric analysis of PCM-enhanced building envelopes, are listed as below:

[10]http://www.hoki.ibp.fhg.de/wufi/WhatsNewIn50_e.html.

- The simulator provides toolboxes and blocksets adequately for setting up almost any complex system configuration.
- The initialization process at the start of any simulation for all parameters of system components is automatic and simulation perimeters can be easily modified.
- Numerical data including any simulation parameters and the numerical results can be easily stored for later processing using MATLAB toolboxes.

Sadasivam and Almeida utilized the MATLAB simulation environment to develop a simple one-dimensional, finite-volume, numerical model for analysis of PCMs (Sadasivam et al. 2011; Almeida et al. 2011). It was using a corrective iterative scheme, based on the enthalpy linearization, which was introduced by Swaminathan and Voller (1993). This METLAB PCM model was compared against ESP-r simulations. For a series of simultaneous simulations performed with the use of PCM and the BESTEST[11] Case 600 model of the lightweight building containing PCM-enhanced structural components. The METLAB simulations have proven to be relatively accurate while using ESP-r results as a benchmark. In another project, Chen et al. (2008b) utilized MATLAB for development of an implicit one-dimensional finite-difference model. This model was later used for thermal performance analysis of the PCM-enhanced wallboard, ceiling, and floor. In this development, the heat capacity method was used. The discretized equation was solved using the Gauss–Seidel iterative method.

6.4.3 PCM Simulations with the Use of ANSYS Fluent and Other CFD Codes

Several commercially available computational fluid dynamic (CFD) tools have been used in the past for numerical analysis of building technologies using PCMs, both in the system scale and for the whole buildings. CFD models are typically used to predict air flow and temperature fields in indoor environments by numerically solving the Navier–Stokes set of partial differential equations for mass, energy, and momentum. These equations are linearized, discretized, and applied to finite volumes in the solver to obtain a detailed numerical solution, including velocity and temperature fields. In PCM applications, CFD analysis is very often combined with TRNSYS whole-building energy simulations—as discussed in Gowreesunker et al. (2013). ANSYS Fluent is one of the general-purpose CFD tools which has been used for thermal analysis of PCM-enhanced building envelopes. It solves the approximate form of the governing equations in order to provide different solution fields for a particular domain.

[11] http://apps1.eere.energy.gov/buildings/tools_directory/software.cfm/ID=85/pagename=alpha_list.

As in all CFD tools, the accuracy of phase-transition simulations with the use of CFD models varies depending on the level of model discretization. However, more refined the model, the longer computer time is required. In case of building envelopes containing PCMs, CFD codes, most often, solve phase-transition problems through the enthalpy porosity method (Susman et al. 2011). This method assumes, in its default state, a linear enthalpy–temperature relationship and ignores enthalpy hysteresis. Susman et al. (2011) and Ye et al. (2011) employed the enthalpy–porosity method to simulate PCM sail and plate units and found reasonable accuracy in temperature predictions for internal building space. PCM sail is a foil or fabric containing PCM hanging from the ceiling and working as a passive heat exchanger. It is usually fabricated with the use of commercially available PCMs and lightweight composite carriers.

An application of the enthalpy–porosity method may yield inaccuracies in dynamic simulations where local envelope and air temperatures are of a prime interest (see Gowreesunker et al. 2013). In order to improve accuracy of transient PCM simulations with the use of CFD commercial packages, Gowreesunker et al. (2013) proposed a new method of simulating phase-change processes in conduction-dominant heat transfer cases, through a use of source terms that considers hysteresis and varying enthalpy–temperature relationship. A series of cyclic simulations of boards containing PCM were used to validate above-described numerical algorithm. According to Gowreesunker et al. (2012), an application of the newly developed method can improve simulation accuracy by about 10 %, over the enthalpy–porosity model. In his work, Gowreesunker experimentally estimated the enthalpy–temperature relationship using DSC testing, since nonlinear enthalpy–temperature function and hysteresis are very important during periodic transient simulations. For dominant conduction heat transfer cases of buildings envelopes using solid boards containing PCM, the governing energy equation was expressed in the following simplified form (Gowreesunker et al. 2012):

$$\frac{\mathrm{d}}{\mathrm{d}t}(\rho H) = \left[\lambda \frac{\mathrm{d}T}{\mathrm{d}x_j}\right] + S_E \tag{6.27}$$

where

S_E represents a user-defined function approximating test-generated freezing and melting curves.

For a commercially available 13-mm-thick clay board, impregnated with 21 % (by volume) with microencapsulated paraffinic PCM, Gowreesunker et al. (2012) developed two separate formulas for melting and freezing:

$$S_{E,\mathrm{melting}} = 12.78\,T^3 - 10932.54\,T^2 + 3117074.92\,T - 296256708.90 \tag{6.28}$$

$$S_{E,\mathrm{freezing}} = -13.02\,T^3 + 11055.07\,T^2 - 3129090.90\,T + 295246573.26 \tag{6.29}$$

In author's opinion, a serious demand for CFD modeling of PCM-enhanced building envelopes may come together with an increasing need for accurate analysis of building shell components containing airspaces and vented air cavities. In these applications, PCM thermal storage is only a single component. Enclosed spaces and air cavities within building envelopes may also use reflective surfaces. Analysis of the above systems is additionally complicated by the fact that a local mass transport usually takes place there as well (Karagiozis 1993; Hens 1996). Today, thermal performance of these dynamic systems is usually approximated using simplified analytical methods and limited testing. Initial laboratory and field testing of similar technologies containing PCM demonstrated great potential for reduction of building thermal loads (Miller and Kośny 2008; Kośny et al. 2010b; Kośny and Kossecka 2013).

6.5 Whole-Building-Scale Energy Modeling

A large number of above-discussed system-scale numerical studies have been focused on understanding the physics behind PCM applications in building envelopes. System-scale models combined with laboratory or test room experiments were very useful in projects investigating temperature distributions or potential heat-flux reductions in PCM-enhanced building envelopes. Some other studies attempted to analyze the energy performance of different PCM products in building applications. However, as mentioned in earlier sections, most often PCM envelope technologies interacted with other components of the building fabric. That is why relatively early, variety of whole-building computer models have been employed for the performance analysis of PCM-enhanced building envelopes. In one of the first whole-building studies performed by the Oak Ridge National Laboratory (ORNL), USA, TRNSYS, a transient model, was modified to enable thermal simulations of the gypsum boards impregnated with paraffin (Tomlinson and Heberle 1990). Later, a single-zone program RADCOOL, combined with DOE-2, a whole-building energy simulation tool, was used by the Lawrence Berkeley National Laboratory (LBNL), USA, for energy analysis in several theoretical studies of the PCM-enhanced gypsum boards (Stetiu and Feustel 1996; Feustel and Stetiu 1997).

During the last three decades, numerous research projects focused on energy savings and peak load decrease associated with PCM building envelope applications have used a number of whole-building energy simulation tools. In Crawley et al. (2005), a comprehensive overview was given of various computational capabilities and features which can be found in 20 commonly used building simulation software tools. Examples of numerical applications are listed in Table 6.3 together with used energy simulation computer programs.

Table 6.3 References for PCM numerical applications in heat transfer tools and whole-building energy programs

Computer program	References
TRNSYS[a]	Stritih and Medved (1994), Stovall and Tomlinson (1995), Koschenz and Lehmann (2000), Ibanez et al. (2005), Ahmad et al. (2006), Kuznik et al. (2010), Poulad et al. (2011)
ESP-r	Heim and Clarke (2004), Schossig et al. (2005), Gelissier (2008), Ponechal (2009), Heim (2010), Fallahi et al. (2012)
EnergyPlus	Almeida et al. (2011), Kośny et al. (2010a, c), Tardieu et al. (2011), Campbell and Sailor (2011), Tabares-Velasco (2012)
RADCOOL[b]	Stetiu and Feustel (1996), Feustel and Stetiu (1997)
CoDyBa[c]	Virgone et al. (2009)
BSim[d]	Rode et al. (2003), Grau and Rode (2007), Rose et al. (2009)
PCM Express[e]	Ubiñas et al. (2009)
WUFI[f]	WUFI (a family of hygrothermal tools with whole-building simulation capabilities) (Shrestha 2012)

[a]http://www.trnsys.com/#1
[b]http://epb.lbl.gov/thermal/docs/chapter3.pdf
[c]http://www.jnlog.com/codyba1_en.htm
[d]http://sbi.dk/en/bsim/about-bsim
[e]http://www.valentin-software.com/en/products/pcm/58/pcm-express
[f]http://www.verlag.fraunhofer.de/wufi/?local=en

Different numerical procedures with diverse levels of complexity have been used in the numerical tools listed above. A number of simplified approaches can be listed here, starting from the following:

(i) the first numerical schemes developed for TRNSYS, which in early 1990s, enabled analysis of PCM building products (Tomlinson and Heberle 1990; Stovall and Tomlinson 1995) and
(ii) empirical models using the equivalent heat transfer coefficient (Ibanez et al. 2005).

In later projects, more advanced, fully implemented finite-difference (Koschenz and Lehmann 2000; Pedersen 2007), finite-element (Stritih and Medved 1994; Ahmad et al. 2006), and control-volume models (Heim and Clarke 2004) have been employed. It is important to notice that there is a great spread in PCM effectiveness predictions coming from numerical studies, today. Simulated energy savings are ranging between almost no noticeable energy benefits (see Pedersen 2007) and 90 % reduction of heating energy demand during the heating season, as predicted by Heim and Clarke (2004). Following this fact, IEA Annex 23 (2011) recommends that before numerical energy predictions generated by thermal and energy models can be fully adopted by building designers and energy code officials, more experimental validation work is necessary to build confidence in the accuracy of energy simulation tools for PCM-enhanced building technologies.

6.5.1 TRNSYS

TRNSYS[12] is the energy simulation software package, developed by the Solar Energy Laboratory at the University of Wisconsin, Madison, USA, and it has been extensively used worldwide in the fields of renewable energy engineering, building energy simulations, passive and active solar designs, for over three decades. TRNSYS is also one of the first commercial numerical models, which has been utilized for whole-building energy performance analysis of the PCM-enhanced building envelopes. At the Oak Ridge National Laboratory, USA, Tomlinson and Heberle (1990) modified the TRNSYS code to incorporate a model of the PCM wallboard. Almost at the same time, Ghoneim et al. (1991) from the University of Wisconsin, Madison, developed a new subroutine for TRNSYS enabling numerical analysis of solar storage walls. Within this project, Ghoneim performed numerical analysis of the following thermal storage mediums: sodium sulfate decahydrate, paraffin, P116-wax, and structural concrete. Following this work, Stritih and Novak (1996) of the University of Ljubljana, Slovenia, analyzed experimentally and with the use of the TRNSYS subroutine a solar wall composed of a transparent thermal insulation material, a thermal storage module/solar collector using paraffinic PCM, and an air channel bringing preheated air to the internal space.

Commercially available since 1975, TRNSYS has a modular internal construction and it has been developed with the use of FORTRAN programming language (Klein et al. 2011). It uses a user-friendly graphical interface and stand-alone components and mathematical modules for a wide variety of applications such as building envelopes, heat pumps, ground-coupled systems, and solar building applications. The TRNSYS code permits the explicit definition of each building component and a variety of control strategies. Each system component in TRNSYS is defined as a type and contains all the relevant mathematical parameters to integrate it into the TRNSYS code. The TRNSYS engine calls the system components based on the input file and iterates at each time step until the system of numerical equations is solved. Each of TRNSYS subroutines (types) can be coupled together to enable the flow of data during the simulation. In order to include weather conditions in building simulations, TRNSYS uses a typical meteorological year (TMY) file, which is run through hourly values.

In TRNSYS, the multi-zone building model, known as the Type 56 is utilized for whole-building thermal simulations. This component model deals with the thermal balances of a building having up to 25 thermal zones. In this module, one air node is assigned to each building zone. Air node represents the thermal capacity of the zone air volume and other heat capacities associated with the air node (i.e., furniture and books). The inputs and outputs of the Type 56 are controlled by a separate preprocessing program, called PREBID. The PREBID program reads building descriptions and generates files that are later used by the Type 56 components during a TRNSYS simulation.

[12]http://sel.me.wisc.edu/trnsys/index.html.

In Type 56, which is a one-dimensional heat balance model, heat transfer through multilayer building shell components is solved with the use of the conduction transfer function (CTF) method (see Mitalas and Stephenson 1971). Keep in mind that, during last decades, the CTF method and the response factor method have been widely used in a large number of the whole-building energy simulation tools (Kossecka and Kośny 1998). Unfortunately, neither of these methods allow transient simulations with the use of materials having non-linear temperature-dependent physical properties. That is why the CTF-based Type 56 module is known as having limitations with building envelopes containing PCMs. This fact has generated an interest in development of external numerical subroutines that enable simulations of the PCM-enhanced building materials. Usually, these subroutines (Types) enable thermal behavior analysis for only a single architectural component. Examples of the currently developed user-defined TRNSYS Types, supporting simulations of the building fabric components containing PCM are presented below. In general, all of these Types have to communicate with Type 56 for whole-building simulations (Ibanez et al. 2005; Kuznik et al. 2010).

6.5.1.1 TRNSYS Type 204

Prior to the development of the Type 204 model in TRNSYS, it was quite difficult to simulate heat transfer processes in buildings with architectural components containing PCM. While in 1980s and 1990s, a majority of research projects were focused on the experimental analysis of building integrated PCMs; more recently, with the development of a new TRNSYS module enabling analysis of the PCMs integrated with architectural components, it was possible to investigate in detail the thermal properties of a wide variety of PCM-enhanced technologies without the need for elaborate experimentation. The Type 204 subroutine was developed in FORTRAN and integrated to TRNSYS by the research team from the Helsinki University of Technology, Finland (Jokisalo et al. 2000; Lamberg et al. 2000; Lamberg 2003). Utilizing the finite-difference method with a Crank–Nicholson scheme, the Type 204 model enables multi-dimensional heat transfer simulations of PCM-enhanced architectural components. In order to account for the changes of the PCM's specific heat capacity, due to temperature variations, the model uses the effective heat capacity method (Minkowych et al. 2006; Stefanescu 2009; Muhieddine et al. 2009). At each node, the conduction, convection, and radiation heat transfer, along with the temperature is calculated. Type 204 allows use of one-, two-, and three-dimensional geometries. Thanks to this, even complex cases of three-dimensional heat transfer can be defined in details (Ahmad et al. 2006). The three-dimensional building envelope can be defined precisely to specify the PCM concentration, overall latent heat storage capacity, and phase-transition points of the PCM used. The properties of the composite building materials used in conjunction with the PCM can also be easily defined. In the Type 204, numerical models should contain detailed geometry descriptions and dimensions, heat transfer coefficients for radiation and convection, followed with physical properties of all building materials and insulations used in conjunction with the PCM. For PCM-enhanced bulk materials and for PCM composites, the Type 204 module in TRNSYS requires the following input parameters that must be entered into the model:

- A number of iterations: This parameter can be given any value between one and infinity and is used primarily to adjust precision.
- PCM melting temperature and crystallization point: Melting point is considered as the initial temperature in which the PCM undergoes phase transition. The crystallization temperature is determined by the point where the PCM changes phase back to a solid.
- Phase-transition range: Unlike in some pure PCMs, which change phase at a distinct temperature, most often phase changes undergo within a temperature range.
- Latent heat, specific heat, and density of PCM.
- If needed, specific heat and density of the PCM carrier.
- Volume Fraction of PCM (representing the overall concentration of PCM in a particular specimen): Unlike in most numerical tools using the overall concentration of PCM characterized as a fraction of weight, this input value must be converted into a volume fraction.

6.5.1.2 TRNSYS Type 222

As mentioned earlier, during whole-building simulations TRNSYS is controlled by the multi-zone building Type 56. Ibanez et al. (2005) of University of Lleida, Spain, developed a new PCM model for TRNSYS 15 in order to control the Type 56 in thermal behavior simulations of buildings containing PCM-enhanced components. It has been originally called the Type 222. However, more often it is called the PCM-active layer controller for the Type 56.

In this work, Ibanez adopted, developed earlier, theoretical approach for modeling of thermally activated constructions (Koschenz and Lehmann 2000). In the numerical approach proposed in Ibanez et al. (2005), the PCM-enhanced component was indirectly represented by a so-called active thermal layer, a procedure already included in the Type 56 for analysis of thermally activated constructions. In this indirect simulation method, the heat conduction through the element is reduced to a one-dimensional form by applying thermal resistance models. The transfer function method of the building model the Type 56 for walls was utilized for solving the one-dimensional heat conduction in PCM assemblies. In this approach, numerical model of the wall containing PCM can be defined as an active layer. The Type 222 was based on heat transfer analogies between thermally activated construction elements and PCM-enhanced building envelopes. In the expert mode, a fictitious equivalent heat transfer coefficient can be defined by the user. Thanks to the application of an active layer containing tubes a circulating heat exchanging fluid, the overall thermal effect of PCMs could be indirectly replicated. In the Type 222, the detailed dimensions of the active layer and heat exchange fluid data are required as input data.

Even though the above approach did not simulate the real heat transfer process through a PCM component, the overall energy impact was quite similar to what would be expected with a wall containing PCM (Ibanez et al. 2005). In each simulation step, Type 222 determined the enthalpy and temperature of the PCM and the variations with the previous simulation step. To properly represent in simulation the heat transfer characteristics of the PCM-enhanced building component, Ibanez utilized a user-defined equivalent heat transfer coefficient. The methodology was initially validated with the use of laboratory experiments and field test data from a prototype room built with concrete panels containing PCM. According to Ibanez et al. (2005), the suitability of their methodology has been proven and Type 222 is a good numerical tool for analysis of PCM applications in buildings. However, more experimental validation work was recommended as well.

6.5.1.3 TRNSYS Type 241

The Type 241 is a PCM Wall Model for TRNSYS 16 which has been developed in the Institute of Thermal Engineering, University of Technology Graz, Austria. It is capable of modeling a PCM-enhanced material as a dynamic layer attached to a wall (Schranzhofer et al. 2006). This is a serious limitation, which makes quite difficult simulations of very common building envelopes containing integral PCM layers, PCM-enhanced insulations, or PCM dispersed inside the wall core material. Since the Type 241 model is coupled with the existing multi-zone building model of TRNSYS 16 (Type 56), the PCM layer is theoretically connected to the wall assembly in the Type 56, through a so-called direct contact zone. Such a fictitious contact zone is characterized by a no-mass air layer of negligible thickness, having close to zero heat transfer resistances. Thermal calculations within the PCM layer are carried out by the Type 241 model using a finite-difference approach. In order to account for the changes of the PCM's specific heat capacity, due to temperature variations, the Type 241 uses the heat capacity method. The PCM wall model (Type 241) has not been fully validated yet against experimental data (Schranzhofer et al. 2006; Castell et al. 2009).

6.5.1.4 TRNSYS Type 260

The Type 260 model for TRNSYS 16 has been developed and validated by a research team, from the Université de Lyon, France. In this numerical procedure a one-dimensional finite-difference scheme is utilized to simulate the thermal performance of an external wall containing a layer of the PCM-enhanced material (Kuznik et al. 2010). The PCM Type 260 is associated with the multi-zone building model (Type 56), as well as with the "Slab on Grade Type"—as specified in the TRNSYS standard component library.[13] During coupled Type 56 and Type 260 simulations, the interior wall surface temperature is calculated by the Type 56 and later used as an input by the PCM Type 260. This numerical procedure uses solar heat flux (short wave radiation), ambient temperature, and a view factor (long wave radiation) as the external condition inputs. The global conductance of the layers situated between the thermal zone and the PCM layer and all material layers which are situated between the PCM and the exterior surface are parameters within the Type 260. Up to 4 exterior material layers are allowed, for which thickness, thermal conductivity, density, specific heat, and a number of mesh nodes have to be defined. Transient characteristics of the PCM layer are specified through either effective, or apparent heat capacity curve, and thermal conductivities in solid and liquid phases. The Type 260 was validated using experimental data from the earlier test room experiment (Kuznik and Virgone 2009). Composite boards containing 60 % of microencapsulated PCM were tested at that time with the use of the MICROBAT test apparatus, which was composed of two identical test cells. Each test cell was a cubical enclosure with an internal dimension of about 0.50 m. Good agreement between numerical predictions and test results was reported by Kuznik et al. (2010).

6.5.1.5 TRNSYS Type 399

Currently developed new PCM model for TRNSYS 17 (Type 399) allows simulations of different types of passive and active building envelopes, including PCM-enhanced walls. It has been developed by Dentel and Stephan (2013) from the Institute for Energy and Building, of the Georg Simon Ohm University of Applied Sciences, Nurnberg, Germany. To allow PCM analysis, the Type 399 is capable to model temperature-dependent thermal characteristics. It also offers ability to simulate two separate enthalpy curves for PCM melting and freezing. Transient thermal simulations of building envelopes are performed in the Type 399 algorithm with the use of the Crank–Nicolson method and elimination procedure, solving the heat conduction equation. The discretization scheme is one-dimensional. The amount of nodes depends on the thickness of each layer. In the TYPE 399 model, the enthalpy changes are expressed as an invertible function of the temperature; therefore, two different data files with a temperature-dependent heat capacity of the PCM are used. The approximation of the phase change with a hysteresis was developed based on work of Glück (2006). As for January 2014, accuracy of his new numerical procedure has not been evaluated against the experimental data yet.

[13] http://www.trnsys.com/assets/docs/03-ComponentLibraryOverview.pdf.

6.5.1.6 TRNSYS Type 1270

Simple materials ongoing phase transition can be simulated using the Type 1270 controller, which was developed in 2012 by Saleh Al Saadi at the University of Colorado, Boulder, USA, and published in the TRNSYS Individual Component Libraries.[14] This algorithm is designed to interact with the TRNSYS multi-zone building model—Type 56. The Type 1270 makes use of the Type 56 "BOUNDARY" wall concept to allow the user to place a PCM layer wherever is needed within a wall structure. In this case, the "standard" parts of the wall are simulated in the Type 56, while the PCM layer is modeled externally by the Type 1270.

6.5.2 ESP-r—University of Strathclyde, Glasgow, UK

ESP-r, a whole-building energy simulation tool, is often a selection of choice in case of thermal analysis of building components and for whole-building energy simulations. This software has a built-in capability to simulate PCMs with notable hysteresis. For this purpose, ESP-r uses two different specific heat profiles (during melting and solidification). ESP-r is an open-sourced whole-building performance energy tool developed in 1974 by the Energy Systems Research Unit (ESRU), University of Strathclyde, UK.[15] ESP-r was originally designed to work in the Unix operating system, yet today it can be run on Windows computers as well. It uses a finite-difference central solver discretizing the problem domain in a controlled volume scheme and solves the corresponding conservation equations for mass, momentum, and energy. ESP-r can integrate the effect of a variety of factors including, climatic conditions, occupancy schedules, air leakage distribution, external shading, internal energy gains, building mechanical systems, and many others. ESP-r has a predefined database of materials and construction layers that can be used directly or modified. Since ESP-r allows a detailed parametric analysis of a number of factors effecting the energy and environmental performance of buildings, for decades, it has been widely utilized to simulate energy consumption in multi-zone buildings, thermal behavior of building components, air ventilation, space-conditioning systems, etc. Today, ESP-r is best-known as a research tool, even though it has been also used for numerous consulting and teaching applications.

Before 2000, a non-PCM part of the ESP-r program was extensively validated (Strachan 2000). Since 2000, with added PCM module, ESP-r can simulate phase-changing materials using a concept of special materials (Kelly 1998). This module enables the introduction of a phase transition to the energy balance equation via material substitution. Special materials are used in ESP-r as means of modeling active building components that have the ability to change their thermo-physical

[14]http://www.trnsys.com/tess-libraries/individual-components.php.

[15]http://www.esru.strath.ac.uk/Programs/ESP-r.htm.

properties in response to some external factors (i.e., temperature or radiation). In ESP-r, the special material functions can be applied to a particular node within a multilayer construction. Any node defined as a special material is then subjected to a time variation in its basic thermo-physical properties. ESP-r assumes one-dimensional heat transfer across the PCM layer(s). As stated by Heim and Clarke (2004), the differential equation of transient heat conduction with variable thermo-physical properties is as follows:

$$\frac{\partial}{\partial t}\rho(T)h(T) = \nabla \cdot [k(T)\nabla T(r^{\rightarrow},t)] + q(r^{\rightarrow},t) \tag{6.30}$$

where T is the temperature, ρ the density, h the enthalpy, k the conductivity, and q the heat-generation rate.

ESP-r offers a number of the special material files SPMCMP53 through SPMCMP56 enabling modeling of the PCM-enhanced building envelopes. PCM numerical models in ESP-r represent different levels of complexity in describing thermal behavior of PCMs during phase transition, ranging from a simplified single enthalpy curve model, to more advanced numerical methods incorporating the effect of hysteresis. A general assumption is that within the phase-transition temperature range, the PCM-enhanced material can be either in a solid, liquid or a mushy, two-phase state. In the simpler SPMCMP53 model, PCM performance can be simulated by assigning phase-changing properties to the construction layer that corresponds to a PCM impregnated board. The thermal modeling is based on two numerical methodologies: the control-volume approach and the apparent heat capacity method. Whenever the temperature of the node representing the layer is within the phase-change limits, the thermo-physical properties of the material are replaced with the properties of the PCM. The heat capacity is substituted by the effective heat capacity method. More advanced SPMCMP56 model was developed by Geissler (2008), based on Hoffmann's (2006) theoretical work. This model is capable of taking into account the PCM hysteresis effect. Material data input requires PCM density, conductivity, and phase-change temperatures. In addition, in order to describe PCM thermal properties using mathematical formula, this model uses the temperature-dependent heat storage characteristics of PCMs both for the melting and for solidification processes. In this method, the stored/released latent heat, *LH(T)*, during phase change, is calculated using the following formula:

$$LH(T) = \int_{T_1}^{T_2} T^f \left(\frac{a + cT + dT^2}{1 + b + eT^2}\right) \mathrm{d}T \tag{6.31}$$

where *a, b, c, d, e*, and *f* are curve-fitting parameters approximating specific heat capacity of the PCM as a function of temperature during phase change. T_1 is the onset melting/solidification temperature and T_2 is the temperature where melting/solidification ends. Outside those two limits (T_1 and T_2), the PCM stores/releases energy only in the form of sensible heat. Within the limits, the heat capacity of the PCM is a function of temperature.

Since introduction of the PCM module in 2000, the ESP-r program has been used worldwide by different research and engineering groups for analysis of PCM-enhanced building technologies. As in the cases of other similar building simulation tools, most of the research work has been focused on PCM-enhanced gypsum boards (Heim and Clarke 2004; Schossig et al. 2005; Castell et al. 2009; Sigalas 2011). In 2006, Fraunhofer ISE, Freiburg, Germany, evaluated the accuracy of the ESP-r building and plant models during a series of laboratory-scale experiments, performed on the commonly operated office room. The simulation study was carried out for a ceiling containing 20 % of microencapsulated PCM (Kalz et al. 2006, 2007). In 2012, Fraunhofer CSE, Boston, USA, performed numerical and experimental study on different types of PCM-enhanced fiber insulations. In this work, Fallahi et al. (2012) validated the ESP-r special material file—SPMCMP56, against the field test data collected earlier by the Oak Ridge National Laboratory, USA, during a series of field test experiments performed in a costal climate of South Carolina, USA, on wood-framed walls containing blown PCM-enhanced cellulose (Kośny 2008). An excellent agreement was reported between recorded and simulated heat flows.

6.5.3 EnergyPlus—Lawrence Berkeley National Laboratory (LBNL), Berkeley, CA, USA

EnergyPlus is an energy simulation software package developed by the U.S. Department of Energy. It enables modeling of heat transfer in buildings while providing for consideration of a large range of HVAC equipment. Its extensive building design options make it a very powerful software and ideal for building energy studies. The EnergyPlus program is a collection of many program modules that work together to calculate the energy required for heating and cooling a building using a variety of systems and energy sources. While originally based on the most popular features and capabilities of earlier-developed whole-building energy tools such as BLAST or DOE-2, EnergyPlus includes many innovative simulation capabilities such as time steps of less than an hour, modular systems and plants integrated with heat balance-based zone simulation, multi-zone air flow, thermal comfort, water use simulations, natural ventilation, and photovoltaic systems.

EnergyPlus models follow fundamental heat balance principles very closely in almost all aspects. However, the simulation of building surface constructions has relied on a conduction transfer function (CTF) method inherited from BLAST. This fact leads to a number of program restrictions such for example: constant material properties or fixed values of simulation parameters. Since building envelope and insulation technologies move today toward using advanced dynamic components, such as switchable materials, PCMs, or insulations with variable thermal conductivity, it becomes necessary to also use more advanced numerical methods in thermal and energy analyses. Accordingly, a conduction finite-difference (CondFD) method has been incorporated into EnergyPlus. It doesn't fully replace a relatively fast CTF solution algorithm, which still can be used for simpler material configurations. However, it complements the CTF method in cases where EnergyPlus is utilized to simulate PCMs or other materials of variable physical properties. It is also possible to use the finite-difference algorithm where very small time steps are required.

Originally EnergyPlus was based on the FORTRAN language. Recently, it has been converted into the C++ language. Two following input files are used to run the simulation: (i) the IDF-file (EnergyPlus input data file), which contains all the information about the location, geometry, materials, structure and all the HVAC equipment in the building, and (ii) the weather file.

For the finite-difference model, EnergyPlus offers for user selection two different computational options. The first scheme is referred to as semi-implicit Crank–Nicholson, and it is based on the Adams–Moulton solution method. In time of introduction of the EnergyPlus version 7, a second, fully implicit, scheme was added. Although these two numerical options differ in their fundamental heat transfer methodologies, they share nearly all the same supporting subroutines for material properties, data storage, solution schemes, and spatial discretization algorithms. In addition, an implicit numerical module utilizing the enthalpy method was later introduced in EnergyPlus by Pedersen (2007), in order to enable PCM simulations.

Please keep in mind that EnergyPlus can simulate PCMs only with the use of the conduction finite-difference (CondFD) solution algorithm. CondFD discretizes building envelope into several nodes and uses an implicit finite-difference scheme to numerically solve the appropriate heat transfer equations. For the PCM algorithm, the CondFD method is coupled with the enthalpy–temperature function, which accounts for enthalpy changes during the phase-transition process (Pedersen 2007). The enthalpy–temperature function is used to develop an equivalent specific heat at each time step. The resulting model is a modified version of the enthalpy method. In the CondFD algorithm, the automatic element discretization depends on the space discretization constant, the thermal diffusivity of the material, and the time step. The presented below formula[16] (Eq. 6.32) describes the calculation method for the fully implicit scheme used in EnergyPlus.

$$C_p \rho \Delta x \frac{T_i^{j+1} - T_i^j}{\Delta t} = k_W \frac{\left(T_{i+1}^{j+1} - T_i^{j+1}\right)}{\Delta x} + k_E \frac{\left(T_{i-1}^{j+1} - T_i^{j+1}\right)}{\Delta x} \tag{6.32}$$

[16]EnergyPlus Engineering Reference: The Reference to EnergyPlus Calculations. 2011, Ernest Orlando Lawrence Berkeley National Laboratory. p. 1130—available at http://apps1.eere.energy.gov/buildings/energyplus/pdfs/engineeringreference.pdf.

where

$$k_E = \frac{\left(k_{i-1}^{j+1} + k_i^{j+1}\right)}{2}$$

$$k_E = \frac{\left(k_{i-1}^{j+1} + k_i^{j+1}\right)}{2}$$

$$k_i = k\left(T_i^{j+1}\right)$$

where

k_W thermal conductivity for interface between i node and i + 1 node
k_E thermal conductivity for interface between i node and i − 1 node

For temperature-dependent thermal conductivity, specific heat (C_p), and density (ρ) of a material, time step is denoted as Δt and finite-difference layer thickness as Δx. The following node numbering notation is used:

i a number of node being simulated,
$i + 1$ adjacent node to interior of construction,
$i - 1$ adjacent node to exterior of construction,
$j + 1$ a new temperature at end of time step, and
j temperature at the end of previous time step

In several experimental and numerical studies focused on PCM-enhanced building board products, EnergyPlus capability of simulating PCMs was utilized to analyze their thermal and energy performance in the whole-building scale (see Kuznik and Virgone 2009; Zhuang et al. 2010; Tardieu et al. 2011). In general, the authors reported a relatively good agreement between the simulation results and the test data. It was concluded that the algorithm incorporated into EnergyPlus can well reflect thermal behavior of the PCM-enhanced boards in whole-building applications.

In a different group of PCM projects, thermal performance data generated during material and system-scale testing and thermal simulations were utilized for development of whole-building EnergyPlus models. During one of the projects, a dynamic heat flow meter apparatus (HFMA) measurements, combined with thermal and energy simulations, were used for whole-building performance analysis of a complex array of the PCM pouches (Kośny et al. 2010a). In this work, the system-scale and whole-building-scale simulations were used together to develop energy performance data of the wall configuration containing PCM. At first, a series of steady-state and dynamic experiments with the use of the HFMA were performed to measure thermal properties of the analyzed PCM system imbedded in an insulation assembly. In addition, differential scanning calorimetry (DSC) test data was utilized to analyze phase-transition characteristics of the PCM. Next, the HFMA test results and the DSC data were combined with a 3-D finite-difference model to determine the effective heat capacity of the system. This multitask experimental–numerical procedure permitted the heat storage evaluation of a complex PCM-enhanced building component and its response to changing temperatures. In order to enable whole-building EnergyPlus simulations, a simplified 1-D computer model was developed using the equivalent wall procedure (Kossecka and Kośny 1997; ASHRAE RP-1145 2001).

The second example project focused on walls containing PCM-enhanced cellulose fiber insulation. In this study, material-scale simulations and dynamic HFMA testing were used together to develop material characteristics necessary for the whole-building EnergyPlus model. At first, a series of steady-state and dynamic experiments with the use of the HFMA were performed to measure thermal properties of the PCM-enhanced cellulose insulation. In addition, microencapsulated PCM was tested with the use of the differential scanning calorimetry (DSC) apparatus. Next, the HFMA test results and the DSC data were used to estimate the percent-content of PCM in thermal insulation (Kossecka and Kośny 2009; Kośny et al. 2009b). This procedure enabled development of the PCM wall data for the whole-building Energy Plus model (Kośny et al. 2010c).

The conduction finite-difference and PCM algorithms in EnergyPlus were validated by Tabares-Velasco et al. (2012). In this study, experimental data from whole room testing was used for comparisons (Kuzmik et al. 2010). Theoretical approach used in this process was similar to what is recommended by the ASHRAE Standard 140, which requires analytical verification, comparative testing, and empirical validation (ASHRAE 2004). In addition, experimental data generated during dynamic hot-box testing of the wall assembly containing PCM-enhanced boards were used for comparisons (Haavi et al. 2011). Several key limitations and numerical problems within the EnergyPlus PCM model were identified and as a result a number of model improvements were implemented into the program version 7.1. Accordingly, Tabares-Velasco et al. (2012) recommend that when using the EnergyPlus PCM model, (i) time steps less than 3 min should be used for whole-building simulations that included PCM components, (ii) the default CondFD node spacing values can be used with acceptable monthly and annual results, (iii) when subhourly energy performance analysis is required, users should use node spacing values about one-third of the default size (equal to using a space discretization constant in EnergyPlus of about 0.3–0.5). Finally, it was concluded that since the EnergyPlus PCM model uses only a single enthalpy curve, accuracy issues can arise when modeling PCMs with notable hysteresis. In practice, the above statement can be translated into the additional recommendation of using EnergyPlus simulations only in cases of building technologies containing paraffinic PCMs.

6.5.4 RADCOOL—Lawrence Berkeley National Laboratory (LBNL), Berkeley, CA, USA

RADCOOL is a whole-building thermal simulation model which has been used as one of the first whole-building numerical applications in several theoretical studies of PCM-enhanced gypsum boards, performed by Lawrence Berkeley National

Laboratory (LBNL), USA (Stetiu and Feustel 1996; Feustel and Stetiu 1997). It is a single-zone computer code, which is written in the SPARK, a simulation tool (Buhl et al. 1990). SPARK is a simulation environment that generates source and executable code for solving nonlinear differential–algebraic equation systems. It is intended to automate the generation of modular, easy to maintain simulation programs. RADCOOL program allows performance analysis of building materials containing PCM (Stetiu et al. 1995). RADCOOL is based on the finite-difference approach. In numerical analysis, each building component can be divided into several layers. This allows modeling of simple one-dimensional multilayer systems. In order to simulate phase-change processes, RADCOOL uses a unique formula for the enthalpy (*h*), which also describes the enthalpy as a function of the temperature (*T*) for the whole temperature range found in buildings. It is based on original work of Egolf and Manz (1994), who formulated the following description for the specific enthalpy:

$$h(T) = c_{p1}T + \eta_1 \quad T \leq T_m \tag{6.33}$$

$$h(T) = c_{p1}T_m + (h_2 - h_1) + c_{p2}(T - T_m) - \eta_2 \quad T > T_m \tag{6.34}$$

where

$$\eta_n = \left(\frac{h_2 - h_1}{2}\right) \mathrm{e}^{\left[-2\left(\frac{T - T_m}{n}\right)\right]} \quad n \in [1, 2]$$

The enthalpy can be described as a hyperbolic function and its derivative (Feustel 1995). The following equation for the specific enthalpy was used at that time:

$$h(T) = c_{p,\mathrm{const}}T + \frac{h_2 - h_1}{2} \times \left\{1 + \tanh\left[\frac{2\beta}{\tau}(T - T_m)\right]\right\} \tag{6.35}$$

A similar formula can be derived for the specific heat capacity.

$$c_p(T) = c_{p,\mathrm{const}}T + \frac{h_2 - h_1}{2} \times \frac{\frac{2\beta}{\tau}}{\cosh^2\left[\frac{2\beta}{\tau}(T - T_m)\right]} \tag{6.36}$$

Because of its limitations, RADCOOL had to be used in conjunction with a whole-building energy code having capability to simulate performance of the space-conditioning system. In several LBNL studies focused on energy performance of buildings containing gypsum boards and plasters using PCM, initial DOE-2 whole-building simulations were performed to obtain anticipated temperatures and air flows for the space-conditioning system. Next, RADCOOL was used to model the thermal performance of a simple building containing walls and ceilings utilizing PCM-enhanced wallboards (Stetiu and Feustel 1996; Feustel and Stetiu 1997).

6.5.5 CoDyBa

CoDyBa is a software used to analyze dynamic energy, thermal, and hygrothermal performance of building elements when subjected to external climatic conditions (Noel et al. 2001). The tool is aimed to conduct studies of heating and cooling strategies, air-conditioning or ventilation options, and insulating materials to be installed. CoDyBa can be used to investigate the thermal, hygrothermal, and energy performance of building components and buildings of almost any type and size. In addition, it can perform peak load calculations necessary for building shell optimization and space-conditioning system design. It also permits estimation of the instant heating or cooling power needed to maintain a given set point temperature, or to calculate the indoor temperatures when the heating or cooling system is insufficient.

CoDyBa is based on a modular internal format enabling analysis of relatively complex buildings. Through the graphical interface, a user can construct a model of the building's geometry using basic elements such as thermal zones, building envelopes, internal space schedules, and mechanical systems. With the program interface, users can define thermal zones, construction components such as walls windows, or roof, add mechanical systems, and specify internal loads. After completion of the data input process, CoDyBa can perform thermal and energy calculations using the climate, internal loads and their schedules (lighting, people, or internal equipment), and internal airspace parameters (i.e., set points for temperature, humidity, or ventilation rates) as main control parameters for simulations. Since CoDyBa has no limit on number of thermal zones, it can simulate buildings of almost any size. The only real limit is the time used to input data.

CoDyBa produces input summary reports that may view on-screen or stored as rtf files. A number of results graphs may be viewed and printed. The following graphical outputs are available (as hourly time function):

- Building: sensible and latent heat powers,
- Thermal zones: sensible and latent powers, indoor air temperatures, resultant temperatures (average of air and wall temperatures), relative humidity,
- Building exterior surfaces: absorbed direct and diffuse solar fluxes, surface temperatures,
- Building interior surfaces: surface temperature (useful for thermal comfort evaluation),
- Windows: transmitted direct and diffuse solar fluxes, sunlit fraction, direct and diffuse transmission coefficient, global transmission coefficient g,
- Mechanical systems: power profiles and energy consumption.

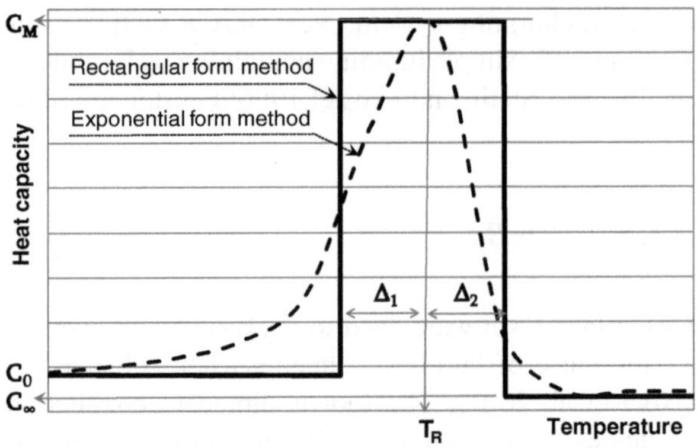

Fig. 6.9 Two heat capacity methods used by CoDyBa for PCM simulations

In addition, CoDyBa can be used for forecasting the dynamic thermal behavior of a building, which may include a performance analysis of building components containing PCM. According to Virgone et al. (2009), this numerical capability has been validated through comparisons against experimental data (using measurements on simple material configurations) with satisfactory agreement. In PCM modeling, one-dimensional analysis of the temperature within the building envelope component is defined by the Fourier equation. The enthalpy h exchanged by the PCM is considered based on the heat capacity $C(T)$ of the material, as a function of temperature.

Heat capacity can be simulated with the use of two simplified methods presented in Fig. 6.9. The first one method describes heat capacity change during phase transition as a simple step function using rectangular shape of the heat capacity $C(T)$ function. The second, more advanced method uses the following exponential formulae for the heat capacity $C(T)$, where T_R is the temperature of the peak of the heat capacity curve:

$$C = C_0 + (C_M - C_0)e^{-\left(\frac{T_R - T}{\Delta_1}\right)^2} \quad \text{for; T} \leq \text{T}_\text{R} \tag{6.37}$$

$$C = C_\infty + (C_M - C_\infty)e^{-\left(\frac{T_R - T}{\Delta_2}\right)^2} \quad \text{for; T} > \text{T}_\text{R} \tag{6.38}$$

Next, the latent heat of the phase-transition process can be computed using the equation below:

$$L = \frac{\sqrt{\pi}}{2}[\Delta_1(C_M - C_0) + \Delta_2(C_M - C_\infty)] \tag{6.39}$$

According to Virgone et al. (2009), the above model represents a compromise between speed of calculations and overall precision of numerical predictions. Comparing to the first rectangular function approach, the second model utilizing the "exponential" expression for heat capacity, enables significantly better approximation of the direct measurement results. In addition, in CoDyBa simulations, separate thermal conductivities can be considered for melted and solid PCM.

6.5.6 BSim

The integrated whole-building numerical tool called BSim was developed by the Danish Building Research Institute, of Aalborg University, Denmark. This is a one-dimensional computer model for analysis of heat and moisture transport. BSim is a successor of several earlier—validated numerical components (Lomas et al. 1997; Rode et al. 2002; Grau and Rode 2006; Kalyanova et al. 2009; Scarpa et al. 2009). The core of the BSim program is the Tsbi5 module serving for analysis of energy and indoor conditions in buildings. The BSim program includes a number of advanced numerical tools for analysis of, e.g., heat transfer in building components, whole-building energy consumption, indoor climate conditions, heat and moisture transport, daylight conditions, and natural ventilation. In BSim simulations, a building is divided into a number of thermal zones which are separated from each other, from the exterior conditions, or from any thermal zone by different types of constructions. The heat balance for each zone couples to the heat transport through all adjacent constructions. In the BSim numerical model, the dynamic behavior of a building is described in a discreet form. The Tsbi5 module is the thermal simulation core of the BSim program package. Tsbi5 makes it possible to perform a dynamic simulation of the indoor climate, energy, and moisture conditions in buildings. This can contribute to the selection of heating, cooling, and ventilating systems at the design stage of new buildings and refurbishment of existing buildings and installations. By using of the moisture module, the program user can analyze hygrothermal behavior of the building interior, moisture accumulation in construction materials, or can evaluate condensation risk at the building shell surfaces of interest.

BSim can perform analysis of a multi-zone building with consideration of heat exchange through opaque building envelopes, solar heat gains through windows, performance of space-conditioning system, steady-state moisture balance and local condensation risks. Building geometry can be either created with the use of the graphic editor, or just imported from CAD drawings. BSim contains both the graphic model editor and a graphic user interface, supporting several applications included in the program package. A unique feature of the BSim program is its capability of analysis of materials with phase transition. It is, for instance, possible to make precise predictions of dynamic energy performance of the building in case of the PCM application in the building construction components. Rose et al. (2009) used BSim to analyze energy performance of the building containing PCM-enhanced gypsum boards. This case study has presented an example showing how the BSim PCM module can be utilized for optimizing indoor climate in a room with walls containing PCM-enhanced gypsum boards.

6.5.7 PCM Express

PCM Express is a whole-building energy simulation tool for buildings using PCMs. The main framework of the PCM Express program was developed by Valentin Energie software GmbH, Berlin, Germany, in connection with the research project focused on development of a user-friendly planning and simulation program for analysis of "Active PCM Storage Systems in Buildings." This project was cosponsored by the German Ministry for Business and Technology and performed in collaboration with the Fraunhofer ISE in Freiburg, Germany, and other industrial partners (Zöller et al. 2008). The major goal of this development program was to support building designers by facilitating a reliable decision-making tool for configuring the PCM-enhanced building components, which was expected to accelerate market adoption of PCM building applications. This includes the most commonly used applications of PCMs in building fabric components. Since, depending on the situation, the use of PCMs can be either justified by the increased level of occupational comfort, or by economic considerations, the PCM Express tool enables analysis based on both above-listed criteria, through the use of different internal menus, predefined usage schedules, and adapted presentation of the results.

In PCM Express, PCM can be a part of the floor, ceiling, or wall construction. To describe the impact of the phase transitions within materials containing PCM, it is necessary to properly represent their transient characteristics and accurately simulate PCM's charging and discharging. In PCM Express, all layers of the building envelope component are simulated as a node in the mathematical node-edge-model. Non-PCM materials are defined with a constant heat capacity. PCMs are defined with the use of the temperature-dependent enthalpy curve.

The mathematical models and numerical algorithms in the PCM Express program were developed by the Fraunhofer ISE. During PCM Express simulations, each room is denoted by a single node representing the inside air. The heat flow between the nodes is computed as a function of the thermal conductivity and is represented by a fictitious edge connecting two layers. The outside layer is linked over an edge with the exterior environment. Irradiation through the windows is distributed to the inside air nodes and the inner layer of the building fabric. All air nodes are linked with space-conditioning systems such as heating, cooling, and ventilation. In order to numerically represent building components containing integral hydronic systems, each layer can be defined as an active layer, e.g., a plaster layer with imbedded plastic heat exchangers.

One of the numerical studies that used PCM Express was performed by Ubiñas et al. (2009), in order to predict and compare the interior temperature behavior of a room using PCM for heat storage. For several Spanish climates, the authors simulated thermal and energy performance of two test rooms, with and without PCM. The thermal performance of the test rooms was evaluated on an hour to hour basis, zones with diverse combinations of facade glazing ratio and window shadow factors. The results of the study indicated that in all considered Spanish climates, the addition of PCM to the building fabric helped in maintain the interior thermal comfort and in shaving of high and low internal temperature peaks. It was demonstrated that PCM's ability to thermally stabilize building interior is more noticeable in summer.

6.5.8 WUFI

WUFI represents a family of numerical tools for modeling the heat and moisture transfer within building structures. WUFI, which in German stands for "*Wärme und Feuchte instationär*", was developed by the Fraunhofer IBP, Holzkirchen, Germany (Kunzel 1995). Coupled transient heat and moisture transport simulations can be performed in WUFI using user-defined time steps and within arbitrary simulation time periods. For this purpose, the finite-volume method is employed. The only limitation is the hourly weather data, which has to be interpolated for smaller time steps. Initially, the main area of WUFI's applications focused on forecasting possible moisture damage or humidity-dependent heat losses caused by indoor climate and natural weather effects. The WUFI algorithm for hygrothermal calculations considers thermal transport, vapor diffusion, and liquid flow. It is based on theoretical equations for heat and mass transport through a defined section of the building element. In time, numerical capabilities of the WUFI models have been gradually expanded to form a complex software package for modeling of complete buildings, including heating systems and ventilation. WUFI Plus is an interior room climate model which focuses on calculating the thermal behavior of the building taking into consideration hourly outdoor climatic data, interior thermal loads, various temperature schedules, and ventilation strategies. Building materials containing PCM can be simulated in WUFI 5.0 and WUFI Plus using temperature depended enthalpy option. Numerical descriptions of the building components and building geometry can be produced by using wizards. Next, boundary conditions can be specified and whole-building simulation conducted. Simulation results such as the indoor temperature are graphically depicted and users are provided with support in interpreting the results. Time-dependent simulations that incorporate all significant boundary conditions also help when making designing decissions regarding the building operation, the ventilation strategies, the dynamic effects caused by thermal storage, and strategies to prevent space overheating in summer.

In Baghban et al. (2010), a transient heat transfer problem of an exterior building wall containing PCM and aerogel insulation has been simultaneously simulated with the use of COMSOL 4.0 and WUFI 5.0. The results confirmed the advantage of using high-performance insulation materials in new construction systems. Numerical analysis showed that, using a layer of concrete containing PCM behind the aerogel in the exterior wall can be a good idea for saving energy in the building envelope. PCM can absorb the passive solar energy which passes through the aerogel layer during the daytime and releases it at night with a low heat loss due to existence of aerogel insulation. After the series of COMSOL 4.0 and WUFI 5.0. simulations, the authors indicated a need for incorporation of the hysteresis effect in thermal and energy simulations of building envelope systems containing PCMs.

Results of WUFI Plus simulations are discussed by Spaeh (2012) for the wood-framed wall containing PCM-enhanced gypsum board. It was analyzed as an alternative system of providing energy-efficient improvements for buildings in hot and mild climates. This wall configuration was compared against conventional low-mass wood-framed walls using conventional fiberglass and foam-sheathing insulations. Transient WUFI simulations were used in analysis of thermal energy flows and interior temperature fluctuations of buildings with increased insulation values and PCM thermal storage. The findings showed that installation of PCM in building enclosures near the interior surface can have significant benefits for buildings in the southern U.S. regions. These benefits include increased energy efficiency, reduced surface temperature fluctuations, stabilized operative temperatures, reduced humidity fluctuations, and a reduced likelihood of moisture condensation.

A series of thermal simulations were performed by Korjakins et al. (2013) to study the optimal configuration of the PCM-enhanced material inside the building. A numerical study utilizing WUFI Plus and EnergyPlus whole-building simulation tools was focused on the building plaster containing PCM. The authors investigated various fusion temperature ranges of PCM in order to optimize night and day performance in summer conditions characteristic for Latvian summers. The PCM effect on the whole-building energy performance during the heating season was investigated as well. The results demonstrated that an application of PCMs in Latvian climatic conditions may help with natural cooling of the building by stabilizing the internal room temperature variations during summer days. This is a subject to sufficient precooling conditions during the night. Correspondingly, PCM usage was found useful in stabilizing indoor temperature during the heating season.

6.6 Modeling of PCM-Enhanced Building Envelopes—Future Challenges

Current developments in areas of building envelopes and PCM heat storage technologies have generated a demand for numerical tools to support the decisions of architects, builders, or mechanical system designers in choosing optimum configuration of building components. That is why before PCM-enhanced building envelopes become routinely used standard building systems in different climatic zones, computer models need to be able to accurately predict thermal and energy performance of buildings containing PCM. This will allow a development of a new generation of building construction standards and code energy requirements, which will support wider application of these technologies. Finally, a larger number of building applications will increase production volume in this area, reducing prices.

However, to get to this point, computer models used for analysis of PCM-enhanced building envelopes need to undergo a serious transition improving overall confidence of their users. We all understand that development and validation of numerical models supporting analysis of PCM-enhanced building systems is challenging due to the nonlinear nature of the heat transfer phenomena in PCM-enhanced building envelope components. In addition, thermal analysis and design of PCM-enhanced building envelopes is a relatively difficult task, because PCMs used in these applications, need to go through a complete phase transition during 24-h time periods in order to be fully effective. It is also widely known that, in case of the PCM-enhanced building envelopes, exterior climatic conditions (such as ambient temperature and solar radiation) are not always"fully cooperative", staying within the range required to enable full phase-transition processes (Kośny et al. 2012a). Other technical issues such as lack of proper material characteristics for PCM-enhanced building materials, or complex multi-dimensional heat transfer in great majority of building envelopes, add complexity to this problem—see Kośny and Kossecka (2002). Lastly, in large number of published studies, it has been assumed that the entire heat transported through the PCM-enhanced building envelopes is available at approximately the same temperature for both, the melting and solidification processes. However, this assumption only applies to the paraffinic PCMs. That is why, for proper designing of PCM systems, it is crucial that this temperature range, or separate temperature ranges for melting and freezing, are precisely known and adequately used in numerical work.

All of the above challenges cannot be used as an excuse for an application of simplistic numerical models leading to poor overall accuracy of computer simulations. Unfortunately, according to the IEA Annex 23 (2011), today's confidence in available simulation tools is too low to justify their use in performance predictions of buildings containing PCM. Moreover, it has been found that overall thermal behavior of PCM-enhanced materials and systems used in building envelopes is poorly known, which by itself generates essential unknowns in any numerical analysis. In addition, in case of many numerical models, simulation results can be considered as suspicious, since they have not been rigorously enough validated against experimental results. It is also known that in many performance studies, materials' characteristics used in modeling have been too simplified to realistically describe the complete physics needed to predict long-term performance. As a result, serious steps need to be taken now to raise standard requirements for PCM model development and benchmarking, with hope that this process will improve the confidence of the future users. The following sections will discuss several crucial problems that should be addressed in the future by the scientific community, to improve overall confidence in results of thermal and energy simulations of the PCM-enhanced building envelope technologies.

6.6.1 Material Characteristics

One of the key problems, faced even before starting the modeling process, is associated with the characterization of PCMs themselves. In building applications, arrays of the PCM pouches and composite PCM materials are two favored methods for exterior thermal envelopes. Gypsum boards and plasters remain primarily selection of choice for other inner building fabric components. In these forms,

PCMs can be incorporated into a building using similar techniques as used for any building sheathing systems or thermal insulations, which would provide a seamless integration. Unluckily, thermal characteristics in case of a large number of these materials are rather difficult to characterize.

One of the most important factors effecting the accuracy of thermal modeling of PCM-enhanced building systems is quality of enthalpy data used in numerical analysis. Typically, PCM modelers use enthalpy data developed for pure PCMs during DSC measurements. However, this type of physical characteristics can be a source of significant inaccuracies. Please keep in mind that PCM-enhanced building products are seldom high-quality, pure substances. Most commonly, PCM-enhanced building envelopes utilize PCM products which are complex mixtures with fire retardants, nucleation additives, thickening agents, and thermal conductivity enhancers. The above facts definitely limit usefulness of DSC testing procedures for development of thermal characteristics, necessary for thermal simulations of PCM-enhanced building envelopes. It is strongly recommended that only physical characteristics developed for analyzed PCM-enhanced building products are used in numerical analysis. The T-history method or dynamic heat flow meter apparatus testing can be utilized for this purpose. As mentioned earlier, both of them enable testing of non-homogenous products containing PCM.

The second very important factor effecting the accuracy of PCM simulations is a need for usage of separate enthalpy data for melting and freezing. There are three following major reasons for this constraint:

- Frequently measured notable differences between melting and freezing energies
- Hysteresis effect in large number of commonly used PCMs
- Different thermal conductivity values for melted and solid PCMs.

IEA Annex 23 (2011) reported that melting and freezing enthalpies can easily differ by more than 15 % in the case of composite PCMs. This fact is difficult to explain from the perspective of the energy conservation law. However, according to IEA Annex 23, there is a hypothesis that some energy might be stored mechanically by the deformation of the matrix as a consequence of the PCM dilatation. Additionally, in some PCMs, the wide temperature ranges for phase-transition processes weaken the division between latent and sensible heat, making difficult precise estimation of the latent heat–temperature correlation. It is also good to remember that, in some cases, PCMs may stay in supercooled stage during the entire thermal cycle. Finally, in most PCMs, thermal conductivity values are different for liquid and solid phases. All these facts make relatively difficult to distinguish between sensible and latent heat. More accurate laboratory measurement procedures are still in development to help in better understanding above-described physical processes.

In addition, from almost all commonly used PCM compounds, only paraffins do not demonstrate notable hysteresis between the cooling and heating processes. The hysteresis phenomenon is not fully explained yet, due to a complex interaction between the nucleation process, progressive dissolution, glass transition, or metastable crystalline phases—IEA Task 23 (2011). In case of numerical analysis of the PCM-enhanced envelopes, it is critical to precisely know onset temperatures for melting and solidification. In whole-building simulations, both these parameters determine a number of days with either a full or a partial phase-transition (see Kośny et al. 2012b). For PCMs with notable hysteresis, an application of simplified models using a single enthalpy curve (typically for melting) may lead to significant errors in energy performance analysis based on incorrect assumptions regarding the PCM's regeneration potential during the night.

6.6.2 Understanding of Long-Term Performance Characteristics

Since different configurations of PCM-enhanced building envelopes can be located in different parts of the building, these technologies have to be often individually designed in order to survive different sets of environmental factors specific for each considered location . The most common environmental factors effecting long-term performance of building envelopes include the following:

- High and low peak temperatures,
- Daily temperature swings,
- UV exposure,
- Hygrothermal processes, and
- Mechanical stresses.

Although some building shell components may be designed to perform according to a single required task, the modern building envelopes must usually fulfill many functions. First and foremost, building envelopes serve as environmental barriers and prime construction components of buildings. They must resist all environmental loads such as wind, rain, snow, dynamic and static service and seismic loads, effectively protecting building interior from exterior environment. That means, future numerical models focused on long-term energy performance of the PCM-enhanced building envelope systems have to combine lifetime performance characteristics of all building envelope components including PCM products.

Long-term performance of building materials as estimated for the length of a service lifetime is usually considered as one of the key building component characteristics. At the same time, aging factors for a large number of building materials such as some thermal insulations, roof finish materials, concretes, ceramic masonry, and wood framing have been already thoroughly investigated and they are well described in the engineering literature. It is also well known that aging-based deterioration of performance characteristics may influence thermal performance of all building envelopes, including those containing PCMs. A good example of these processes is aging of some plastic foams, caused by release of the blowing agent to surrounding atmosphere. This process increases foam thermal conductivity making older foams less insulating, which is followed by higher overall heat conduction through the building envelope component. It is good to remember that an increase of heat conduction can be also caused by hygrothermal processes in building envelopes. Moisture accumulation in thermal insulations, layers of porous materials, or sheathing boards increases their thermal conductivity, which impairs their overall thermal performance. It is recommended that in long-term energy performance analysis of PCM-enhanced envelopes, all above-described processes, associated with material aging and long-term impact of moisture accumulation, were included in numerical models.

Finally, PCMs in building envelopes represent integral parts of the building structure, in contrary to PCMs used in heat storage containers, heat exchangers, hydronic tubing, etc. The major difference is that in building envelopes, PCM systems cannot be freely replaced or repaired in case of their failure. They have to remain fully functional as long as building is in operation.

Considering that building life span is usually over 50 years, this fact imposes serious long-term durability requirements on the PCM-enhanced building envelope technologies.

In Europe, the main long-term durability criteria for PCMs were introduced in 2004. As defined in the RAL quality mark,[17] they are as follows:

- Phase-transition heat as a function of temperature,
- The cyclic reproducibility of the heat storage process, and
- The thermal conductivity of the PCM.

Nevertheless, long-term performance deterioration processes in large number of PCMs and PCM-enhanced products are widely acknowledged today, and well investigated for basic PCMs, they have not been, however, routinely incorporated yet into numerical performance predictions of building envelope components. The main reason is that specific long-term performance characteristics of PCM-enhanced building envelopes have not been fully investigated so far. More experimental and theoretical work is needed in this area.

6.6.3 Validation of Computer Models

The validation of PCM modeling algorithms is also a challenging task. Keep in mid that older numerical algorithms enabling PCM simulations haven't been originally developed as thermal analysis tools for building applications and therefore they haven't been validated for building applications. That is why it is critical today that in case of PCM-enhanced building envelopes, computer models were validated using experimental data both from system-scale experiments and from the whole-building testing. According to Castell et al. (2009), there are many numerical models for PCM-enhanced building envelopes available in the literature. However, most of them have not been experimentally validated.

[17]http://195.20.235.12/pdf/RAL-GZ_896.pdf.

Please also keep in mind that thermal loads generated by specific building envelope components such as walls, roofs, or windows, represent only a fraction of a total building load. In regards to Northern American residential buildings, a fraction of the whole building thermal loads is usually well below 25 % in case of fenestration and below 15 % for roofs or opaque walls. That is why, very often, the whole-building energy savings generated by PCM-enhanced building components can be below 5 % in the whole building scale, which is well below accuracy estimates for most of the currently used whole-building energy tools. That means, experimental validation of whole-building computer models, used for analysis of PCM-enhanced building envelopes, cannot be only performed based on the whole-building energy consumption analysis. Their validation process has to be more detailed and should include local temperature profiles and heat fluxes for building envelope components containing PCM. It is also highly recommended that all numerical algorithms used for PCM simulations were separately validated using results of the system-scale experiments, before their inclusion into whole-building energy models.

Please consider that improper validation of whole-building numerical tools, or sometimes an absence of program validation, usually leads to conflicting energy performance predictions, which yield a fall of confidence in numerical analysis. According to IEA Annex 23 (2011), in case of a large number of earlier numerical models used for PCM analysis, experimental validation was usually performed. In addition, analytical solutions for selected periodic problems had been often used to validate these models (Dutil et al. 2011). Unfortunately, this is not a common practice today. In many recent publications, an accuracy of numerically generated results is only qualitatively discussed, using comparisons with graphical results taken from other referenced publications. Interestingly enough, among more than 250 numerical models reviewed by Dutil et al. (2011), in only one case, the authors stated that the results were not "in good agreement with those found in the references." The above situation is a consequence of a widely reported lack of clear standard requirements and accuracy recommendations regulating development and validation of computer models used for analysis of PCM-enhanced building envelope systems.

A lack of respected, widely used, and well-validated whole-building energy models, capable to simulate PCM-enhanced building envelopes, leads to serious problems with justification of energy savings potential for building materials containing PCMs. It is important to mention that these facts affect in the same way both governmental energy policies for the buildings' sector, as well as manufactures of PCM-enhanced building products. That is why it might be beneficial to consider a collaborative international effort focused on development of special research projects focused on experimental validation of selected, most popular computer tools for whole-building energy analysis of PCM-enhanced building envelopes. These projects could be jointly financed by governments and industry in countries of largest market potential for the PCM building product applications.

6.6.4 Required Quality and Complexity of System-Scale and Whole-Building Computer Models

As stated in earlier sections, PCM-enhanced building envelopes are integral components of the building fabric and they are exposed to the same group of physical processes as the other parts of building envelope. That means, all computer models developed for analysis of the PCM-enhanced building envelopes need to be able to analyze these processes as well. Below, there are listed the most important factors effecting accuracy of thermal and energy performance modeling of building envelopes:

1. Thermal bridging and multi-dimensional heat transfer
2. Air leakage and air ventilation through internal cavities within building enclosures
3. Hygrothermal processes
4. Variable physical characteristics of materials.

Typically, system-scale thermal performance analysis of building envelopes requires at least consideration of multi-dimensional heat transfer effects generated by structural components.

In addition, due to common complex geometries of arrays of CPM containers (which are sometimes used in building envelopes), heat flow in their vicinities is not one-dimensional. When taking into account additional thermal bridging caused by structural and attachment components, it is obvious that, heat transfer in PCM-enhanced envelopes may have a strong three-dimensional character (Kośny et al. 2010a; Childs and Stovall 2012).

As air leakage through the building enclosure can have a significant impact on heating and cooling loads, it is crucial to include this process in whole-building analysis of PCM-enhanced building envelopes. The good news is that most of the widely used whole-building energy models such as EnergyPlus, ESP-r, or TRNSYS enable analysis of air leakage effects. The only problem is a usage of realistic levels of air leakage in whole-building energy analysis. More difficult to analyze are thermal effects generated within internal ventilated cavities. Several advanced roof and wall configurations containing vented air cavities combined with an application of PCMs have been developed and tested in the past (Kośny et al. 2010b). Numerical representation of thermal processes taking place in the above structures is quite challenging and usually requires CFD analysis combined with system-scale performance testing, before development of the computer algorithm (Stovall and Karagiozis 2004).

It is already well known that moisture transport processes within building envelopes may significantly affect the overall energy balance of the building (see Desta and Roels 2010). Bear in mind that in comparison with the older buildings, currently used building envelope components, due reduced transmission losses, are exposed to notably lower levels of energy transport. In these circumstances, an importance of potential moisture issues tends to rise not only because energy conservation requirements, but also because potential long-term durability concerns. The most important moisture sources are wetting during the construction phase, wind-driven rain water leakage, local moisture condensation, and moisture accumulation due to vapor diffusion and convection. In case of PCM-enhanced building envelopes, where very dynamic heat transfer processes take place, moisture accumulation problems may occur prevalent. Very often, these problems may be associated with the envelope cladding, weather resistive barriers, installation details, or just climatic conditions. That is why, a development of hygrothermal models capable to analyze PCMs would be very important for the future long-term performance predictions of the PCM-enhanced building envelopes. Several hygrothermal simulation tools have been developed in the last decades to analyze the moisture response of building envelopes (Pedersen 1990; Karagiozis 1993; Künzel 1994; Grunewald 1997). However, at the present time only, the WUFI family of hygrothermal models offers a capability for numerical analysis of PCM-enhanced building envelopes.[18]

Last but not least, variable physical characteristics of many materials used to construct building envelopes may play important role in accuracy of thermal and energy performance simulations. Some of these materials may have temperature-dependent physical properties in a similar way to PCMs. There is also a second group of materials having switchable physical characteristics. Best-known applications from this group are switchable roofing membranes and electrochromic or thermo-chromic windows with variable optical properties.[19,20] It is also well known that aging-based deterioration of performance characteristics of materials may influence long-term energy performance of all building envelopes, including those containing PCMs. Since PCMs can be integral parts of these technologies, it is recommended that, in such cases, all of the above factors related with variable physical characteristics are included into the system-scale and whole-building-scale computer models.

References

Agyenim F, Hewitt N, Eames P, Smyth M (2010) A review of materials, heat transfer and phase change problem formulation for latent heat thermal energy storage systems (LHTESS). Renew Sustain Energy Rev 14:615–628

Ahmad M, Bontemps A, Salle H, Quenard D (2006) Thermal testing and numerical simulation of a prototype cell using light wallboards coupling vacuum insulation panels and phase change materials. Energy Build 38(2006):673–681

Alderman RJ, Yarbrough DW (2007) Use of phase-change materials to enhance the thermal performance of building insulations. In: Proceedings of the 29th international thermal conductivity conference, Birmingham, 24–27 June 2007

[18]http://www.pcm-ral.de/en/products/planungs-und-simulationsprogramme.html.

[19]http://energy.gov/articles/making-smart-windows-smarter.

[20]http://newscenter.lbl.gov/news-releases/2013/08/14/raising-the-iq-of-smart-windows/.

Alexiades V, Solomon AD (1993) Mathematical modeling of melting and freezing processes. Hemisphere Publishing Corporation, New York

Almeida F, Zhang D, Fung AS, Leong WH (2011) Comparison of corrective phase change material algorithm with ESP-r simulation. In: Proceedings of the building simulation 2011, 12th conference of international building performance simulation association, Sydney

AL-Saadi SN, Zhai Z (2013) Modeling phase change materials embedded in building enclosure: a review. Renew Sustain Energy Rev 21:659–673

Antonopoulos KA, Tzivanidis C (1997) Numerical solution of unsteady three-dimensional heat transfer during space cooling using ceiling-embedded piping. Energy 22(1):59–67

Antonopoulos KA, Koronaki EP, Tzivanidis C (1998) Effects of building thermal mass during space cooling using ceiling-embedded piping. In: Proceedings of international conference ECOS '98, Nancy France

ASHRAE RP-1145 (2001) Modeling two and three-dimensional heat transfer through composite wall and roof assemblies in hourly simulation programs. Project report prepared by Enermodel Engineering Ltd. for American society of heating, refrigeration and air-conditioning engineers, Inc. (ASHRAE), Atlanta

ASHRAE (2004) ANSI/ASHRAE standard 140-2004. In: Standard method of test for the evaluation of building energy analysis computer programs. ASHRAE, Atlanta, p 151

Athienitis AK, Liu C, Hawes D, Banu D, Feldman D (1997) Investigation of the thermal performance of a passive solar test-room with wall latent heat storage. Build Environ 32 (5):405–410

Baghban MH, Hovde PJ, Gustavsen A (2010) Numerical simulation of a building envelope with high performance materials. In: Proceedings of the 2010 COMSOL conference, Paris

Balcomb JD, Kosiewicz CE, Lazarus GS, McFarland RD, Wray WO (1983) Passive solar design handbook vol.3, Los Alamos National Lab. ISBN 0-89553-106-2

Basu B, Date A (1988) Numerical modeling of melting and solidification problems—a review. Sadhana 13:169–213

Beasley DE, Clark JA (1984) Transient response of a packed bed for thermal energy storage. Int J Heat Mass Transf 27(9):1659–1669

Bedecarrats J, Castaing-Lasvignottes J, Strub F, Dumas J (2009a) Study of a phase change energy storage using spherical capsules; part I: experimental results. Energy Convers Manage 50 (10):2527–2536

Bedecarrats J, Castaing-Lasvignottes J, Strub F, Dumas J (2009b) Study of a phase change energy storage using spherical capsules; part II: numerical modeling. Energy Convers Manage 50 (10):2537–2546

Bentz DP, Turpin R (2007) Potential applications of phase change materials in concrete technology. Cement Concr Compos 29(2007):527–532

Bony J, Citherlet S (2007) Numerical model and experimental validation of heat storage with phase change materials. Energy Build 39(10):1065–1072

Brillouin M (1930) Sur Quelques Problèmes non Résolus de la Physique Mathématique Classique Propagation de la Fusion. Annales de l'institut Henri Poincaré 1(3):285–308

Brousseau P, Lacroix M (1998) Numerical simulation of a multi-layer latent heat thermal energy storage system. Int J Energy Res 22(1998):1–15

Buhl WF, Erdem E, Nataf J-M, Winkelmann F, Moshier MA, Sowell EF (1990) Advances in SPARK. In: Proceedings of the 3rd international conference on system simulation in buildings, Lawrence Berkeley laboratory report LBL-29419, Rev. 1, Liege, 3–5 Dec 1990

Cabeza LF, Castellón C, Nogués M, Medrano M, Leppers R, Zubillaga O (2007) Use of microencapsulated PCM in concrete walls for energy savings. Energy Build 39(2007):113–119

Caldwell J, Kwan YY (2004) Numerical methods for one-dimensional Stefan problems. Commun Numer Methods Eng 20:535–545

Campbell KR, Sailor DJ (2011) Phase change materials as thermal storage for high performance homes. In: Proceedings of the 2011 ASME international mechanical engineering congress & exposition, IMECE 2011, Denver, 11–17 Nov 2011

Cao Y, Faghri AA (1990) Numerical analysis of phase-change problems including natural convection. J Heat Transf 112:812–816

Carslaw HS, Jaeger JC (1959) Conduction of heat in solids. Clarendon Press, Oxford

Castell A, Medrano M, Castellón C, Cabeza LF (2009) Analysis of the simulation models for the Use of PCM in buildings. In: Proceedings of Effstock 2009—the 11th international conference on thermal energy storage, Stockholm, 14–17 June 2009

Castell A, Martorell I, Medrano M, Pérez G, Cabeza LF (2010) Experimental study of using PCM in Brick constructive solutions for passive cooling. Energy Build ENERG BLDG 42(4):534–540

Castellón C, Günther E, Mehling H, Hiebler S, Cabeza LF (2008) Determination of the enthalpy of PCM as a function of temperature using a heat-flux DSC—a study of different

Chen CR, Sharma A (2006) Numerical investigation of melt fraction of PCMs in a latent heat storage system. J Eng Appl Sci 1(2006):437–444

Chen CR, Sharma A, Tyagi SK, Buddhi D (2008a) Numerical heat transfer studies of PCMs used in a box-type solar cooker. Renew Energy 33(2008):1121–1129

Chen C, Guo H, Liu Y, Yue H, Wang C (2008b) A new kind of phase change material (PCM) for energy-storing wallboard. Energy Build 40:882–890

Childs KW, Stovall TK (2012) Use of phase change material in a building wall assembly: a case study of technical potential in two climates. In: Proceedings of international high performance buildings conference, Purdue University, West Lafayette, 16–19 July 2012

Christian JE, Kośny J (1995) Toward a national opaque wall rating label. In: Proceedings thermal performance of the exterior envelopes VI, ASHRAE ISBN 1-883413-29-X, Dec 1995

Clapeyron BPE, Lamé MM (1831) Mémoire sur la solidification par Refroidissement d'un globe liquid. Ann Chim Phys 47:250–256

Comini G, Del Guidice S, Lewis RW, Zienkiewicz OC (1974) Finite element solution of non-linear heat conduction problems with special reference to phase change. Int J Numer Methods Eng 8:613–624

Constantinescu M, Anghel EA, Buixaderas E, Pavel PM, Popa VT (2013) Shape-stabilized nano composite elements based on phase change materials (PCM)-Epoxy for hauses with low energy consumption. In: Proceedings of the 2nd international conference on sustainable energy storage in buildings, Trinity College, Dublin, Ireland, June 19–21, 2013

Costa M, Buddhi D, Oliva A (1998) Numerical simulation of a latent heat thermal energy storage system with enhanced heat conduction. Energy Convers Manage 39(3/4):319–330

Crank J (1981) How to deal with moving boundaries in thermal problems. In: Lewis RW, Morgan K, Zienkiewicz OC (eds) Numerical methods in heat transfer, Wiley, New York, pp 177–200

Crank J (1984) Free and moving boundary problems. Clarendon Press, Oxford

Crawley DB, Hand JW, Kummert M, Griffith BT (2005) Contrasting the capabilities of building energy performance simulation programs. A joint report, US DOE, University of Strathclyde UK, University of Wisconsin, Madison, USA and National Renewable Energy Renewable Laboratory, USA

Dalhuijsen A, Segal A (1986) Comparison of finite element techniques for solidification problems. Int J Numer Methods Eng. 23:1807–1829

Darkwa K, O'Callaghan PW (2005) Simulation of phase change drywalls in a passive solar building. Appl Therm Eng 26:853–858

Dentel A, Stephan W (2013) TRNSYS TYPE 399, Phase change materials in passive and active wall constructions: model description and implementing into TRNSYS. Georg Simon Ohm University of applied sciences, institute for energy and building. Version 1.5, Germany, May 2013. http://www.transsolar.com/__software/download/de/ts_type_399_en.pdf

Desta TZ, Roels S (2010) Experimental and numerical analysis of heat, air, and moisture transfer in a lightweight building wall. In: Proceedings of DOE, ASHRAE, ORNL, conference—thermal envelopes XI—thermal performance of the exterior envelopes of buildings, Clearwater, Dec 2010

Dhanusiya G, Rajakumar S (2013) Thermal analysis of PCM based building wall for cooling. Int J Sci Eng Technol Res (IJSETR) 2(4)

Douglas J (1957) A uniqueness theorem for the solution of a Stefan problem. Proc Am Math Soc 8:402–408

Drake JB, Geist GA, Morris MD, Solomon AD, Martin J, Tomlinson J (1987) Design of PCM enhanced passive solar structures, ORNL-6281, Oak Ridge National Laboratory, project report

Dutil Y, Rousse DR, Salahb NB, Lassue S, Zalewski S (2011) A review on phase-change materials: mathematical modeling and simulations. Renew Sustain Energy Rev 15(1):112–130

Egolf PW, Manz H (1994) Theory of phase change materials with and without mushy regions. Int. J Heat Mass Transf 37:2917–24

Evans GW (1951) A note on the existence of a solution to a problem of Stefan. Q Appl Math 9:185–193

Eyres NR, Hartree DR, Ingham J, Jackson R, Sarjant RJ, Wagstaff JB (1946) The calculation of variable heat flow in solids. Philos Trans Roy Soc London Ser A, Math Phys Sci 240:1–57

Fallahi A, Shukla N, Kośny J (2012) Numerical thermal performance analysis of PCMs integrated with residential attics. In: SimBuild-2012 conference, Madison

Fang Y, Medina M (2009) Proposed modifications for models of heat transfer problems involving partially melted phase change processes J ASTM Int 6(9). ISSN: 1546-962X

Feustel HE (1995) Simplified numerical description of latent storage characteristics for phase change wallboard. Lawrence Berkeley laboratory report LBL-36933, Berkeley

Feustel HE, Stetiu C (1997) Thermal performance of phase change wallboard for residential cooling application. Indoor environment program energy and environment division Lawrence Berkeley national laboratory, University of California Berkeley, CA 94720, USA. Report—LBL-38320 UC-1600

Fox L (1975) What are the best numerical methods? In: Ockendon JR, Hodgkins WR (eds) Moving boundary problems in heat flow and diffusion. Clarendon Press, Oxford, pp 210–241

Friedman A (1964) Partial differential equations of parabolic type. Prentice-Hall, New Jersey

Furzeland RM (1980) Comparative study of numerical methods for moving boundary problems. IMA J Appl Math 26:411–29

Gelissier A (2008) SPMCMP56 subroutine in ESP-r source standard code. http://espr.trac.cvsdude.com/espr/browser/branches/Joe_Clarke/src/esrubld/spmatl.F?rev=4529

Glück B (2006) Dynamisches Raummodell zur wärmetechnischen und wärmephysiologischen Bewertung. Rud. Otto-Meyer Umweltstiftung, Hamburg, Germany

Ghoneim AA, Klein SA, Duffie JA (1991) Analysis of collector-storage building walls using phase-change materials. Sol Energy 47:237–242

Gowreesunker BL, Tassou SA, Kolokotroni M (2012) Improved simulation of phase change processes in applications where conduction is the dominant heat transfer mode. Energy Build 47:353–359

Gowreesunker BL, Tassou SA, Kolokotroni M (2013) Coupled TRNSYS-CFD simulations evaluating the performance of PCM plate heat exchangers in an airport terminal building displacement conditioning system. Build Environ 65:132–145

Grau K, Rode K (2006) A model for air flow in ventilated cavities implemented in a tool for whole-building hygrothermal analysis. IEA annex 41 meeting, Lyon, Oct 2006

Grau K, Rode C (2007) A model for advection heat and moisture flows implemented in a program for whole-building hygrothermal simulation. IEA Annex 41 meeting, April 2007, Florianopolis, Brazil

Grunewald J (1997) Diffusiver und konvektiver stoff- und energietransport in kapillarporösen baustoffen. Ph.D. Thesis, Institut für Bauklimatik, Technische Universität Dresden, Dresden

Günther E, Mehling H, Hiebler S (2007) Modeling of subcooling and solidification of phase change materials. Modell Simul Mater Sci Eng 15(8)

Günther E, Hiebler S, Mehling H, Redlich R (2009) Enthalpy of phase change materials as a function of temperature: required accuracy and suitable measurement methods. Int J Thermophys 30:1257–1269

Haavi T, Gustavsen A, Cao S, Uvsløkk S, Jelle BP (2011) Numerical simulations of a well-insulated wall assembly with integrated phase change material panels—comparison with hot box experiments. In: Proceedings of the international conference on sustainable systems and the environment, Sharjah

Halford CK, Boehm RF (2007) Modeling of phase change material peak load shifting. Energy Build 39:298–305

Hashemi HT, Sliepcevich CM (1967) A numerical method for solving two-dimensional problems of heat conduction with change of phase. Chem Eng Prog Symp Ser 63: 34–41

Hawes DW, Banu D, Feldman D (1990) Latent heat storage in concrete. Sol Energy Mater 21:61–80

Hawes DW, Feldman D (1992) Absorption of phase change materials in concrete. Solar Energy Mat Solar Cells 27(2):91–101

Hawlander M, Uddin M, Zhu H (2002) Encapsulated phase change materials for thermal energy storage: experiments and simulation. Int J Energy Res 26(2):159–171

Heim D (2010) Isothermal storage of solar energy in building construction. Renew Energy 35:788–96

Heim D, Clarke JA (2004) Numerical modeling and thermal simulation of PCM–gypsum composites with ESP-r. Energy Build 36(2004):795–805

Hens H (1996) IEA annex 24: heat, air and moisture transfer through new and retrofitted insulated envelope parts. Task 1: modelling. Acco, Leuven, Belgium

Hill JM (1987) One-dimensional Stefan problems: an introduction. Longman Scientific Technical, Harlow

Hoffmann S (2006) Numerische und experimentelle Untersuchung von Phasenübergangsmaterialien zur Reduktion hoher sommerlicher Raumtemperaturen‖. PhD Thesis. Bauhaus-Universität Weimar, Germany

Hu H, Argyropoulos SA (1996) Mathematical modelling of solidification and melting: a review. Model Simul Mat Sci Eng 4:371–96

Ibanez M, Lazaro A, Zalba B, Cabeza LF (2005) An approach to the simulation of PCMs in building applications using TRNSYS. Appl Therm Eng 25(2005):1796–1807

Idelsohn S, Storti M, Crivelli L (1994) Numerical methods in phase-change problems. Arch Comput Methods Eng 1:49–74

IEA Annex 23, Dutil Y, Rousse D, Lassue S, Zalewski L, Joulin A, Virgone J, Kuznik F, Johannes K, Dumas J-P, Bédécarrats J-P, Castell A, Cabeza LF (2011) Modeling phase change materials behaviour in building applications: selected comments. In: Proceedings of world renewable energy congress, Linköping, Sweden, 8–13 May 2011

Jayalath A, Mendis P, Aye L, Ngo T (2012) Thermal performance of concrete with PCM. In: Proceedings of 2012 ICSBE—international conference on sustainable built environment, Kandy, Dec 2012

Jokisalo J, Lamberg P, Siren K (2000) Thermal simulation of PCM structures with TRNSYS. In: Proceedings of the 8th international conference on thermal energy storage, Stuttgart, Germany

Kalyanova O, Heiselberg P, Felsmann C, Poirazis H, Strachan P, Wijsman A (2009) An empirical validation of building simulation software for modeling of double-skin facade. In: Proceedings of IBPSA BS2009, University of Strathclyde, Glasgow, 27–30 July 2009, pp 1107–1114

Kalz D, Pfafferott J, Schossig P, Herkel S (2006) Thermally activated building systems using phase-change-materials. In: Proceedings of the EuroSun 2006 conference, international solar energy society—ISES, Glasgow, 27–30 June 2006

Kalz D, Pfafferott J, Schossig P, Herkel S (2007) Thermoaktive Bauteilsysteme mit integrierten Phasenwechselmaterialien - eine Simulationsstudie. Bauphysik 29(1):27–32

Karagiozis A (1993) Overview of the 2D hygrothermal heat-moisture transport model LATENITE. Institute for Research in Construction, National Research Council, Canada

Katsourinis D, Founti M, Romero-Sanchez MD, Lopez-Buendia AM (2010) On the computational modelling of thermal energy storage in natural stone treated with PCMs. In: Proceedings of the 9th IIR conference on phase change materials and slurries for refrigeration and air conditioning, Sofia, Bulgaria, Oct 2010

Kedl RJ (1990) Conventional wallboard with latent heat storage for passive solar applications. 25th Intersociety energy conversion engineering conference, Reno, NV, 12–17

Kelly NJ (1998) Towards a design environment for building-integrated energy systems: the integration of electrical power flow modelling with building simulation, PhD Thesis, Glasgow: University of Strathclyde, UK

Kendrick C, Walliman N (2007) Removing unwanted heat in lightweight buildings using phase change materials in building components: simulation modelling for PCM plasterboard. Archit Sci Rev 50(3):265–273

Kissock KJ, Hannig JM, Thomas IW (1998) Testing and simulation of phase change wallboard for thermal storage in buildings. In: Morehouse, Hogan (eds). Proceedings of 1998 international solar energy conference, June 14–17, Albuquerque, N. M. USA, American Society of Mechanical Engineers, New York

Kissock KJ, Limas S (2006) Diurnal load reduction through phase-change building components. ASHRAE Transactions 112(1):509–517

Knoll DA, Keyes DE (2004) Jacobian-free Newton–Krylov methods: a survey of approaches and applications. J Comput Phys 193:357–397

Klein SA, Beckman WA, Mitchell JW, Duffie JA, Duffie NA, Freeman TL, Mitchell JC, Braun JE, Evans BL, Kummer JP, Urban RE, Fiksel A, Thornton JW, Blair NJ, Williams PM, Bradley DE, McDowell TP, Kummert M, Duffy MJ (2011) TRNSYS 17—A TRaNsient SYstem simulation program, user manual. University of Wisonsin-Madison, USA

Korjakins A, Kazjonovs J, Bajāre D, Ozoliņš D, Jakovičs A (2013) Simulation of PCM behaviour in buildings for latvia climate conditions. In: Proceedings of sustainable energy storage in buildings conference: sustainable energy storage in buildings, Ireland, Dublin, 19–21 June 2013. Dublin, Ireland, pp 211–217

Koschenz M, Lehmann B (2000) Handbuch Thermoaktive Bauteilsysteme. TABS, EMPA Abteilung Energiesysteme/Haustechnik, CH-8600, Düsseldorf, Germany

Kośny J (2008) Field testing of cellulose fiber insulation enhanced with phase change material. Oak Ridge National Laboratory report-ORNL/TM-2007/186, Oak Ridge, Sept 2008

Kośny J, Christian JE (1995) Reducing the uncertainties associated with using the ASHRAE ZONE method for R-value calculations of metal frame walls. ASHRAE Trans 101:2

Kośny J, Kossecka E (2000) Computer modeling of complex wall assemblies—some accuracy problems. In: Proceedings of the international building physics conference, Eindhoven, The Netherlands

Kośny J, Kossecka E (2002) Multidimensional heat transfer through complex building envelope assemblies in energy simulation programs. Energy Build 34:445–454

Kośny J, Kossecka E (2013) Understanding a potential for application of phase-change materials (PCMs) in building envelopes. In: Proceedings of the ASHRAE 2013 winter conference, ASHRAE transactions, vol 119. Part 1. DA-13-001, Dallas, TX, 26–30 Jan 2013

Kośny J, Kossecka E, Desjarlais AO, Christian JE (1998) Dynamic thermal performance of concrete and masonry walls. In: Proceedings of DOE, ASHRAE, ORNL conference—thermal envelopes VII, Clearwater, Florida, Dec 1998

Kośny J, Yarbrough DW, Syed AM (2007a) Performance of thermal insulation containing microencapsulated phase change material. In: ITCC29 and ITES17 conference, Birmingham, 24–27 June 2007

Kośny J, Yarbrough DW, Miller W, Childs P, Syed AM (2007b) Thermal performance of PCM-enhanced building envelope systems. In: Proceedings of X conference—thermal performance of the exterior envelopes of buildings, Clearwater, Florida, Dec 2007

Kośny J, Yarbrough DW, Miller WA (2008a) Use of PCM-enhanced insulations in the building envelope. J Build Enclosure Des (Summer/Fall)

Kośny J, Yarbrough DW, Miller WA (2008b) Experimental analysis of dynamic thermal performance of building shell components containing phase change materials. Building Enclosure Technology and Environment Council (BETEC) of the National Institute of Building Sciences BEST1 Conference, 10 June 2008

Kośny J, Stovall TK, Yarbrough DW (2009a) Dynamic heat flow measurements to study the distribution of phase-change material in an insulation matrix. In: Proceedings of the 2009 international thermal conductivity conference (ITCC) and the international thermal expansion symposium (ITES), Pittsburgh, Aug 29–Sept 2 2009

Kośny J, Kossecka E, Yarbrough DW (2009b) Use of a heat flow meter to determine active PCM content in an insulation. In: Proceedings of the 2009 international thermal conductivity conference (ITCC) and the international thermal expansion symposium (ITES), Pittsburgh, Aug 29–Sept 2 2009

Kośny J, Yarbrough DW, Riazzi T, Leuthold D, Smith JB, Bianchi M (2009c) Development and testing of ignition resistant microencapsulated phase change material. In: Proceedings of the Effstock 2009—the 11th international conference on thermal energy storage, Stockholm, Sweden, 14–17 June 2009

Kośny J, Stovall T, Yarbrough DW (2009d) Dynamic heat flow measurements to study the distribution of phase-change material in an insulation matrix. In: Proceedings of the 2009 international thermal conductivity conference (ITCC) and the international thermal expansion symposium (ITES), Pittsburgh, Aug 29–Sept 2 2009

Kośny J, Stovall T, Shrestha S, Yarbrough D (2010a) Theoretical and experimental thermal performance analysis of complex thermal storage membrane containing bio-based phase-change material (PCM). In: Proceedings of DOE, ASHRAE, ORNL conference -thermal envelopes XI—thermal performance of the exterior envelopes of buildings, Clearwater, Florida, Dec 2010

Kośny J, Miller W, Zaltash A (2010b) Dynamic thermally-disconnected building envelopes—a new paradigm for walls and roofs in low energy buildings. In: Proceedings of DOE, ASHRAE, ORNL conference—thermal envelopes XI—thermal performance of the exterior envelopes of buildings, Clearwater, Florida, Dec 2010

Kośny J, Yarbrough D, Miller W, Shrestha S, Kossecka E, Lee E (2010c) Numerical and experimental analysis of building envelopes containing blown fiberglass insulation thermally enhanced with phase change material (PCM). In: Proceedings of the 1st Central European symposium on building physics, Cracow, Poland, Sept 2010

Kośny J, Biswas K, Miller WA, Kriner S (2012a) Field thermal performance of naturally ventilated solar roof with PCM heat sink. Solar Energy 86(9):2504–2514

Kośny J, Kossecka E, Brzezinski B, Tleoubaev A, Yarbrough D (2012b) Dynamic thermal performance analysis of fiber insulations containing bio-based phase change materials (PCMs). Energy Build 52:122–131

Kośny J, Fallahi A, Shukla N (2013) Cost analysis of simple phase change material-enhanced building envelopes in Southern US Climates, DOE building America report, submitted to the US department of energy building technology program, Jan 2013. http://www.eereblogs.energy.gov/buildingenvelope/file.axd?file=2013%2F1%2Fcost_analysis_simplephase_change_material-enhanced_building_envelopes_southernclimates.pdf

Kossecka E, Kośny J (1996) Relations between structural and dynamic thermal characteristics of building walls. In: Proceedings of 1996 international symposium of CIB W67 energy and mass flow in the life cycle of buildings, Vienna, 4–10 Aug 1996, pp 627–632

Kossecka E, Kośny J (1997) Equivalent wall as a dynamic model of the complex thermal structure. J Therm Insul Build Envelopes 20:249–268

Kossecka E, Kośny J (1998) Effect of insulation and mass distribution in exterior walls on the dynamic thermal performance of a whole buildings. In: Proceedings of DOE, ASHRAE, ORNL conference—thermal envelopes VII, Clearwater, Florida, Dec 1998

Kossecka E, Kośny J (2001) Modelling three-dimensional heat transfer through composite wall assemblies. In: Proceedings of the VIII conference building physics in theory and practice, Technical University of Lodz, Poland, June 2001, pp 313–320

Kossecka E, Kośny J (2008) Hot box testing of building envelope assemblies; a simplified procedure for estimation of minimum time of the test. J Test Eval (JTE) 36(3). ISSN: 0090-3973

Kossecka E, Kośny J (2009) Dynamiczna Metoda Pomiaru Zawartosci Materialu Fazowo-Zmiennego w Izolacji (Dynamic method of analyzing PCM content in thermal insulations). In: Proceedings of the XII conference building physics in theory and practice, £ódz, Poland, 23–26 June 2009. ISSN 1734-4891

Kossecka E, Kośny J (2010) Thermal balance of a wall with PCM-enhanced thermal insulation. In: Proceedings of the 1st central European symposium on building physics, Cracow, Poland, Sept 2010

Künzel H (1994) Simultaneous heat and moisture transport in building components. One- and two-dimensional calculation using simple parameters. Ph.D. Thesis, Fraunhofer institute of building physics, Universität Stuttgart, Germany

Kunzel H (1995) Simultaneous heat and moisture transport in building components. Ph.D. thesis. Wiley IRB Verlag, Stuttgart

Kuznik F, Virgone J (2009) Experimental investigation of wallboard containing phase change material: data for validation of numerical modeling. Energy Build 41(5):561–570

Kuznik F, Virgone J, Noel J (2007) Optimization of a phase change material wallboard for building use. Appl Therm Eng 28:1291–1298

Kuznik F, Virgone J, Johannes K (2010) Development and validation of a new TRNSYS type for the simulation of external building walls containing PCM. Energy Build 42(7):1004–1009

Lamberg P (2003) Mathematical modeling and experimental investigation of melting and solidification in a finned phase change material storage. Helsinki University of Technology. Department of Mechanical Engineering, Helsinki, Finland

Lamberg P, Jokisalo J, Sirén K (2000) The effects on indoor comfort when using phase change material with building concrete products. In: Proceedings of healthy buildings, SIY indoor air information OY, vol 2. pp 751–756

Lamberg P, Lehtiniemi R, Henell AM (2004) Numerical and experimental investigation of melting and freezing processes in phase change material storage. Int J Therm Sci 43(Compendex): 277–287

Lazaro A et al. (2013) Intercomparative tests on phase change materials characterization with differential scanning calorimeter. Appl Energy 109:415–420

Lee T, Hawes DW, Banu D, Feldman D (2000) Control aspects of latent heat storage and recovery in concrete. Solar Energy Mater & Solar Cells 62:217–237

Lee AS, Hittle DC (2005) Encapsulated paraffin wax in floor tiles for thermal energy storage. In: Proceedings of the clima 2005 conference, Lausanne, Switzerland

Lemmon E (1981) Multidimensional integral phase change approximations for finite element conduction codes. In: Lewis RW, Morgan K, Zienkiewicz OC (eds) Numerical methods in heat transfer. Wiley, New York, pp 201–213

Liu H, Awbi HB (2008) Performance of phase change material boards under natural convection. Build Environ 44:1788–1793

Lomas KJ, Eppel H, Martin CJ, Bloomfield DP (1997) Empirical validation of building energy simulation programs. Energy Build 26(1997):253–275

Lunardini VJ (1981) Heat transfer in cold climates. Van Nostrand Reinhold, New York

Medina MA, Stewart R (2008) Phase-change frame walls (PCFWs) for peak demand reduction, load shifting, energy conservation and comfort. In: Proceedings of 16th symposium on improving building systems in hot and humid climates, Plano, TX, 16–17 Dec 2008

Miller W, Kośny J (2008) Next generation roofs and attics for residential homes. 2008 ACEEE summer study on energy efficiency in buildings, Pacific Grove, California, 17–22 Aug 2008

Miller W, Kośny J, Shrestha S, Christian J, Karagiozis A, Kohler C, Dinse D (2010) Advanced residential envelopes for two pair of energy-saver homes. In: Proceedings of 2010 ACEEE summer study on energy efficiency in building, Pacific Grove, California, 15–20 Aug 2010

Minkowycz WJ, Sparrow EM, Murthy JY (eds) (2006) Handbook of numerical heat transfer, 2nd edn. Wiley, New York

Mitalas GP, Stephenson DG (1971) Calculation of heat conduction transfer functions for multi-layer slabs. National research council of Canada, division of building research, ASHRAE 1971, Atlanta

Morgan K, Lewis RW, Zienkiewicz OC (1978) An improved algorithm for heat conduction problems with phase change. Int J Numer Methods Eng 12:1191–1195

Muhieddine M, Canot E, March R (2009) Various approaches for solving problems in heat conduction with phase change. Int J Finite 6

Noel J, Roux J-J, Schneider PS (2001) CoDyBa, a design tool for buildings performance simulation. In: Proceedings of the 7th international IBPSA conference, Rio de Janeiro, Brazil, 13–15 Aug 2001

Ogoh W, Groulx D (2010) Stefan's problem: validation of a one-dimensional solid-liquid phase change heat transfer process. In: Proceedings of the COMSOL conference, 2010 Boston, USA

Ozısık MN (1993) Heat conduction, 2nd edn. Wiley, New York

Patankar SV (1980) Numerical heat transfer and fluid flow. Hemisphere Publishing Corporation, New York

Pedersen CR (1990) Combined heat and moisture transfer in building constructions. Ph.D. Thesis, Technical University of Denmark, Report no. 214

Pedersen CO (2007) Advanced zone simulation in EnergyPlus: incorporation of variable properties and phase change materials (PCM) capability. In: Proceedings of the 10th international IBPSA conference, Beijing, China, pp. 1341–1345

Petrie TW, Childs KW, Childs PW, Christian JE, Shramo DJ (1997) Thermal behavior of mixtures of perlite and phase change material in a simulated climate. In: Graves RS, Zarr RR (eds) Insulation materials: testing and applications: third volume, ASTM STP 1320. ASTM—American Society for Testing and Materials, West Conshohocken, pp 180–194

Pham QT (1985) A fast, unconditionally stable finite-difference scheme for heat conduction with phase change. Int J Heat Mass Transf 28:2079–2084

Pham QT (2006) Modeling heat and mass transfer in frozen foods: a review. Int J Refrig 29:876–888

Poirier DJ (1988) On numerical methods used in mathematical modeling of phase change in liquid metals. J Heat Transf 110:562–570

Ponechal R (2009) An experimental study and simulations of phase change materials in an office thermal environment. Slovak J Civ Eng 3:24–29

Poulad ME, Fung AS, Naylor D (2011) Effects of convective heat transfer coefficient on the ability of PCM to reduce energy demand. In: Proceedings of building simulation 2011: 12th conference of international building performance simulation association, Sydney, 14–16 Nov

Regin AF, Solanki SC, Saini JS (2008) Heat transfer characteristics of thermal energy storage system using PCM capsules: a review. Renew Sustain Energy Rev 12:–2438–2458

Regin AF, Solanki S, Saini J (2009) An analysis of a packed bed latent heat thermal energy storage system using PCM capsules: numerical investigation. Renew Energy 34(7):1765–1773

Riza M (2007) Thermal characteristics of eutectic mixture of capric-lauric acids as phase change material (PCM) in gypsum board. Jurnal Teknologi Proses, Media Publikasi Karya Ilmiah Teknik Kimia 6(1):31–38. ISSN 1412-7814

Rode C, Grau K, Sørensen LC, Christoffersen LD (2002) Application of a computer model for integrated hygrothermal building analysis. In: Proceedings of 11th symposium for building physics, Dresden, Germany, Sept 2002

Rode C, Salonvaara M, Ojanen T, Simonson C, Grau K (2003) Integrated hygrothermal analysis of ecological buildings. In: Proceedings of the 2nd international building physics conference, Leuven, Belgium, 14–18 Sept 2003

Romanowska A, Jablonski M, Klemm P (1998) Kompozyt Ceramiczny o Podwyzszonej Akumulacji Ciepła. In: Proceeding of the IV polish engineering and technological conference Energodom '98, Krakow, Poland, pp 164–171 (in Polish)

Romero-Sánchez MD, Guillem-López C, López-Buendía AM, Stamatiadou M, Mandilaras I, Katsourinis D, Founti M (2012) Treatment of natural stones with phase change materials: experiments and computational approaches. Appl Therm Eng 48:136–143

Rose J, Lahme A, Christensen NU, Heiselberg P, Hansen M, Grau K (2009) Numerical method for calculating latent heat storage in constructions containing phase change material. In: Proceedings of the 11th international IBPSA conference, Glasgow, Scotland, pp 400–407

Rozanna D, Salmiah A, Chuah TG, Medyan R, Thomas Choong SY, Sa'Ari M (2005) A study on thermal characteristics of phase change material (PCM) in gypsum board for building application. J Oil Palm Res 17:41–46

Rubinstein LI (1971) The Stefan problem. Amer Math Soc (Translated from Russian)

Sadasivam S, Almeida F, Zhang D, Fung AS (2011) An iterative enthalpy method to overcome the limitations in ESP-r's PCM solution algorithm. ASHRAE Trans 117:100–107

Salyer IO, Sircar AK, Kumar A (1995) Advanced phase change materials technology: evaluation in lightweight solite hollow-core building blocks. In: Proceedings of the 30th intersociety energy conversion engineering conference, Orlando, FL, USA. 217–224

Scarpa M, Grau K, Olesen BW (2009) Development and validation of a versatile method for the calculation of heat transfer in water-based radiant systems. In: Proceedings of IBPSA BS2009, University of Strathclyde, Glasgow, 27–30 July 2009, pp 688–695

Schossig P, Henning HM, Raicu A, Haussmann T (2003) Mikroverkapselte phasenwechselmaterialien in baustoffen. In: Proceedings of 13th symposium thermische solarenergie OTTI Technologie-Kolleg, Staffelstein, Germany. 489–494

Schossig P, Henning HM, Gschwander S, Haussmann T (2005) Micro-encapsulated phase-change materials integrated into construction materials. Sol Energy Mater Sol Cells 89(2–3):297–306

Schranzhofer H, Puschnig P, Heinz A, Streicher W (2006) Validation of a TRNSYS simulation model for PCM energy storages and PCM wall construction elements. In: Proceedings of ECOSTOCK 2006—10th international conference on thermal energy storage. Pomona, NJ

Shamsundar N, Sparrow EM (1975) Analysis of multidimensional conduction phase change via the enthalpy model. J Heat Transf 97:333–340

Sharma A, Sharma SD, Buddhi D, Won LD (2006) Effect of thermo physical properties of heat exchanger material on the performance of latent heat storage system using an enthalpy method. Int J Energy Res 30(2006):191–201

Sharma A, Tyagi VV, Chen CR, Buddhi D (2009) Review on thermal energy storage with phase change materials and applications. Renew Sustain Energy Rev 13(2009):318–345

Shrestha M (2012) PCM application-effect on energy use and IA temperature. Norwegian university of science and technology, Faculty of architecture and fine arts. M.Sc. thesis—sustainable architecture, 01 July 2012

Shukla NC, Fallahi A, Kośny J, Harasim S, Blair C (2012) Aerogel for thermal insulation of interior wall retrofits in cold climates. The building enclosure science and technology (BEST), AIA conference, Atlanta, April 2012

Sigalas G (2011) Computational optimization of passive use of phase change materials in lightweight low-energy houses. Graduation project report for the sustainable energy technology master program, Department of the built environment, Eindhoven University of Technology, The Netherlands. Sept 2011

Solomon AD (1979) An easily computable solution to a two-phase stefan problem. Sol Energy 23 (6):525–528

Solomon AD, Wilson DG, Alexiades V (1984) The quasi-stationary approximation for the stefan problem with a convective boundary condition. Int J Math Matm Sci 7(3):549–563

Solomon AD, Wilson DG, Alexiades V (1986) An approximate analysis of the formation of a buoyant solid sphere in a supercooled melt, ORNL-6212 Oak Ridge National Laboratory, project report

Solomon A, Alexiades V, Wilson DG (1990) The mathematical modeling of melting and freezing processes. Hemisphere Press

Spaeh GC (2012) Thermal storage with PCM vs. Thermal insulation in warm and hot climates. In: Proceedings of the 27th RCI international convention and show, Dallas, TX, 15–20 March 2012

Stefan J (1889) Uber Einige Probleme der Theorie der Warmeleitung. Sitzungsber Wiener Akad Math Naturwiss Abt 98(1889):473–484

Stefanescu DM (2009) Science and engineering of casting solidification, 2nd edn. Springer, Berlin

Stetiu C, Feustel HE (1996) Phase change wallboard as an alternative to compressor cooling in California residences. In: Proceedings of the ACEEE 1996 summer study on energy efficiency in buildings, vol 10. Washington, D.C., American Council for an Energy-Efficient Economy, pp 157–163

Stetiu C, Feustel HE, Winkelmann FC (1995) Development of a model to simulate the performance of hydronic radiant cooling ceilings. ASHRAE Trans 101(2):730–743

Stovall TK, Karagiozis AN (2004) Airflow in the ventilation space behind a rain screen wall. In: Proceedings of IX conference—thermal performance of the exterior envelopes of buildings, Clearwater, Florida, Dec 2004

Stovall TK, Tomlinson JJ (1995) What are the potential benefits of including latent storage in common wallboard? J Sol Energy Eng 117

Strachan PA (2000) 'ESP-r: Summary of validation studies. ESRU technical report, University of Strathclyde, Glasgow, UK

Stritih U, Butala V (2010) Experimental investigation of energy saving in buildings with PCM cold storage. Int J Refrig 33(8):1676–1683

Stritih U, Medved S (1994) Use of phase change materials in the wall with TIM. Strojniski Vestnik/J Mech Eng 40:155–160

Stritih U, Novak P (1996) Solar heat storage wall for building ventilation. Renewable Energy 8:268–71

Susman G, Dehouche Z, Cheechern T, Craig S (2011) Tests of prototype PCM 'sails' for office cooling. Appl Therm Eng 31:717–726

Swaminathan CR, Voller VR (1993) On the enthalpy method. Int J Numer Meth Heat Fluid Flow 3(1993):233–244

Tabares-Velasco PC (2012) Energy impacts of nonlinear behavior of PCM when applied into building envelope. In: Proceedings of the ASME 2012 6th international conference on energy sustainability and 10th fuel cell science, engineering and technology conference, San Diego, California, 23–26 July 2012

Tabares-Velasco PC, Christensen C, Bianchi M (2012) Verification and validation of energyplus phase change material model for opaque wall assemblies. Build Environ 54:186–196. NREL Report No. JA-5500-54019, August 2012

Tardieu A, Behzadi S, Chen JJ, Farid M (2011) Computer simulation and experimental measurements for an experimental PCM-impregnated office building. In: Proceedings of the building simulation 2011: 12th conference of international building performance simulation association, vol. 1, pp 56–63

Tikhonov AN, Samarskii AA (1963) Equations of mathematical physics. Pergamon Press, UK

Tomlinson J, Heberle D (1990) Analysis of wallboard containing a phase change material. In: Proceedings of the 25th intersociety energy conversion engineering conference, Published by IEEE, vol 4. Reno, NV, USA, 12–17 Aug 1990

Ubiñas RE, Cronemberger J, Vega Sánchez S, García Santos A (2009) Performance of passive application of PCM in Spain. In: Proceedings of: I Congreso Internacional de Investigación en Edificación, Madrit, Spain, 24–26 June

Verma P, Varun P, Singal SK (2008) Review of mathematical modeling on latent heat thermal energy storage systems using phase-change material. Renew Sustain Energy Rev 12:999–1031

Virgone J, Noël J, Reisdorf R (2009) Numerical study of the influence of the thickness and melting point on the effectiveness of phase change materials: application to the renovation of a low inertia school. In: Proceedings of the 11th international IBPSA conference. Glasgow, Scotland, 27–30 July 2009

Voller VR (1990) Fast implicit finite-difference method for the analysis of phase change problems. Numer Heat Transf Part B Fundam 17:155–169

Voller VR (1997) An overview of numerical methods for solving phase change problems. In: Minkowycz WJ, Sparrow EM (eds) Advances in numerical heat transfer. Taylor & Francis, UK, pp 341–380

Voller VR, Cross M (1983) Application of control volume enthalpy methods in the solution of Stefan problems. In: Lewis RW, Morgan K, Johnson JA, Smith R, Smith WR (eds) Computational techniques in heat transfer, vol 1. Pineridge Press, Swansea, pp 245–275

Voller VR, Swenson JB, Kim W, Paola C (2006) An enthalpy method for moving boundary problems on the earth's surface. Int J Numer Methods Heat and Fluid Flow 16(6):41–54

Ye WB, Zhu DS, Wang N (2011) Numerical simulation on phase-change thermal storage/release in a plate-fin unit. Appl Therm Eng 31:3871–3884

Zalba B, Marın JM, Cabeza LF, Mehling H (2003) Review on thermal energy storage with phase change: materials, heat transfer analysis and applications. Appl Therm Eng 23:251–283

Zeng X, Faghri A (1994a) Temperature-transforming model for binary solid–liquid phase-change problems. Part I: Mathematical Modeling and Numerical Methodology. Numer Heat Transf Part B Fundam 25:467–80

Zeng X, Faghri A (1994b) Temperature-transforming model for binary solid-liquid phase-change problems. Part II: numerical simulation. Numer Heat Transf Part B: Fundam 25:481–500

Zhang Y, Jiang Y, Jiang Y (1999) A simple method, the t-history method, of determining the heat of fusion, specific heat and thermal conductivity of phase-change materials. Measur Sci Technol 10:201–205

Zhang D, Li Z, Zhou J, Wu K (2004) Development of thermal energy storage concrete. Cem Concr Res 34(Compendex):927–34

Zhanhua M, Yuwen Z (2006) Solid velocity correction schemes for a temperature transforming model for convection phase change. Int J Numer Meth Heat Fluid Flow 16:204–225

Zhou G, Zhang Y, Wang X, Lin K, Xiao W (2007) An assessment of mixed type PCM-gypsum and shape-stabilized PCM plates in a building for passive solar heating. Sol Energy 81:1351–1360

Zhou G, Zhang Y, Kunping L, Xiao V (2008) Thermal analysis of a direct-gain room with shape-stabilized PCM plates. Renew Energy 33:1228–1236

Zhuang C-L, Deng A-Z, Chen Y, Li S-B, Zhang H-Y, Fan G-Z (2010) Validation of Veracity on simulating the indoor temperature in PCM light weight building by EnergyPlus. In: Proceedings of the 2010 international conference on life system modeling and intelligent computing, and 2010 international conference on intelligent computing for sustainable energy and environment: part I, LSMS/ICSEE'10. Springer, Berlin, pp 486–496

Zöller A, Jung M, Schmidt M, Brenner T, Gatzka B, Schossig P, Haussmann T (2008) Abschlussbericht zum Verbundforschungsvorhaben, Aktive Latentspeichersysteme für Gebäude. Fraunhofer ISE, Gemeinsamer Bericht aller Projektteilnehmer in den Teilprojekten 0327370G-K, 0327370S

Chapter 7
Basic Building Envelope Products Containing PCM and Related Patents

7.1 Introduction

For variety of climatic conditions, PCM-enhanced building envelope products are widely considered as prospective building technologies which can help in close future to achieve peak-hour load and HVAC energy consumption reductions. As discussed in previous chapters, notable energy savings can be accomplished in residential and commercial buildings, by means of variety of applications utilizing external envelopes, as well as internal building fabric technologies having capability of sensible and latent heat accumulation. Building components with phase transition ability are allowing large amounts of heat to be stored or released due to the latent heat of fusion or solidification. It has been also demonstrated that with the market arrival of different forms of PCMs implemented in gypsum boards, plaster, concrete, thermal insulation, building membranes, or other board products, latent heat storage can control the building thermal response even in case of lightweight structures.

The heat capacity of lightweight construction buildings can be significantly increased by including latent heat storage materials into the building envelope surfaces in addition to the interior building components. The use of PCMs in construction materials in a passive way is relatively not new. There were several moderately successful attempts in the 1970s and 1980s to utilize different types of organic and inorganic PCMs to reduce peak loads and heating and cooling energy consumption (Balcomb 1979; Hawes et al. 1990; Salyer et al. 1985). These investigations focused on impregnating concrete, gypsum, or ceramic masonry with salt hydrates or paraffinic hydrocarbons.

Several researchers have investigated methods for impregnating gypsum wallboard and other architectural materials with PCMs (Salyer and Sircar 1989; Shapiro et al. 1987; Babich et al. 1994; Banu et al. 1998). First application of this type tested by Oak Ridge National Laboratory, USA, were PCMs installed directly on interior building surfaces (Drake 1978; Tomlinson and Herbele 1990; Tomlinson et al. 1992;

J. Kośny, *PCM-Enhanced Building Components*, Engineering Materials and Processes, DOI 10.1007/978-3-319-14286-9_7

Stovall and Tomlinson 1995). Masonry blocks or other ceramic building materials impregnated with a PCM have been also used in building construction for decades.

An incorporation of PCMs in concrete or ceramic masonry can be a complex technological process. Please notice that great majority of PCMs used in building envelope applications need to go through a complete phase transition during 24-h time periods in order to be fully effective. This is why it is crucial that the temperature in the location where PCM is installed is circulating above and below the PCM functional temperature range. In addition, there are several durability concerns known for PCM-enhanced masonry and ceramic products. Some of these concerns include volume changes during melting and freezing, slow heat transfer rates of inorganic PCM products, problems of PCM leakage, and adverse effects on the physical properties of the PCM carrier materials. One of the simplest methods for addition of PCM to the concrete or gypsum mixtures is the usage of microencapsulated PCM. Different manufacturing techniques, followed with thermal performance of gypsum wallboards and concrete blocks containing PCM, have been discussed by Kudhair and Farid (2004), Zalba et al. (2003), Hauer et al. (2005), and Cabeza et al. (2008). In the past, variety of PCMs of different phase transition characteristics had been considered for building applications. However, today, after several decades of laboratory testing and full-scale trials, the most popular are microencapsulated paraffinic hydrocarbons. Also, PCM-enhanced gypsum boards are widely considered as the most popular building materials using latent heat thermal storage.

In addition, in early 2000s, an innovative concept of the PCM-enhanced building envelope insulation was proposed by Oak Ridge National Laboratory, USA. It is based on incorporating the microencapsulated PCM into different types of fiber insulations and plastic foams (Kośny et al. 2006, 2007). This application is counter intuitive, because it deliberately restrains energy transport between the PCM and both the conditioned space and the exterior environment (Khudhair and Farid 2004). However, the following research demonstrated that PCMs significantly modify thermal characteristics of such insulations and thus control the temperature profile and overall heat exchange within the building envelope components. In some configurations, this can reduce the net energy transported through the interior envelope surfaces; in others, it changes the time when the peak energy crosses that boundary. In most air-conditioning operations, nighttime precooling is a well-established energy savings method. In case of PCM application, the time delay can prove especially valuable, in economic terms, when utilities offer different peak-time and off-peak-time electricity rates. From the energy consumption perspective, the air-conditioning system can operate more efficiently (less costly) during the time of the shifted space-conditioning loads. Using a similar load shifting approach, Zhang and Medina of University of Kansas, USA, developed a thermally enhanced wood frame wall that integrated a paraffinic PCM via macro-encapsulating (Zhang et al. 2005).

Today, great majority of market and economic studies concentrate mostly on PCM manufacturing, microencapsulation, and basic packaging. Production of PCM-enhanced building envelope products is still considered as a gray sphere between chemical and construction materials' industries. After decades of laboratory testing and field trials of different PCMs, organic PCM applications represent high market share in terms of value, mostly due to their high availability and an excellent durability, while inorganic PCMs represent more concentrated consumption in terms of volume. Traditionally, a widespread usage in various building applications can be observed for paraffinic PCMs with convenient melting points, chemical stability, and capability for microencapsulation. Current developments in area of bio-based PCMs, derived from plant or animal base, contribute added advantage of lower flammability and high biodegradability over the organic products.

The main growth drivers for the PCM market worldwide are energy savings, occupants' comfort, temperature control, environment friendliness, and wide range of other applications. It is estimated that the building and construction sector represents the largest share of the PCM market today, followed by space-conditioning applications. According to Markets and Markets Research,[1] the global PCM market is estimated to grow from $460 million in 2013 to $1150 million by 2018, at a compound annual growth rate (CAGR) of 20.1 % for the same period. It is also anticipated that reducing price of PCM and R&D for novel materials will create new opportunities and will drive the market in the future. The highest growth is observed in Europe where most of PCM companies have been already well established due to relatively high energy cost and several government incentive programs in energy storage areas. North America is another market with significant R&D potential, where a number of major players exist. Growing number of PCM applications, constantly increasing cost of energy, and regulations imposed by governments worldwide are attracting the companies. In line with the mentioned above Markets and Markets Research report, Europe is expected to remain the largest PCM market (in general) in terms of both volume and value by 2018. However, it is also believed that the great growth potential for building applications exists also in Asiatic countries (China, India, Korea, Japan), Australia, and mid-eastern region.

One of the major barriers reported today for extensive application of the latent heat technologies in buildings is their relatively high cost. The most effective way for cost reductions seems to be an application of less costly source PCMs, combined with the overall system optimization. A variety of successful applications have been described in technical literature for PCM-enhanced building envelope systems. This chapter is mainly focused on most popular building construction products containing PCM. Associated patents are discussed as well.

7.2 PCM–Concrete Blends and Ready Plaster Blends Containing PCM

A large number of experiments worldwide demonstrated that in buildings where PCM is imbedded into the interior plaster wall finish, it is possible to effectively manage building space-conditioning and associated energy consumption. In these

[1]http://www.marketsandmarkets.com/PressReleases/advanced-phase-change-material-pcm.asp.

situations, due to large heat exchange surfaces, internal space temperature can be effectively controlled by PCM with the functional phase-change temperature range close to the room set-point temperature. This means that significantly less energy can be required to maintain expected thermal comfort level. It has been already proven that the heat accumulation capability of a typical plaster wall finish can be significantly increased by adding microencapsulated PCM. In passive applications, PCM phase transition processes are initiated by changes of the interior space temperature. However, in active systems, the PCM can be activated by using dynamic ventilation strategies or through an application of integrated hydronic technologies, which usually include capillary tube mats or pipe heat exchangers.

In addition, work performed by Fraunhofer IBP, Germany, demonstrated that PCM can be also added to the exterior wall stucco, in order to control wall surface temperature.[2] This effect can be used for the reduction of the early morning surface condensation, leading to development of an esthetical microbial growth (Fitz et al. 2006). A series of field experiments followed with numerical hygrothermal analysis led to the conclusion that the potential to solve the exterior wall condensation problem looks promising.

Several concrete and stucco ready-mix products are available right now. This paragraph lists a number of well known, commercially available products of this type.

7.2.1 Weber.Mur Clima® Plaster Mix[3]

Weber.Mur Clima® dry gypsum plaster mix is manufactured in Germany by Saint-Gobain Weber. This preformulated product is recommended for interior plaster applications. Weber.Mur Clima® contains microencapsulated PCM for passive temperature control, mineral lightweight aggregates, chemical additives, and adhesives for better workability and easier site application. The total heat capacity of this product can be easily controlled by varying the application thickness. Table 7.1 shows basic physical properties of this product.

7.2.2 Maxit-Clima[4] Gypsum Plaster Mix

Maxit-Clima gypsum plaster containing PCM is manufactured by Maxit Deutschland GmbH—a German manufacturer of construction material. Maxit Gypsum plaster contains around 20 % by weight of microencapsulated PCM. It is usually

[2]http://cse.fraunhofer.org/Portals/55819//docs/buildings-xi/kuenzel%20workshop%202.pdf.

[3]http://www.sg-weber.de/fassade-wand/produkte/gipsprodukte/webermur-clima-26.html.

[4]http://www.maxit-kroelpa.de/.

Table 7.1 Physical properties of the Weber.Mur Clima®—dry gypsum plaster containing PCM

Physical property	Value
Compressive strength	≥1.5 N/mm^2
Thermal conductivity	>0.3 W/mK
Water vapor diffusion resistance factor μ	10
Shift range of the PCM (clima 23/26)	About 23 °C/ca. 26 °C
Enthalpy of fusion of the PCM	About 100 J/g
Enthalpy of fusion (Ts = 23–26 °C)	18 kJ/kg
Thermal conductivity $\lambda_{(10\ °C)}$	0.19 W/mK
Density	About 0.90 g/cm^3

Table 7.2 Physical properties of the Klima 544 TYNK—cement/lime plaster containing PCM

Physical property	Value
Compressive strength	1.8 N/mm^2
Thermal conductivity	0.47 W/mK
Water vapor diffusion resistance factor μ	14
Shift range of the PCM	26 °C
Enthalpy of fusion of the PCM	About 100 J/g
Enthalpy of fusion	18 kJ/kg

applied with up to 15-mm-thick layer. In addition to passive applications, the Maxit-Clima plaster can be also activated by hydronic water-based systems. Latent storage capacity of the Maxit-Clima plaster is around 70 Wh/m^2. This PCM-enhanced product is available in three phase transition ranges: 21, 23, and 26 °C with approximate heat of fusion of around 18 kJ/kg, and density of about 1340 kg/m^3.

7.2.3 Klima 544 TYNK[5] Cement–Lime Plaster Mix

A similar ready-mix plaster product for internal applications, called Klima 544 TYNK, was offered in the past by the building materials manufacturer CSV from Poland. Dry plaster mix Klima 544 TYNK is based on a proprietary blend of cement, lime, a mixture of mineral aggregates and fillers, chemical additives, and plasticizers for better workability. About 20 % by weigh of PCM is used for local heat storage and temperature control. It is good to notice, here that a large number of PCM-enhanced concrete and plaster ready-mix products use the same type of microencapsulated paraffinic PCM manufactured by BASF. Physical properties of the plaster made with use of the Klima 544 TYNK are presented in Table 7.2.

[5]www.csv.pl.

7.2.4 *ENERCIEL*

Interior Decorative Coating Featuring Organic PCM Winco Technologies ENERCIEL decorative coating is manufactured by WINCO Technologies SAS in Tregueux, France. ENERCIEL is a ready-to-use decorative coating recommended for interior walls and ceiling applications. ENERCIEL contains 50 % by weight of the bio-based microencapsulated PCM (enthalpy—184 J/g) developed by MCI Technologies and the remaining 50 % of proprietary additives to optimize application and its long-term durability. ENERCIEL latent heat storage capacity is 51 Wh/m^2 for 3-mm thick layer and can be adjusted with different thicknesses (for example 102 Wh/m^2 for 6-mm and 153 Wh/m^2 for 9-mm). ENERCIEL has been designed to absorb heat in the temperature range of 23 °C–26 °C and to diffuse heat in the range of 22 °C–18 °C.

7.3 PCM-Enhanced Gypsum Board and Panel Products

Energy storage capabilities of internal surfaces in buildings can be enhanced by applications of different types of wall/ceiling/floor board or panel products containing PCM. The main purpose of these applications is direct capture of dynamic thermal loads (such as, for example, solar energy transmitted through the fenestration) and indirect improvement of human comfort by decreasing the frequency of internal air temperature swings and maintaining temperature closer to the desired temperature for a longer period of time. Several decades of research and development work worldwide demonstrated that wall and ceiling board products can be successfully manufactured to contain in most cases up to 30 % PCM by weight, thus enabling these common building materials to serve as effective thermal energy storage devices. It is good to mention that there are also available several similar composite products with PCM loads higher from 50 %. A large group of PCM-enhanced board products is commercially available today for building applications. This section presents examples of commonly used wall and ceiling boards of that type.

7.3.1 *Knauf PCM Smartboard®*

Smartboard® is the interior plasterboard product manufactured by Knauf, Germany. It is commercially available for drywall construction applications with around 30 % mass fraction of microencapsulated paraffinic PCM manufacture by BASF. Latent heat capacity of the 15-mm-thick Smartboard® product is around 90 Wh/m^2, with two available melting points: 23 and 26 °C. Table 7.3 presents physical properties of the Smartboard® gypsum plaster board.

During last decade, Smartboards® have been successfully installed in a number of buildings mostly in Europe. These include the 3 Liter-Haus in Ludwigshafen,

Table 7.3 Physical properties of the Knauf Smartboard® containing PCM

Physical property	Value
Specific heat	1.2 kJ/kg K
Thermal conductivity at 20 °C	0.20 W/mK
Thermal conductivity at 35 °C	0.19 W/mK
Shift range of the PCM	23 or 26 °C
Enthalpy of fusion of the PCM	About 110 J/g
Latent heat capacity ΔH	330 kJ/m^2

Table 7.4 Physical properties of the National Gypsum ThermalCORE® gypsum board containing PCM

Product picture	Physical property	Value
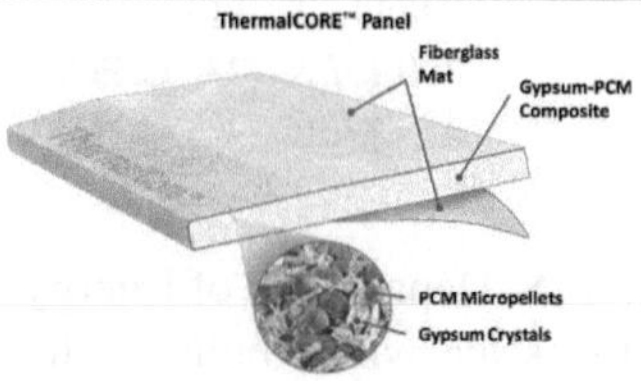	Specific heat	1.2 kJ/kg K
	Board density	800 kg/m^3
	Flame spread ASTM E84	Class B
	Combustibility per ASTM E136	Combustible
	Shift range of the PCM	23 or 26 °C
	Enthalpy of fusion of the PCM	About 110 J/g
	Latent heat capacity ΔH	251 kJ/m^2

Badenova office building in Offenburg, hotel and office complex in Berlin, Haus der Gegenwart in Munich, and the Sonnenschiff Passivhaus Bürokomplex in Freiburg.[6] The above German installations have successfully demonstrated the use of PCM-enhanced gypsum board product as a component in a larger design goal of reducing building space-conditioning energy consumption to a near-zero level.

7.3.2 National Gypsum ThermalCORE®

In 2009, National Gypsum, USA, was the first company to introduce the PCM-enhanced gypsum boards to the North American construction material market. National Gypsum ThermalCORE® panels use Micronal PCM manufactured by BASF. The PCM micropellets absorb and distribute heat energy as the wax melts and solidifies as room temperatures fluctuate around the PCM melting point. Two board options are available, with PCM changing phase at 23 or 26 °C. Table 7.4 summarizes thermal properties of the ThermalCORE® gypsum boards. Unlike other PCM-enhanced gypsum boards which use paper facing, ThermalCORE® has fiberglass skins, but otherwise looks no different than standard drywall. California's

[6]http://www.micronal.de/portal/load/fid443847/BASF_Micronal_PCM_Brochure%202009_English.pdf.

Table 7.5 Physical properties of the Saint-Gobain Rigips Alba®Balance plaster panels containing PCM

Product picture	Physical property	Value
	Board-specific heat	28.3 kJ/m^2 K
	Board weight	25 kg/m^2
	Board density	1000 kg/m^2
	Board flexural strength	8.8 kg/cm^2
	Compressive strength	3.9 N/mm^2
	Share strength	1.4 N/mm^2
	Thermal conductivity λ	0.27 W/mK
	Water vapor diffusion resistance μ	5–10
	PCM phase transition temperatures	23 or 26 °C
	Enthalpy of fusion of the PCM	About 110 J/g
	Latent heat capacity ΔH	300 kJ/m^2 (23 °C PCM) 330 kJ/m^2 (26 °C PCM)

Emerging Technologies Coordinating Council and the U.S. Department of Energy's National Renewable Energy Laboratory performed full-scale whole-building testing on this technology.[7]

7.3.3 Alba®Balance Plaster Panels

Interlocking PCM-enhanced Alba plaster panels are manufactured in Switzerland by Saint-Gobain Rigips. These drywall panels are about 25 mm thick and contain about 3 kg of microencapsulated PCM per m^2. The Alba®Balance panels can be used as interior finish boards in stud walls, wall paneling, and suspended ceilings.[8] Two phase transition options of 23 and 26 °C are available with latent heat of 300 and 330 kJ/m^2, respectively. According to manufacture, the whole board has the total heat storage capacity of about 866 kJ/m^2 for temperature range between 10 and 30 °C. Table 7.5 summarizes thermal properties of the Alba®Balance boards.

7.4 PCM-Enhanced Composite Board Products

PCM-enhanced composite boards or panels are usually manufactured with the use of waste wood, cellulose fibers, or fiber-reinforced mineral composites. Composite wood-based products, also called man-made wood, include a range of derivative

[7]http://www.thermalcore.info/product-info.htm.

[8]http://www.rigips.ch/download/rigips_alba_balance_infobro_de_low.pdf.

wood products which are manufactured by binding the wood strands, particles, fibers, or veneers together with most often adhesives to form composite materials. Plywood and oriented strand board are probably the best-known wood-based composites used for construction purposes today. Other recycled wood/plastic composites, called often man-made lumber, typically consist of a 50/50 mix of wood fibers from recovered saw dust and waste plastics that include high-density polyethylene, PVC, and others. These materials can be formed into both solid and hollow profiles, boards, panels, etc. Composite wood-based boards are used in variety of applications, from home construction to commercial buildings, transportation, and industrial products. They generally exhibit low moisture absorption and high resistance to decay, insects, and UV-ray damage. Because manufacturing processes of these man-made building products usually include a number of powder components (i.e., fillers or fire retardants), these products are excellent PCM carriers containing often over 50 % by weight of the microencapsulated PCM. It is good to mention that hollow composite products can be used to directly enclose PCM without a need for encapsulation.

7.4.1 Wood Fiber-Based Board Products Containing PCMs

This category of building materials is listed separately because of its great potential for PCM applications in buildings. Even though there are no available yet composite PCM board products produced by major building material manufacturers, it is a public secret that a large number of companies are experimenting with integration of PCMs to their composite wood or fiber products. One of the best-known PCM-enhanced composite board products was manufactured in Europe by Rubitherm. Figure 7.1 shows the Rubitherm® FB (fiber board), which comprises a novel type of the heat storage composite in which a PCM is bound within a fiber-based carrier.[9] The RUBITHERM® FB in which approx. 65 % of PCM is contained provides high latent heat storage capacities of over 119 kJ/Kg[10] with working temperatures of 35 and 79 °C.

Today, plywood and oriented strand boards are the most commonly used wood-based composite boards worldwide. All these products can be excellent carriers for microencapsulated PCMs. Plywood is manufactured from sheets of cross-laminated veneer and bonded under heat and pressure with durable, moisture-resistant adhesives. By alternating the grain direction of the veneers from layer to layer, or "cross-orienting," panel strength and stiffness in both directions are maximized. Oriented strand boards (OSBs) are wood structural panels manufactured from rectangular-shaped strands of wood that are oriented lengthwise and then arranged in layers, laid up into mats, and bonded together with moisture-resistant, heat-cured adhesives.

[9]http://www.rubitherm.de/english/pages/04c_glossary_03.htm.

[10]http://www.rubitherm.com/english/pages/02c_latent_heat_fibre_boards.htm.

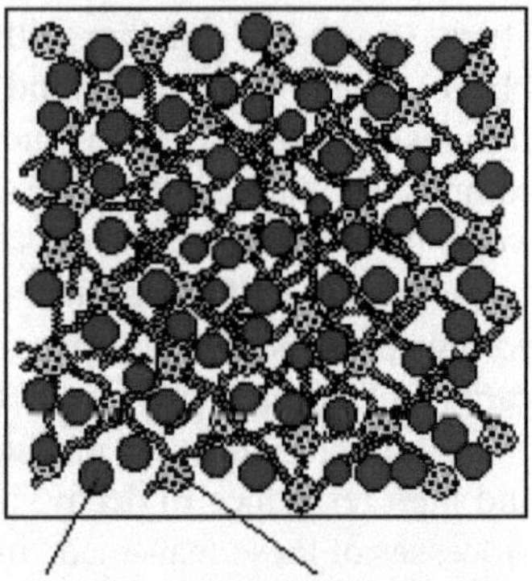

Fig. 7.1 PCM-enhanced fiber boards manufactured by Rubitherm, Germany, and schematic of their cross-linked internal structure—*Source* http://www.rubitherm.de/english/pages/04c_glossary_03.htm

The individual layers are cross-oriented to provide strength and stiffness to the panel. Produced in huge, continuous mats, OSB is a solid panel product of consistent quality with no laps, gaps, or voids.

Another group of composites with great potential for the PCM-enhanced building envelope applications are wood–cellulose–plastic composites. Keep in mind that composite products containing recycled wood, cellulose, or plastic are one of the prime uses for recycled plastic trash bags, waste wood fibers, and cellulose, where wood and cellulose fibers are used as a structural reinforcement. Manufacturers claim that products produced with recycled wood/cellulose/plastics, etc. are more durable than conventional pressure-treated wood and wood-based products. Alternatively, it is also possible to manufacture similar engineered products from the bamboo waste, rye straw, wheat straw, rice straw, hemp stalks, kenaf stalks, or sugar cane residue. In these cases, composites contain no actual wood but other vegetable fibers.

7.4.2 RACUS® PCM Wallboard—A Multi-layer Composite Sandwich System Containing PCM

The RACUS® PCM Wallboard is a novel, passively working, internal sheathing board that offers all attributes of the conventional gypsum boards such as high fire, water, and mechanical impact resistance, with the ability to add thermal energy storage capability to the place of its application. As shown in Fig. 7.2, the RACUS® PCM wallboard comprises of an aluminum honeycomb core with cavities filled

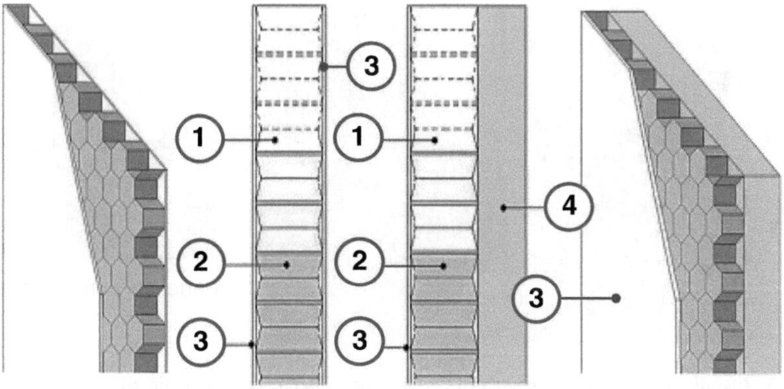

Fig. 7.2 Multilayer PCM-enhanced sandwich board with aluminum honeycomb core—*Source* http://www.datumphasechange.com/index.php?racus-pcm-wallboard. *1* Aluminum Honeycomb, *2* RACUS® PCM infill, *3* Magnoboard®, *4* PIR or Phenolic Foam

Table 7.6 Basic physical properties of the RACUS® PCM wallboards

PCM weight content (kg/m^2)	1.5	4.0
Total panel thickness and dimensions	22 mm, panel size: 1.2 m × 2.4 m	
Sandwich components	6-mm Magnoboard—10-mm aluminum honeycomb filled with PCM—6-mm Magnoboard	
Latent heat storage capacity (kg/m^2)	165	440

with microencapsulated PCM. For use on internal partition applications, the aluminum–PCM layer is sandwiched between two layers of the magnesium oxide (MgO) and magnesium chloride (MgCl)-based construction board product[11] called Magnoboard®. For external wall applications, foam insulation layer can be used on one side. The MgO-based board does not contain any organic solvents, formaldehyde, asbestos, oils, or other toxic substances that can have a negative environmental impact. According to manufacture, the RACUS® PCM Wallboard outperforms cement-based boards and regular gypsum plasterboard. This is due to its superior performance characteristics and environmental benefits, including 100 % water-resistant, high fire resistance, excellent acoustics, and exterior and interior applications.

The RACUS® PCM Wallboards are manufactured by the Datum Phase Change Limited in Clavering, UK. Two PCM weight load levels are offered: 1.5 and 4.0 kg/m^2. Table 7.6 summarizes basic physical properties of the RACUS® PCM Wallboards.

[11]http://www.mgoltd.com/January2009.pdf.

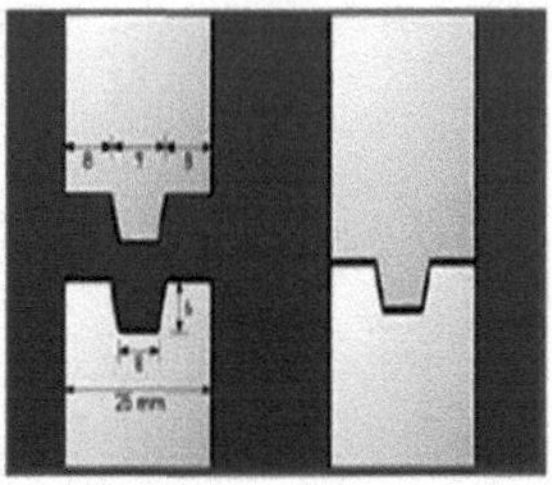

Fig. 7.3 Installation of the ThermaCool® wallboards (*left*). A schematic of tongue and groove connection between the wall board panels is presented on *right—Source* http://www.datumphasechange.com/index.php?home

7.4.3 ThermaCool® Wallboards and Mineral Ceiling Tiles—Manufactured by Datum Phase Change Ltd., UK

Datum Phase Change Ltd. is a UK manufacturer specializing in incorporating PCM into building products under the brand name ThermaCool®. PCM-enhanced wallboards and ceiling tiles are offered for use in both new building construction and retrofit of commercial and residential buildings. ThermaCool® panels are easy to install and are cost competitive comparing to similar wall facing products. They offer a lightweight thermal mass enhancement for lightweight wood- or steel-framed structures. They can be also used in retrofitted buildings, in situations where the original heat accumulation effects are tempered by added levels of thermal insulation. Figure 7.3 shows installation of the ThermaCool® wallboards on the light-gage steel-framed wall. The size of the boards is 0.6×1.0 m, and thickness is 25 mm. The ThermaCool® wallboards contain microencapsulated organic PCM of melting point at 23 °C. The weight of a single board is around 25 kg/m^2. According to the manufacture specs, the total heat capacity of the ThermaCool® wall board is around 157 Wh/m^2.

The ThermaCool® wallboards and ceiling tiles incorporate organic microencapsulated PCM. Datum Phase Change Ltd.—the board manufacturer—claims that, due to unique hygrothermal properties of the ThermaCool® products, they do not promote the growth of mold and mildew and help in humidity level control in the surrounding environment, to improve the occupants' comfort. Table 7.7 lists basic physical properties of the ThermaCool® ceiling tiles.

Datum Phase Change Ltd. is also manufacturing a very unique metal ceiling tile, which incorporates composite PCM core technology with architectural metal-faced ceiling tiles. This product is recommended for projects where an increased thermal

Table 7.7 Basic dimensions and physical properties of the ThermaCool® ceiling tiles

Type of ceiling tile	Standard tile	High thermal mass tile

Panel dimensions (mm)	600 × 600	600 × 600
Panel thickness (mm)	16	16
Panel weight (kg/m^2)	11.6	12.5
PCM melting temperature (°C)	23	23
Total heat capacity (10–30 °C) (Wh/m^2)	85	122

Source http://www.datumphasechange.com/index.php?specification-and-technical-1

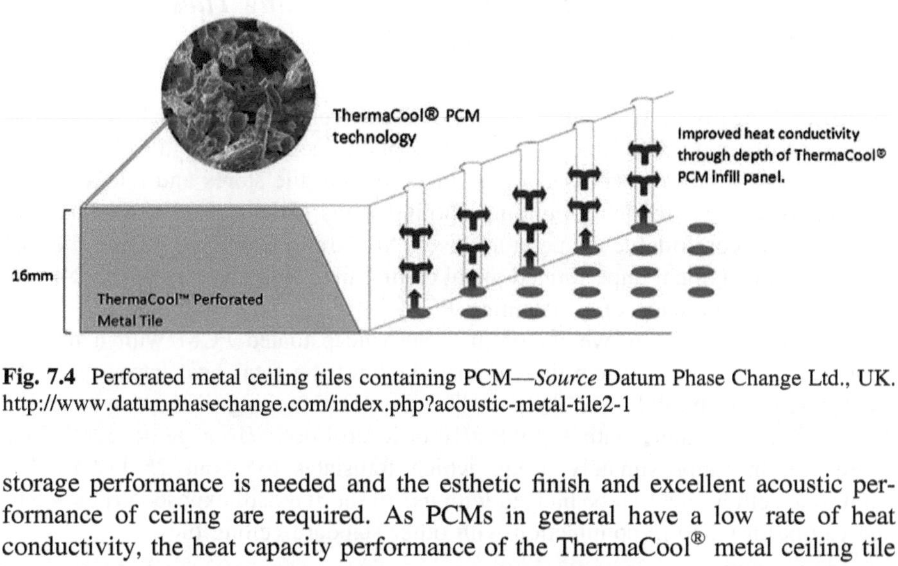

Fig. 7.4 Perforated metal ceiling tiles containing PCM—*Source* Datum Phase Change Ltd., UK. http://www.datumphasechange.com/index.php?acoustic-metal-tile2-1

storage performance is needed and the esthetic finish and excellent acoustic performance of ceiling are required. As PCMs in general have a low rate of heat conductivity, the heat capacity performance of the ThermaCool® metal ceiling tile is accelerated through perforations in both the metal tile and infill panel—see Fig. 7.4. This significantly increases the heat exchange rate through the ceiling surface compared to that of a standard plain metal ceiling tile incorporating PCM. For the temperature range of 10–30 °C, the total heat capacity of this ceiling system is around 122 Wh/m^2. Allowing the unrestricted heat flow through the ceiling, panel perforations help in more efficient utilization of available PCM's heat storage capability. This technology is a winner of prestigious 2012 CIBSE Journal Passive Product of the Year Award.[12]

[12]http://www.datumphasechange.com/index.php?acoustic-metal-tile2-1.

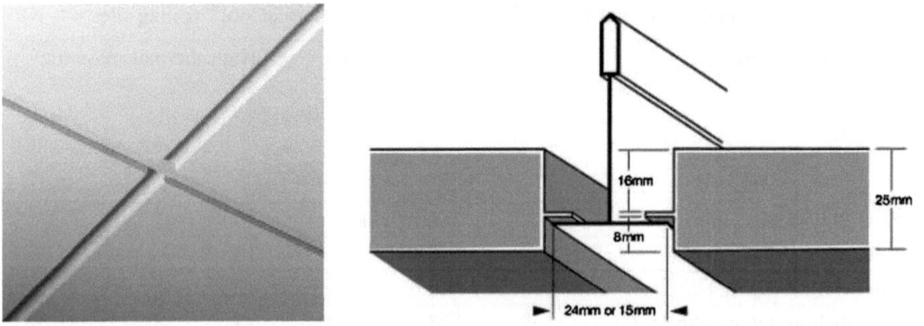

Fig. 7.5 CoolZone—metal ceiling tiles containing microencapsulated PCM—*Source* Armstrong, UK. http://www.armstrong.com/commclgeu/eu1/uk/gb/coolzone-application.html

7.4.4 CoolZone—PCM-Enhanced Metal Ceiling Tiles Manufactured by Armstrong, UK

In 2012, Armstrong, UK, introduced CoolZone ceiling tiles which incorporate microencapsulated PCM into Armstrong's plain metal ceiling tiles.[13] Organic phase-changing material encased in the metal ceiling tile stores and releases large amounts of energy, reducing building cooling loads and, in passive applications, maintaining a comfortable temperature level. Armstrong CoolZone ceiling tiles are based on conventional unperforated metal ceiling tiles with addition of the 24-mm-thick dynamic thermal core containing PCM—see Fig. 7.5. The CoolZone infill contains about 25 % by weight of the microencapsulated PCM with a melting temperature of 23 °C. According to manufacture, the total heat capacity of the CoolZone tile is around 136.2 Wh/m^2. The CoolZone ceiling tiles are available in 600 × 600 mm modules with Tegular 8/16 or MicroLook 8/16 edge details.[14] Each board weights approximately 9 kg, which translates to about 25 kg/m^2. The CoolZone ceiling tiles can be hanged with use of the standard exposed grid systems and can be installed in conjunction with other standard ceiling tiles.

7.4.5 Ecco Building Boards (EBBs)—Unfired Clay Building Boards Containing Microencapsulated PCM

For centuries, in many climates worldwide, clay has been widely accepted as an inexpensive and relatively durable material serving as structural components and at the same time adding thermal mass to buildings. The EBBs are utilizing this old

[13]http://www.architectsjournal.co.uk/specification/product-anatomy/-armstrong-ceilings-debuts-energy-saving-coolzone-system-at-ecobuild/8627739.article.

[14]http://www.armstrong.com/content2/commclgeu/files/71628.pdf.

construction practice. They are original alternative to the widely used today gypsum plasterboard interior finish. The EBB board technology offers practicality of the conventional drywall, combined with unique properties of clay. According to EBB, a series of field experiments demonstrated that the 16-mm-thick unfired clay boards add significantly to the thermal mass of a structure and passively control the relative humidity of the building interior, by acting as a moisture buffering system. The EBB's boards contain about 3 kg/m^2 of microencapsulated PCM with melting point at 23 °C. Their overall weight is around 15 kg/m^2. The EBB clay boards have a total heat storage capacity of 110 Wh/m^2 for a small 5 K temperature change over the melting point.

The EBB board consists of a 4-mm-thick fiber board backing sandwiched with a 10-mm paper honeycomb structure filled with a mixture of clay and PCM microcapsules. The EBB boards are finished with a thin layer of clay plaster to allow plaster coatings to be applied. PCM-enhanced EBB boards are recommended to be used in wall and ceiling applications.

7.4.6 PCM-Enhanced Composite Energain® Boards Manufactured by DuPont™

In 2006, DuPon introduced 5.26-mm-thick Energain® boards, which contain paraffinic PCM with melting point around 22 °C. It is an easy-to-install 1.0 × 1.2-m. wall and ceiling panel system utilizing polymer-based shape-stabilized PCM, faced with two layers of aluminum foil. The Energain® boards are usually installed behind the plasterboard in walls or ceilings—see Table 7.8 for physical properties. In passive applications, the Energain® boards work by absorbing ambient heat as room temperature rises (at around 22°), storing it until the temperature drops again (at around 18°), and then releasing it back into the room. According to manufacturer, in buildings with air-conditioning DuPont™, Energain® can reduce cooling energy

Table 7.8 Basic dimensions and thermal characteristics of the Energain® boards containing paraffinic PCM

Board dimensions	1000 × 1198 × 5.2 mm
Board area weight	24.5 kg/m^2
PCM % weight content	60 %
PCM melting point	21.7 °C
Latent heat (0–30 °C)	70 kJ/kg
Total heat storage capacity (0–30 °C)	140 kJ/kg
Solid phase thermal conductivity	0.18 W/mK
Liquid phase thermal conductivity	0.14 W/mK

costs by up to 35 % and help to reduce heating bills in the winter by up to 15 %.[15] Comparing Energain® boards with traditional thermal mass solutions such as concrete, a research study performed by Lyon University, France, has demonstrated that 5 mm of the Energain® boards shows thermal response similar to 20–40 mm of concrete, depending on set-point temperature.[16]

7.4.7 EcoCore Floor Panels Containing PCM

In 2012, Tate Access Floors, Inc., a U.S. manufacturer and distributor of raised access flooring and in-floor cooling solutions, introduced its new EcoCore Phase Change Panel.[17] Tate Access Floors, Inc. is headquartered south of Baltimore in Jessup, Maryland, USA, and is a member of the Kingspan Group of companies. Tate's EcoCore Phase Change Panel utilizes a cement-based composite containing microencapsulated PCM of melting point around 23 °C. This heat-absorbing material is enclosed within the 610 × 610 × 41-mm welded steel floor panels—see Fig. 7.6. Each panel contains about 1.1 kg of organic PCM, and it can absorb about 51 Wh of energy during the phase-change process. By embedding the PCM in cement and then fully encapsulating it in a steel-welded enclosure, the phase-changing core is protected from the external environment. EcoCore Phase Change Panels are recommended for use in office buildings. The PCM-enhanced EcoCore panel is the recipient of a 2012 Product Innovation Award from Architectural Products Magazine.[18]

Fig. 7.6 EcoCore raised floor system containing microencapsulated PCM—*Source* Tate Access Floor Inc

[15]http://energain.fr/Energain/fr_FR/index.html.

[16]http://energain.co.uk/Energain/en_GB/assets/downloads/documentation/references/ref_Lyon_office_building.pdf.

[17]http://www.tateinc.com/products/ecocore.aspx.

[18]http://tateinc.com/press_ecocore_award.aspx.

7.4.8 Dynamic Ecom4 Floor Tiles Containing PCM

In 2011, Spanish tile manufacturer Ceracasa[19] introduced a new type of floor tiles containing PCM. With PCM melting point around 22 °C, the Ecom4 tiles help maintaining a floor surface temperature close to the interior set-point temperature. Ceracasa partnered with the Polytechnic University of Valencia, Spain, to develop and test this new technology. The joined research demonstrated that in case of a 100 m^2 floor application, the Ceracasa's Ecom4 Tiles can reduce space-conditioning energy use by about 16 %.[20] The Ecom4 tiles are suitable for both floor or wall installations.

7.4.9 DELTA®-COOL 21—Floor Application of Composite Containers with Inorganic PCM

Dörken GmbH & Co. KG—German manufacturer of innovative building materials —developed a series of building products containing inorganic PCMs. In October 2009, one of their technologies—DELTA®-COOL 21—was demonstrated as a floor application during the US Department of Energy Solar Decathlon.[21] A research team from the University of Waterloo, Ontario, Canada, designed and constructed the North House, a compelling, marketable solar-powered home that made use of the latest in high-performance building envelope technologies, system integration, mobile communication technology, sustainable engineering, and original architectural design. In this building, the interior floor assembly employed DELTA®-COOL 21, a PCM system encased in flat composite profiles that were installed under the hardwood floor finish—see Fig. 7.7. The thermal mass of DELTA®-COOL 21 contributes to energy savings as well as to improved comfort inside the building by reducing the effect of ambient day/night temperature swings. The US Department of Energy's Solar Decathlon is a biennial international competition and public event where 20 university teams from around the world compete to design, build, and operate the most attractive, effective, and energy-efficient solar-powered house.

7.5 PCM-Enhanced Mats, Membranes, and Arrays of PCM Containers

As described in earlier chapters, PCMs can be used in building envelopes in many different forms including variety of different construction materials, including foils, membranes, and mats, which can be either attached to or just be carried by the

[19]http://www.ceracasa.com/.

[20]http://inhabitat.com/energy-saving-ecom4-tiles-could-shave-16-off-your-power-bill/.

[21]http://www.cosella-dorken.com/bvf-ca-en/projects/pcm/northhouse.php.

Fig. 7.7 PCM-enhanced flooring technology presented during the 2009 Solar Decathlon in Washington, DC, USA—*Source* Dörken GmbH & Co. KG

building structure. Concentrated PCM applications have been tested as thermal mass components of different builfing envelope configurations in Northern American and European buildings for at least 40 years. Arrays of PCM containers are allowing high thermal storage density per area of application. They are usually utilizing plastic mats, membranes, or foils, which are excellent carriers for PCM. Besides, these systems offer bonus benefit of dual functions, performing often as reflective insulations or air/moisture barriers. Unfortunately, nothing goes for free and the above applications require careful selection of used materials, excellent hygrothermal design, identification of PCM location, and cautious estimation of PCM amount. The section below presents several examples of these technologies.

7.5.1 BioPCmat™—Foil-Based Array of PCM Containers

Phase Change Energy Solutions of Asheboro, NC, USA, offers a wide range of thermal energy storage products for buildings that reduce energy consumption, shift peak demand, and enhance occupant comfort.[22] BioPCmat™ is a foil-based array of PCM containers—see Fig. 7.8. It works in the daytime by absorbing heat from solar and internal gains and then releasing that heat during the night in order to reduce cooling and heating loads along with energy consumption, while enhancing comfort in the process. PCMs used by phase-change energy solutions are soy-based blends of chemicals that change from liquid to solid and back again at different temperatures (i.e., 22.8, 24.4, 26.1, and 29 °C), allowing the material to absorb and release heat.

BioPCM™ is the trade name for family of proprietary bio-based PCMs. Based on different PCM loads, BioPCmat™ products are offered with the following heat storage capacity categories[23]: M27 (307 kJ/m^2), M51 (580 kJ/m^2), and M91 (1033 kJ/m^2). Lengths of these mats are 1.22 and 2.44 m. Only a single width of 0.42 m is available (typical in North America spacing between wall studs and roof rafters). When used in buildings, the BioPCmat™ products are expected to reduce

[22]http://www.phasechange.com/index.php/en/.

[23]http://www.phasechange.com/index.php/en/about/biopcm-biopcmat-thermastix.

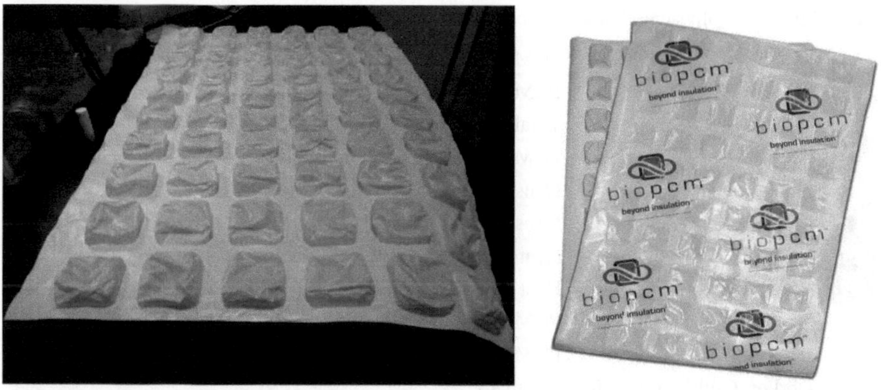

Fig. 7.8 BioPCmat™—*Source* Phase Change Energy Solutions (http://www.treehugger.com/green-architecture/bio-based-phase-changing-material-adds-instant-thermal-mass.html)

cooling loads and HVAC energy consumption while enhancing the building's comfort level. Most typical applications are wall cavities, ceilings, roofs, and floors. According to manufacture, the performance of the BioPCmat™ products is comparable to that of the thermal mass of up to 0.30-m-thick layer of solid concrete.

7.5.2 DELTA®-COOL 24—Foil-Based Array of PCM Containers

Delta Cool 24 has been developed by Dörken GmbH & Co. KG, Germany, to provide increased thermal comfort in interior spaces, especially for (extra) cooling at high internal heat loads. This PCM technology may be integrated in walls and ceilings to accumulate heat, thus increasing the thermal storage capacity of the building. In this instance, salt hydrates are used as PCMs. Under the impact of heat, the salt hydrates in DELTA®-COOL 24 change their state of aggregation from solid to liquid as they reach their melting temperature, absorbing energy (e.g., ambient heat) in the process. As soon as interior temperatures drop below the solidification level, their phase changes back from liquid to solid, and they emit the previously stored heat. DELTA®-COOL 24 can be packaged in many different forms including PCM pouches, PCM-enhanced composite boards, and foil-based arrays of CPM containers.[24] Table 7.9 shows basic physical characteristics of the DELTA®-COOL 24 products. Maximum service temperature for DELTA®-COOL 24 products is 60 °C.

[24]http://www.globalplasticsheeting.com/Portals/32796/docs/delta-cool.pdf.

Table 7.9 Basic physical characteristics of the DELTA®-COOL 24 building products

PCM melting temperature range	22–28 °C
PCM crystallization temperature	22 °C
Total cooling capacity	25–40 W/m²
PCM phase transition enthalpy	158 kJ/kg = 44 Wh/kg
Density (solid)	1.6 kg/L
Density (liquid)	1.5 kg/L
Heat capacity (solid)	2.7 kJ/kg K
Heat capacity (liquid)	2.2 kJ/kg K
PCM thermal conductivity (solid)	1.12 W/mK
PCM thermal conductivity (liquid)	0.56 W/mK

Source Dörken GmbH & Co. KG, Germany

Fig. 7.9 Plastic membrane containing packets with inorganic PCM—an example of floor application—*Source* Alderman Research, USA

7.5.3 Infinite R—A Triple-Layer Plastic Membrane Containing an Array of PCM Containers

Applied Innovation Group of Macon, GA, USA, manufactures phase-changing building membrane, where inorganic PCM is encapsulated in pouches in multi-layer, heavy-duty nylon/PE foil, which imparts good strength, durability, and integrity.[25] Infinite R is a triple-layer membrane both flexible and tough. The product is encased within several rows of packets sealed within the membrane, such that if there were a puncture, damage would be limited to that particular pocket and would not impact the rest of the material. The finished product comes in 0.40-m-wide plastic sheets which are app. 5 mm thick. Infinite R membrane has staple tabs

[25]http://www.phasechangetechnologies.com/.

(wings) that fold out so the roll can be stapled to wall studs, floor joists, or roof rafters—see Fig. 7.9. It is designed to intercept and store heat (energy) from solar load or internal heat gains, which minimizes the transfer of heat from the conditioned space to the unconditioned space—see: Alderman and Yarbrough (2007).

7.6 PCM-Enhanced Building Insulation

PCM-enhanced thermal insulations have been introduced by the textile industry, and they are used today in sport apparel, protective clothes, shoes, etc. As described in previous chapters, there have been also several successful trials as building applications. In buildings, PCM can be directly blended with thermal insulation or added as separate layers of concentrated product.

7.6.1 PCM-Enhanced Cellulose Fiber Insulation from Advanced Fiber Technologies (AFT), Bucyrus, OH, USA

AFT is one of the first insulation companies offering since 2009 a new generation of dynamic insulations containing PCM. This material was developed during a joint project with the U.S. Department of Energy Oak Ridge National Laboratory (ORNL). It is a 20–25 % blend of the microencapsulated organic PCM and cellulose fiber insulation. As shown in Fig. 7.10a, it does not require any additional equipment to install using conventional blowing technology. Three-year-long laboratory and small-scale field testing, performed by the ORNL's Buildings Technology Center and AFT on PCM-enhanced cellulose, demonstrated superior energy performance comparing with traditional insulations.[26]

Thanks to collaboration with Microtek Laboratories Inc., this new insulation product is using ignition-resistant PCM. To ensure that new insulation is in full concurrence with the existing in the U.S. code requirements, the PCM-enhanced cellulose went through a series of flammability tests passing smolder combustion testing and critical radiant heat-flux test—Fig. 7.10b. When the PCM microcapsules are added at 20 % by weight level to blown cellulose insulation intended for attic floor or wall insulation, the product conforms to ASTM C1485-00—"Standard Test Method for Critical Radiant Flux of Exposed Attic Floor Insulation Using an Electric Radiant Heat Energy Source"—ASTM C1485 (2000).

[26]http://web.ornl.gov/sci/roofs+walls/research/detailed_papers/PCM_enhance/content.html.

Fig. 7.10 PCM-enhanced cellulose insulation: **a** installation process using conventional equipment, **b** ASTM C1485-00 flammability testing of the PCM-enhanced blown cellulose—*Source* AFT. http://www.advanced-fiber.com/dynamicci.pdf

7.7 PCM-Enhanced Window Components

The major objective of using PCM in fenestration systems is to utilize its high latent heat of fusion to reduce the heat gain by absorbing the solar thermal load before it reaches the indoor space. The phase-change cycle may stabilize the indoor building temperature and reduce the heating and cooling loads. An addition of PCM systems to the glazing enables utilization of theoretically unrelated characteristics of construction materials, such as thermal mass and interior space illumination. Both of them can be provided at the same time with use of the PCMs packed in transparent plastic containers. During the sunny days, such technologies can reduce interior space temperature peaks due to their significant heat storage capacity. PCM-enhanced windows are usually based on a thin layer of translucent or transparent PCM combined with a thermally insulating window located on the exterior side. Depending on the building's architecture and orientation, the PCM layer can be installed separately, can be preassembled as directly integrated into an insulating glazing unit, or can be combined with a shading device. Since these dynamic insulated glass technologies have the ability to absorb significant amounts of the exterior thermal loads, they can notably reduce the energy consumption associated with the building space conditioning.

7.7.1 GLASSX Façade System

Almost two decades ago, GLASSX Inc. of Zurich, Switzerland, developed a translucent glass façade system aiming to minimize a building's energy consumption while improving the thermal comfort inside the building. GLASSX

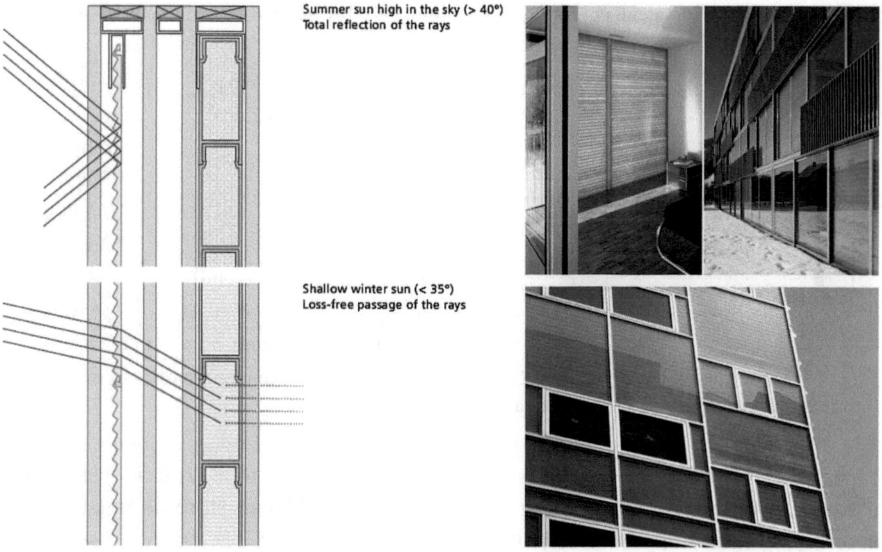

Fig. 7.11 GLASSX—a multi-layer translucent glass façade system containing PCM—*Source* GLASSX Inc., Zurich, Switzerland. http://inhabitat.com/phase-change-glassx-windows-offer-amazing-performance/

façade has unique optical and thermal characteristics and it was licensed to Saint-Gobain[27]—a global building products company. Currently, GLASSX façade fenestration system is manufactured in Austria and Germany. The technology is based on a thin layer of translucent PCM packed into 16-mm-thick transparent plastic containers. The GlassX technology is available now in several dimensional configurations. Figure 7.11 explains configuration of the GlassX Crystal system, which is being marketed in North America. As you can see, outer insulated glazing unit (IGU) has a suspended prismatic filter (like a Fresnel lens) between the panes of glass that reflects higher angle sunlight back out while transmitting low-angle sunlight. This offers a "passive" solar-control mechanism for south-facing glass to keep out most of the high-angle summer sun, while benefiting from the lower angle winter sun. Two separate low-emissivity (low-e) coatings and low-conductivity gas are utilized in two sealed air-spaces in order to enhance thermal insulation characterisitcs of the exterior part of the glazing. Overall, this glazing assembly is about 80 mm thick and weighs around 95 kg/m^2. Glazing modules are available in a maximum height of 2.80 m and a maximum width of 1.50 m.[28] The U-factor is about 0.48 W/m^2 K. The direct-beam light transmission (assuming the sunlight is not blocked by the prism layer) is up to 45 % when the PCM is liquid and up to 28 % when the PCM has crystallized. This glazing system has a heat storage

[27] http://www.glassxpcm.com/.

[28] http://greenspec.buildinggreen.com/blogs/high-tech-glazing-phase-change-material.

Table 7.10 Basic physical characteristics of the DELTA®-COOL 28 thermal mass window component

PCM melting temperature range	26–30 °C
PCM crystallization temperature	26 °C
PCM density (solid)	1.6 kg/L
PCM density (liquid)	1.5 kg/L
Heat capacity (solid)	2.7 kJ/kg K
Heat capacity (liquid)	2.2 kJ/kg K
PCM thermal conductivity (solid)	1.12 W/mK
PCM thermal conductivity (liquid)	0.56 W/mK
Max. operational temperature	60 °C

Source Dörken GmbH & Co. KG, Germany, http://www.energiesparhaus-kfw.de/Daten/Informationsbroschueren/Doerken-Delta-Cool-PCM-Bauteile.pdf

capacity of 1185 Wh/m^2, and it is acting in a similar way to a Trombe wall (glazed thermal storage wall). According to GlassX, the PCM can store as much heat as about 24-cm-thick layer of concrete.

7.7.2 Delta Cool 28 Translucent Heat Storage Containers for Building Fenestration

Dynamic window component designs from Dörken GmbH & Co. KG, Germany, emerge from the variable translucency of some PCM-based products. At the same time, they enhance building thermal mass and provide a source of natural lighting to the interior. Delta Cool 28 system utilizes transparent plastic containers containing salt hydrate. The Delta Cool 28 window thermal mass component has the total heat storage capacity of about 75 Wh/L. Energy state of this system can be visualized as semi-transparent or translucent when used inorganic PCM is melted and milky when PCM is frozen. Working temperature range for PCM in Delta Cool 28 system is between 26 and 30 °C. Table 7.10 shows basic physical properties of the Delta Cool 28 system.

Field performance of semitransparent PCM components from Dorken has been jointly investigated by the glass company Glaswerke Arnold and research institute ZAE Bayern, Germany. A complete window assembly consisted of two glass sheets on the outside and a macro-packaged PCM on the inside—Mehling and Cabeza (2008). Figure 7.12 shows the application of DELTA®-COOL 28 in translucent panels in a glass facade system of a zero energy office building in Kempen, Switzerland.[29] Every second window panel is equipped with the phase-change material in order to reduce solar heating of the interior office space. The solar energy that is stored in the PCM is later released during nighttime, when typically the building would need to be heated to maintain a desirable temperature range.

[29]http://www.cosella-dorken.com/bvf-ca-en/projects/pcm/kempen.php.

Fig. 7.12 Application of the DELTA®-COOL 28 technology in a zero energy office building in Kempen, Switzerland

7.8 Current Patents in Area of PCM-Enhanced Building Envelopes

There are available great variety of intellectual property data focused on PCMs, PCM admixtures, PCM packaging and microencapsulation, PCM manufacturing processes, etc. However, because the PCM-enhanced building envelope technologies are the main objective of this publication, the following section will mostly present a patent work associated with building envelopes, building envelope components, and building envelope materials—all containing PCM. Earlier patents introducing PCM encapsulation, PCM packaging, gypsum board impregnation, etc. will be listed as well, due to their great impact on the historical and recent technology developments. In addtion, the original abstract language from the listed patents is presented below.

7.8.1 Early PCM Encapsulation and Micro-packaging Patents

US 4504402 A
"*Encapsulated phase change thermal energy storage materials*"
Inventors: Johnson C.H. Chen, John L. Eichelberger
Priority Date: Jun 13, 1983

Abstract A thermal energy storage composite pellet-shaped product of about 1/8 to 1 in. in size (3 to 24 mm.), formed of an outer seamless shell member which defines a cavity that permanently encases a phase-change composition, the composition being present in a quantity such that the total volume of the composition, whether in its solid form, liquid form, or both forms as in its transition state, never exceeds the volume of the cavity; and a process for making the product which includes using the phase-change composition in a compacted powder form as cast about which the shell member is formed by roll-coating steps.

US 4797160 A
"*Phase change compositions*"
Inventor: Ival O. Salyer
Priority Date: Aug 31, 1984

Abstract Compositions containing crystalline, straight chain, alkyl hydrocarbons as phase change materials including cementitious compositions containing the alkyl hydrocarbons neat or in pellets or granules formed by incorporating the alkyl hydrocarbons in polymers or rubbers; and polymeric or elastomeric compositions containing alkyl hydrocarbons.

EP 0437331 B1
"*Dry powder mixes comprising phase change materials*"
Inventor: Ival O. Salyer
Priority Date: Jan 09, 1990

Abstract Free flowing, conformable powder-like mix of silica particles and a phase change material (p.c.m.) is disclosed. The silica particles have a critical size of about 7×10^{-3} to about 7×10^{-2} microns and the PCM must be added to the silica in an amount of 80 wt% or less pcm per combined weight of silica and PCM. The powder-like mix can be used in tableware items, medical wraps, tree wraps, garments, quilts and blankets, and in cementitious compositions of the type in which it is beneficial to use a PCM material. The silica-pcm mix can also be admixed with soil to provide a soil warming effect and placed about a tree, flower, or shrub.

US 5565132 A
"*Thermoplastic, moldable, non-exuding phase change materials*"
Inventor: Ival O. Salyer
Priority Date: Jun 06, 1995

Abstract A thermoplastic, moldable, non-exuding phase change material in the form of a composite useful for thermal energy storage. The composite is preferably a solidified melt mixture of a polyolefin resin, and ethylene copolymer, silica particles and a fatty acid, fatty acid ester, primary alcohol or hydrocarbon phase change material. For a microwave heating capability, a microwave absorbing additive may be added as a fifth major ingredient. The composite can be formed into a variety of configurations such as pellets, sheets, rods, tubes, plugs for hollow core cement blocks, films, and fibers, all for thermal energy storage uses.

7.8.2 Early Patens Describing PCM Integration with Building Products or Fabrics

US 4178727 A
"*Heat absorbing panel*"
Inventors: Richard C. Prusinski, Timothy E. Johnson
Priority Date: Feb 01, 1978

Abstract A heat absorbing panel is provided having an outer casing of resinous material intermixed with aggregate material. A central core is provided within the casing. The core includes substantial amounts of a phase change material having a melting point in the range of normal room temperature.

US 4277357 A
"*Heat or cold storage composition containing a hydrated hydraulic cement*"
Inventor: Bryan J. Boardman
Priority Date: Jan 30, 1980

Abstract A polyphase composition for the storage of heat or cold utilizing the latent heat of fusion of a salt hydrate continuous phase intimately intermixed with a hydrated hydraulic cement continuous phase and wherein said continuous phases are optionally in contact with a discontinuous crystalline phase comprising a nucleating component and wherein the composition is enveloped, contained, or packaged within a vapor impermeable material.

US 4747240 A
"*Phase change materials*"
Inventors: Walter E. Voisinet, Deall Daymon, John A. Slyh
Priority Date: Aug 06, 1981

Abstract Phase change materials (PCM's) having a phase change at about 10–30 °C are incorporated like a fine aggregate into building materials.

EP 0306202 A
"*Fiber with reversible enhanced thermal storage properties and fabrics made therefrom*"
Invertors: David P. Colvin, Yvonne G. Bryant
Priority Date: Aug 31, 1987

Abstract A fiber with integral microspheres filled with phase change material or plastic crystals has enhanced thermal properties at predetermined temperatures. The fibers may be woven to form a fabric having the enhanced thermal storage properties, and articles of manufacture may be formed therefrom.

US 4988543 A
"*Impregnation a porous board, changing phases, spraying*"

Inventors: Jean-Francois Houle, Jean Paris
Priority Date: Sep 25, 1989

Abstract A method and an apparatus for impregnating one side of a porous board, such as gypsum boards, with a precise amount of compatible phase change material. A predetermined amount of the phase change material is applied at a predetermined uniform rate to the one surface of the board so that a predetermined amount of the phase change is impregnated. The rate of application is lower than the absorption rate of phase change material. Also, the temperature of the board must be above the melting point of the phase change material during the impregnation. The process and apparatus may be either a continuous, a semi-continuous or a batch operation, and the resulting product is intended as a building construction material.

US 5501268 A
"*Method of energy load management using PCM for heating and cooling of buildings*"
Inventors: Therese K. Stovall, John J. Tomlinson
Priority Date: Jun 28, 1993

Abstract A method of energy load management for the heating and cooling of a building. The method involves utilizing a wallboard as a portion of the building, the wallboard containing about 5 to about 30 wt% a phase change material such that melting of the phase change material occurs during a rise in temperature within the building to remove heat from the air, and a solidification of the phase change material occurs during a lowering of the temperature to dispense heat into the air. At the beginning of either of these cooling or heating cycles, the phase change material is preferably "fully charged". In preferred installations one type of wallboard is used on the interior surfaces of exterior walls, and another type as the surface on interior walls. The particular PCM is chosen for the desired wall and room temperature of these locations. In addition, load management is achieved by using PCM-containing wallboard that form cavities of the building such that the cavities can be used for the air handling duct and plenum system of the building. Enhanced load management is achieved by using a thermostat with reduced dead band of about the upper half of a normal dead band of over three degree. In some applications, air circulation at a rate greater than normal convection provides additional comfort.

US 5755216 A
"*Building products incorporating phase change materials and method of making same*"
Inventor: Ival O. Salyer
Priority Date: Jun 06, 1995

Abstract A building product having thermal energy storage properties in the form of a cementitious hollow core building block having a hollow core or cores and having within the hollow core(s) a composite containing a phase change material.

The composite may be a cementitious plug sized to fit the hollow core(s) and having a phase change material imbibed into the pores and matrix thereof or having polyolefin pellets containing a phase change material or silica particles containing a phase change material incorporated at the wet mix stage of formation of the cementitious plug. The composite may also be a polyolefin plug sized to fit hollow core(s) and having a phase change material contained within the matrix thereof. Finally the composite may be polyolefin pellets containing a phase change material or silica particles containing a phase change material packed in a container sized to fit the hollow core(s).

7.8.3 Patens Associated with PCM-Enhanced Thermal Insulations and Composites

US 5637389 A
"*Thermally enhanced foam insulation*"
Inventors: David P. Colvin, Yvonne G. Bryant
Priority Date: Feb 18, 1992

Abstract A thermal barrier in the form of a base material with a plurality of microcapsules containing a phase change material forms an insulative pad, such as a shoe insole. The microcapsules are surroundingly encapsulated and embedded within the base material. In addition, substantially all of the microcapsules are spaced apart from each other, the space between neighboring adjacent microcapsules containing a phase change material. The microcapsules may be anisotropically distributed to further reduce thermal conductivity of heat through the thermal barrier.

US 5851338 A
"*Skived foam article containing energy absorbing phase change material*"
Inventor: Robert J. Pushaw
Priority Date: Mar 04, 1996

Abstract A composition for saturation of preformed, previously cured foam substrates having an at least partially open cell configuration, includes a polymer binder in which microspheres containing a phase change material are dispersed. Preferred phase change materials include paraffinic hydrocarbons. The microspheres may be microencapsulated. A preferred cured foam substrate is a skived polyurethane or polyether foam of from 20 to 1000 mils in thickness, preferably 20–200 mils in thickness, having up to 6 oz per square yard or more of encapsulated phase change material embedded in a polymer binder. One method of applying the binder with dispersed encapsulated phase change materials is by applying a binder/microsphere dispersion to the upper surface of a previously cured foam sheet with, then drawing a vacuum from the underside of the cured foam sheet to permeate from 20 to 100 %

of the cured foam sheet with the binder/microsphere dispersion. The resulting product is then cured. An alternative method of applying the binder with embedded encapsulated phase change materials using knife over roll technique to an exposed surface of the skived foam is disclosed in which the most preferred viscosity of the uncured polymer binder is from 7000 to 9000 centipoise.

US 6000438 A
"*Phase change insulation for subsea flowlines*"
Inventors: Theodore R. Ohrn
Priority Date: Feb 13, 1998

Abstract A subsea flowline or pipeline with passive insulation having improved transient heat-loss characteristics. The flowline or pipeline has a phase change material insulating the flowline pipe joints and/or pipe sections to greatly extend the cooldown time of produced or processed hydrocarbon fluid mixtures being transported therethrough during shut-in conditions. The phase change material surrounds a carrier pipe of the flowline or pipeline. The phase change material may be either dispersed or encapsulated within standard insulation. Additional insulation and/or a jacket pipe surrounds the phase change material and holds it against the carrier pipe. The phase change material can be of a micro-encapsulated or bulk-encapsulated type.

US 6689466 B2
"*Stable phase change materials for use in temperature regulating synthetic fibers, fabrics and textiles*"
Inventor: Mark H. Hartmann
Priority Date: Sep 21, 2000

Abstract A stabilized phase change composition comprises a phase change material and a stabilizing agent selected from the group consisting of antioxidants and thermal stabilizers. The stabilizing agent provides oxidative or thermal stabilization to the phase change material. The stabilized phase change composition may be used or incorporated in a variety of processes (e.g., melt spinning processes, extrusion processes, injection molding processes, and so forth) to form articles having enhanced reversible thermal properties. Exemplary articles that may be formed include, by way of example and not by limitation, synthetic fibers (e.g., nylon fibers, polyester fibers, polyethylene fibers, polypropylene fibers, and multi-component fibers), fabric materials, textiles, films, foams, sheets, pellets, granules, rods, and injection molded articles.

US 6645598 B2
"*Cell insulation blanket with phase change material, and method of making*"
Inventors: Robert J. Alderman
Priority Date: Jan 04, 2002

Abstract Heat insulation blanket includes a layer of inner cells containing a phase change material, and opposed outer cells containing dead air or other undisturbed gas. When placed in an attic or other insulated area, the phase change material changes phase when the outside atmospheric temperature passes the phase change temperature, resulting in a delay in the transfer of heat between the interior space and the atmosphere because of the heat required for or given up by the PCM during its change of phase. The outer plies of sheet material that form the superposed cells of the cell blanket are formed with a coating of heat reflective material that faces the interior of the cell blanket, whereby the reflective surfaces are protected against deterioration and retain their capacity to reflect radiant heat.

EP 1628110 A1
"*Loose fill insulation product having phase change material therein*"
Inventors: Murray S. Toas, David G. Ober, Michael P. Ellis
Priority Date: Jun 17, 2004

Abstract A thermal insulation product is provided comprising a loose fill insulation for insulating an interior of a hollow or open space in a structure and at least one phase change material dispersed in the loose fill insulation.

US 8091309 B2
"*Insulation containing inorganic fiber and spherical additives*"
Inventor: Christophe Wagner
Priority Date: Jun 17, 2004

Abstract The present invention provides thermal insulation products such as loose fill, batts and boards, such as duct boards and duct liner. The insulation products include randomly distributed inorganic fibers, which are supplemented with at least about five weight percent microspheres, macrospheres, or both, and preferably include hollow microspheres, which boost the insulation value of the fiberglass thermal insulation by at least about 0.5 R.

US 7735327 B2
"*Active thermal insulation system utilizing phase change material and a cool air source*"
Inventors: Keith R. Brower, Burt Dole, Thomas Neal, Daniel J. Shramo
Priority Date: Jul 19, 2006

Abstract An active thermal insulation system is disclosed. The system utilizes a cool air source or ambient air below a threshold temperature in conjunction with a phase change material and/or conventional insulation. In a controlled manner, the cooled air or ambient air facilitates the transition of the phase change material from a substantially liquid state to a substantially solid state allowing the solid phase change material to absorb heat. Cooled air or ambient air may be directed to the phase change material via a gap or plenum capable of introducing the cooled air or ambient air adjacent to the phase change material. A fan, blower or air intake device may be used to direct ambient air to the gap or plenum.

US 20120149795 A1
"*Polyurethane foams containing incorporated phase change material*"
Inventors: Stephan Schleiermacher, Heike Niederelz
Priority Date: Aug 26, 2009

Abstract The invention relates to polyurethane foams with incorporated phase change material, especially for reinforcing the back of deep-drawn films and components.

US 8156703 B2
"*Multiple phase PCM heat insulation blanket*"
Inventor: Robert J. Alderman
Priority Date: Nov 24, 2009

Abstract A PCM heat insulation blanket (10) is positioned in a wall structure of a building for modulating the transfer of heat to and from the controlled internal temperature of the building structure. The PCM heat insulation blanket includes an array of pods (12) with one group of pods 12A containing a first PCM and a second group of pods containing a second PCM 12B having a different phase change temperature than the first group of pods. The pods with different PCMs are substantially uniformly intermixed among each other. The pods are formed of laminated sheets, each of which may include a metal foil that is vapor impermeable and tends to maintain the PCMs in the blanket at approximately the same temperature, and which also functions to reflect radiant heat away from the temperature controlled internal space of the building structure.

US 20130296449 A
"*Polyurethane Gel-Like Polymers, Methods and Use in Flexible Foams*"
Invertors: Bruce W. Peterson, Mark L. Crawford
Priority Date: Feb 26, 2010

Abstract Combinations of open cell flexible foams with polyurethane gel-like polymers, in forms such as layers of different forms and shapes, solid sheets, perforated sheets, and particles, and methods of making the combinations are described using a variety of procedures. Alternatively, the resin to make the polyurethane gel-like polymers may be infused into the polyurethane foams. The open cell flexible foam may partially or wholly comprise polyurethane foam and latex foam.

US 20130295371 A1
"*Thermal Storage Gelatinous Triblock Copolymer Elastomer Particles in Polyurethane Flexible Foams*"
Invertors: Bruce W. Peterson, Mark L. Crawford
Priority Date: Feb 26, 2010

Abstract Combinations of gelatinous elastomer containing one or more phase change materials, known as "phase change gel", and polyurethane foam may be made by introducing at least partially cured phase change gel particles comprising plasticized triblock copolymer resin and/or diblock copolymer resin and one or more phase change materials, into a mixture of polyurethane foam-forming components including a polyol and an isocyanate. The phase change gel can be crosslinked to form a cured gelatinous gel, which is then reduced in size before introduction. After the foam-forming components polymerize to make polyurethane foam, the phase change gel particles are discrete visible particles dispersed throughout the foam. The polyurethane reaction is exothermic and can generate sufficient temperature to at least partially melt the styrene-portion of the triblock copolymer resin thereby extending the crosslinking. The combination of phase change gel and polyurethane foam increases the heat storage capacity and thermal conductivity of the foam.

WO 2013103596 A1
"*Compositions comprising phase change materials and methods of making the same*"
Inventors: Reyad I. Sawafta, Najih Naser
Priority Date: Jan 03, 2012

Abstract In one aspect, compositions are described herein. In some embodiments, a composition comprises a phase change material, a hydrophobic sorption material, and a viscosity modifier. In some embodiments, a composition comprises a foam and a latent heat storage material dispersed in the foam, the latent heat storage material comprising a phase change material and a hydrophobic sorption material.

7.8.4 Patens Associated with PCM-Enhanced Building Envelope Products

EP 1656239 A1
"*Membrane materials with thermo-regulating properties for fabric structures*"
Inventor: Barbara Pause
Priority Date: Jul 21, 2003

Abstract Membrane materials for fabric structures consisting of a woven fabric which is coated on one or both sides with an elastomeric compound comprising finely divided phase change materials such as crystalline alkyl hydrocarbons or salt hydrates. The membrane materials facilitate thermo-regulation due to latent heat absorption and latent heat release in the phase transition range of the phase change material, which enhances the thermal comfort of the enclosed structure, they are applied to and leads to energy savings.

US 20050055982 A1
"*Phase-change structural insulated panels and walls*"
Inventor: Mario Medina
Priority Date: Aug 13, 2003

Abstract A wall, ceiling, or structural insulated panel having macroencapsulated phase-change materials incorporated therein.

US 8333903 B2
"*Phase change material (PCM) compositions for thermal management*"
Inventors: Loic Pierre Rolland, Raymond Joseph Reisdorf
Priority Date: Dec 09, 2004

Abstract The present invention relates to a Phase Change Material (PCM) composition comprising (a) from 20 to 80 wt% of a PCM; and (b) from 20 to 80 wt% of one or more polymers chosen from the group consisting of (b1) Very Low Density Polyethylene (VLDPE) having a density equal or lower than 0.910 g/cm^3 measured according to ASTM 792; (b2) Ethylene Propylene Rubber (EPR) having a density equal or lower than 0.900 g/cm^3 measured according to ASTM 792; (b3) Styrene Ethylene Butadiene Styrene (SEBS) copolymers; and (b4) Styrene Butadiene Styrene (SBS) copolymers. The PCM composition of the present invention can be used in applications where thermal management is needed, like for example in building, automotive, packaging, garments and footwear.

US 7721566 B1
"*Collapsible interconnected panels of phase change material*"
Inventor: Kenneth Wilken
Priority Date: Aug 14, 2006

Abstract An apparatus having a plurality of phase change material-containing wall panels. The wall panels have a thickness and sides. The wall panels define an inner major surface and an outer major surface. The sides of the wall panels are hingedly interconnected to define a closed structure. Two diametrically opposed hinge interconnections are proximate the inner major surface. The remaining hinge interconnections are proximate the outer major surface.

US 7703254 B2
"*Reflective insulation tiles*"
Inventor: Robert J. Alderman
Priority Date: Oct 08, 2007

Abstract Heat insulation tiles are supported in a grid of support bars of a ceiling assembly with the ceiling tiles being in the form of insulation boxes that form a dead air spaces therein. Reflective material is applied to at least one of the interior surfaces of the box that faces the dead air space and a layer of phase change material is applied to the insulation boxes.

US 8156703 B2
"*Multiple phase PCM heat insulation blanket*"
Inventor: Robert J. Alderman
Priority Date: Nov 24, 2009

Abstract A PCM heat insulation blanket is positioned in a wall structure of a building for modulating the transfer of heat to and from the controlled internal temperature of the building structure. The PCM heat insulation blanket includes an array of pods with one group of pods containing a first PCM and a second group of pods containing a second PCM having a different phase change temperature than the first group of pods. The pods with different PCMs are substantially uniformly intermixed among each other. The pods are formed of laminated sheets, each of which may include a metal foil that is vapor impermeable and tends to maintain the PCMs in the blanket at approximately the same temperature, and which also functions to reflect radiant heat away from the temperature controlled internal space of the building structure.

US 8308861 B2
"*Phase change material compositions*"
Inventors: Loic Pierre Rolland, Jacques Andre
Priority Date: May 13, 2010

Abstract A phase change material (PCM) composition comprises a blend of at least one fatty acid ester and at least one ethylene copolymer, wherein the ethylene copolymer comprises at least 28 % of an at least one polar comonomer by weight, based on the weight of the ethylene copolymer. A film or sheet comprises or is produced from the PCM composition and the film or sheet can be a single layer or multilayer structure.

US 8333903 B2
"*Phase change material (PCM) compositions for thermal management*"
Inventors: Loic Pierre Rolland, Raymond Joseph Reisdorf
Priority Date: Jun 20, 2011

Abstract The present invention relates to a Phase Change Material (PCM) composition comprising (a) from 20 to 80 wt% of a PCM; and (b) from 20 to 80 wt% of one or more polymers chosen from the group consisting of (b1) Very Low Density Polyethylene (VLDPE) having a density equal or lower than 0.910 g/cm^3 measured according to ASTM 792; (b2) Ethylene Propylene Rubber (EPR) having a density equal or lower than 0.900 g/cm^3 measured according to ASTM 792; (b3) Styrene Ethylene Butadiene Styrene (SEBS) copolymers; and (b4) Styrene Butadiene Styrene (SBS) copolymers. The PCM composition of the present invention can be used in applications where thermal management is needed, like for example in building, automotive, packaging, garments and footwear.

WO 2013040404 A2
"*Composite construction panels and applications thereof*"
Inventors: Reyad I. Sawafta, Ken Farrish, Byron C. Owens, Peter F. Horwath, James D. Thornsberry.
Priority Date: Sep 16, 2011

Abstract In one aspect, composite construction materials are described herein. In some embodiments, a composite construction panel comprises a substrate layer, a cover layer separated from the substrate layer by one or more spacers, and at least one mat disposed between the substrate layer and the cover layer, wherein the mat comprises at least one phase change material disposed in at least one phase change region.

References

Alderman RJ, Yarbrough DW (2007) Use of phase-change materials to enhance the thermal performance of building insulations. In: Proceedings of the international thermal conductivity conference, ITCC29 and ITES17 conference, Birmingham, AL, June 24–27

ASTM C1485 (2000) Standard test method for critical radiant flux of exposed attic floor insulation using an electric radiant heat energy source. ASTM International, West Conshohocken, PA http://www.astm.org/DATABASE.CART/HISTORICAL/C1485-00.htm

Balcomb JD (1979) Heat storage effectiveness of concrete masonry units. Los Alamos Scientific Laboratory, report No. LA-UR-82-966

Banu D, Feldman D, Haghighat F, Paris J, Hawes D (1998) Energy-storing wallboard: flammability tests. J Mater Civ Eng 10(1998):98–105

Cabeza LF, Medrano M, Castellón C, Castell A, Solé C, Roca J, Nogués M (2008) Thermal energy storage with phase change materials in building envelopes. Contributions to Science 3(4): 501–510

Clearwater FL, Khudhair AM, Farid MM (2004) A review on energy conservation in building applications with thermal storage by latent heat using phase change materials. Energ Conv Manage 45:263–275

Drake JB (1978) PCMSOL—simulation code for a passive solar structure incorporating phasechange materials. U.S. Department of Energy, Oak Ridge National Laboratory. Report ORNL—6359

Fitz C, Hofbauer W, Sedlbauer K, Krus M, Breuer K (2006) Prognoseverfahren zum biologischen befall durch algen pilze und flechten an bauteiloberflächen auf basis bauphysikalischer und mikrobieller untersuchungen. Bauforschung für die Praxis Band 77, Fraunhofer IRB Verlag, 304 Seiten, Germany

Hawes DW, Banu D, Feldman D (1990) Latent Heat Storage in Concrete. Sol Energ Mater 21, 61–80

Hauer A, Mehling H, Schossig P, Yamaha M, Cabeza LF, Martin V, Setterwall F (2005) International energy agency implementing agreement on energy conservation through energy storage. Annex 17 Final Report

Khudhair AM, Farid MM (2004) A review on energy conservation in building applications with thermal storage by latent heat using phase change materials. Energy Convers Manage 45:263–275

Kośny J, Yarbrough DW, Wilkes KE, Leuthold D, Syed AM (2006) PCM-enhanced cellulose insulation thermal mass in lightweight natural fibers. In: Proceedings of ECOSTOCK—IEA International Conference on Thermal Energy Storage. Stockton College of New Jerse, May 31–June 02, 2006

Kośny J, Yarbrough DW, Miller WA, Childs P, Syed AM (2007) Thermal performance of PCM-enhanced building envelope systems. In: Proceedings of X Conference—Thermal Performance of the Exterior Envelopes of Buildings, December 2007

Mehling H, Cabeza LF (2008) Heat and cold storage with PCM. Springer, Series: Heat and Mass Transfer, ISBN: 978-3-540-68556-2

Salyer IO, Sircar A (1989) Development of PCM Wallboard for heating and cooling of residential buildings. Thermal Energy Storage Research Activities Review, U.S. Department of Energy, New Orleans, LA, March 15–17, pp 97–123

Salyer IO, Sircar AK, Chartoff RP, Miller DE (1985) Advanced phase-change materials for passive solar storage applications. In: Proceedings of the 20th Intersociety Energy Conversion Engineering Conference. Warrendale, Pennsylvania, USA, pp 699–709

Shapiro M, Feldman D, Hawes D, Banu D (1987) P.C.M. thermal storage in drywall using organic phase-change material. Passive Sol J 4(4):419–438

Stovall TK, Tomlinson JJ (1995) What are the Potential Benefits of Including Latent Storage in Common Wallboard? J Sol Energ Eng Trans ASME 117:318–325

Tomlinson JJ, Heberle D (1990) Analysis of wallboard containing a phase change material. In: Proceeding of the 25th Intersociety Energy Conversion Engineering Conference, vol 4, Reno NV, USA

Tomlinson J, Jotshi C, Goswami D (1992) Solar thermal energy storage in phase change materials. In: Proceedings of Solar '92: The American Solar Energy Society Annual Conference, June 15–18, Cocoa Beach, FL, pp 174–79

Zalba B, Marín JM, Cabeza LF, Mehling H (2003) Review on thermal energy storage with phase change: materials, heat transfer analysis and applications. Appl Therm Eng 23:251–283

Zhang M, Medina MA, King J (2005) Development of a thermally enhanced frame wall with phase-change materials for on-peak air conditioning demand reduction and energy savings in residential buildings. Int J Energ Res 29(9):795–809

MIX
Papier aus verantwortungsvollen Quellen
Paper from responsible sources
FSC® C105338

If you have any concerns about our products,
you can contact us on
ProductSafety@springernature.com

In case Publisher is established outside the EU,
the EU authorized representative is:
Springer Nature Customer Service Center GmbH
Europaplatz 3, 69115 Heidelberg, Germany

Printed by Libri Plureos GmbH
in Hamburg, Germany